Industrial Development Report 2016

The Role of Technology and Innovation in Inclusive and Sustainable Industrial Development

UNITED NATIONS
INDUSTRIAL DEVELOPMENT ORGANIZATION

For reference and citation, please use: United Nations Industrial Development Organization, 2015. *Industrial Development Report 2016. The Role of Technology and Innovation in Inclusive and Sustainable Industrial Development.* Vienna.

Cover image: iStock.

UNIDO ID/447
Sales Number: E.15.II.B.47
ISBN: 978-92-1-106454-4
eISBN: 978-92-1-057637-6

Contents

Page

Boxes

Page

Figures

Page

Page

Page

Page

Tables

Foreword

Technological change is recognized as one of the main drivers of long-term growth. In the coming decades, radical innovations such as the mobile internet, the Internet of Things and cloud computing are likely to revolutionize production processes and enhance living standards, particularly in developing countries. The Sustainable Development Goal 9 *Build resilient infrastructure, promote inclusive and sustainable industrialization and foster innovation* adopted on 26 September 2015 implies that without technology and innovation, industrialization will not happen, and without industrialization, development will not happen.

It is undebatable that technology makes production processes more efficient, thereby increasing the competitiveness of countries and reducing their vulnerability to market fluctuations. Structural change, i.e. the transition from a labour-intensive to a technology-intensive economy, drives economic upgrading. Low income countries thus acquire the necessary capabilities to catch up and reduce the gap with per capita incomes in high income countries.

Catching up, unfortunately, does not occur frequently. In the last 50 years, only a few countries were successful in rapidly industrializing and achieving sustained economic growth. Technology was always a key driver in these cases and they successfully developed an advanced technology-intensive industry. Though there is clear evidence that technological change contributes significantly to the prosperity of nations, the debate about the underlying factors deterring countries from promoting technology and innovation more intensively continues.

Though technology is linked to sustainable growth, it is uncertain whether it can simultaneously create social inclusiveness and environmental sustainability. The substitution of labour with capital induced by structural change may reduce employment. Technological change also requires the labour force to be prepared to use increasingly complex machinery and equipment, which widens the inequality between highly skilled and unskilled workers in terms of wage distribution. Industrialization has historically been accompanied by increasing pollution and the depletion of natural resources. Economic growth also entails a rise in the use of inputs, materials and fossil fuels, which generate environmental pollution and degradation, especially in low income countries.

The Lima Declaration approved during the 15th session of UNIDO's General Conference clearly states that "Poverty eradication remains the central imperative. This can only be achieved through strong, inclusive, sustainable and resilient economic and industrial growth, and the effective integration of the economic, social and environmental dimensions of sustainable development". UNIDO strongly promotes paths of economic growth and industrialization that reconcile all relevant dimensions of sustainability.

The *Industrial Development Report 2016* addresses a challenging question: under which conditions do technology and innovation achieve inclusive and sustainable industrial development (ISID)? The main finding of this report is that technology can simultaneously serve all three dimensions of sustainability. Rapid inclusive and sustainable industrialization can be achieved provided that policymakers resolutely facilitate and steer the industrialization process, which requires sound policies and avoiding the mistakes other countries have made in the past.

From an economic point of view, globalization and the fragmentation of production at international level have facilitated the diffusion of new technologies through the intensification of trade in sophisticated manufacturing goods. However, this diffusion of technology has in many cases not translated into concrete growth opportunities due to the lack of technological capabilities and the capacity of countries to promote innovation systems. Innovation needs to be supported by appropriate interventions that strengthen the process from technology invention to adoption by firms as was the case in benchmark countries such as China and the Republic of Korea.

From a social point of view, industrialization contributes to the improvement of many indicators such as the Human Development Index and the poverty rate. Even though technology and automation generally improve people's working conditions, the number of jobs may decrease as a result, with workers being replaced by machines. A key point highlighted in this report is that technological change itself can mitigate this effect. New technologies also generate new markets, for example the waste and recycling industry, reduce the prices of consumer goods and provide opportunities for new investments with higher levels of profitability. Most importantly, the expansion of new technologically-intensive industries absorbs those workers who have lost their jobs to machines.

From an environmental point of view, there is a natural tendency of firms to seek efficiency in the use of resources. Entrepreneurs tend to maximize profits by minimizing the use of inputs through process innovations. During the structural change process, the transition from medium tech industries towards high-tech industries is beneficial from a macro perspective, as it implies a lower level of environmental pollution. Despite these positive dynamics, the current trend of technological change does not guarantee that we will follow a sustainable path in the future. Global concerted action is indispensable to reduce greenhouse gases and to stimulate the creation and diffusion of environmentally friendly technological progress.

It gives me great pleasure to present this report as Director-General of UNIDO. I am particularly pleased that the *Industrial Development Report 2016* emphasizes the critical need for international cooperation to promote technological change and achieve ISID, and that it reaffirms the commitment of my Organization to fulfil its unique mandate in support of this effort. I am grateful to the UNIDO staff and the international experts who joined hands to produce this report, and look forward to seeing it become a key component in the development debate.

LI Yong
Director General, UNIDO

Acknowledgements

The *Industrial Development Report 2016* was prepared under the overall guidance of LI Yong, Director-General of the United Nations Industrial Development Organization (UNIDO). It is the result of two years of intense research efforts, fruitful discussions and close collaboration between the members of a cross-organizational team headed by Ludovico Alcorta, Director of the Research, Statistics and Industrial Policy Branch at UNIDO. This lengthy and at times arduous endeavour was coordinated by Nicola Cantore, Industrial Development Officer (UNIDO), who played an instrumental role in the successful completion of the report. The other members of the UNIDO core team, without whom this report would not have been possible, were Michele Clara, Smeeta Fokeer, Nobuya Haraguchi, Alejandro Lavopa, Ascha Pedersen, Miriam Weiss and Shohreh Mirzaei Yeganeh. The report was a collaboration with the United Nations University—Maastricht Economic and Social Research Institute on Innovation and Technology (UNU—MERIT), in particular Adam Szirmai and Bart Verspagen; and with Paula Nagler and Wim Naudé from the University of Maastricht and the Maastricht School of Management, who were part of the core team. Many of the concepts introduced and elaborated in the report were developed during IDR core team meetings and at workshops at UNIDO headquarters in Vienna in February and April 2015.

The invaluable contributions and insightful comments received greatly enhanced the overall quality of the report—these included experts from UNU—MERIT, namely Ibrahima Kaba, Mary Kaltenberg, Neil Foster-McGregor and Simone Sasso. Other contributing experts were Charles Fang Chin Cheng, University of New South Wales; Valentina De Marchi, University of Padova; Teresa Domenech, University College of London; Elisa Giuliani, University of Pisa; Arjan de Haan, International Development Research Centre; Jojo Jacob, Grenoble Ecole de Management; Raphael Kaplinsky, Open University; Florian Kaulich, University of Vienna; Michael Landesmann, Johannes Kepler University; Carolina Lennon, Vienna University of Economics and Business; Giovanni Marin, National Research Council of Italy; Isabella Massa, Overseas Development Institute; Roberta Rabellotti, University of Pavia; Cornelia Staritz, Austrian Foundation for Development Research; Robert Stehrer, Vienna Institute for Economic Studies; Fiona Tregenna, University of Johannesburg; as well as Juergen Amann and Gary Gereffi from Duke University; Thomas Gries, Rainer Grundmann and Margarete Redlin from the University of Paderborn; and Marianna Gilli, Massimiliano Mazzanti and Francesco Nicolli from the University of Ferrara. Deepest gratitude is also due to Xiaolan Fu from Oxford University, and John Weiss, Emeritus Professor at the University of Bradford, who thoroughly reviewed numerous drafts of the report and significantly improved several sections of the report.

The report further benefited from constructive comments by members of the IDR Advisory Board at UNIDO, specifically Stefano Bologna, Guillermo Lorenzo Castella, Mohamed-Lamine Dhaoui, Sam Hobohm, Steffen Kaeser, Bernardo Calzadilla Sarmiento, Stephan Sicars and Nilgun Tas, by members of the UNIDO Publications Committee Jacek Cukrowski, Frank Hartwick and Patrick Nussbaumer, as well as other UNIDO colleagues including Manuel Albaladejo, Ralph Luken, Valentin Todorov and Shyam Upadhyaya. Profound appreciation is also extended to Taizo Nishikawa, Deputy to the Director-General at UNIDO, for the exceptional support he provided throughout the entire production process.

The authors of the report were backed by a talented and indispensable team of research assistants and interns at UNIDO including Juan Carlos Castillo, Emi Mima, Stefano Olivari, Francis Ostermeijer and Sheng Zhong.

UNIDO staff members Debby Lee, Fernando Russo and Iguaraya Saavedra, without whom a smooth production process would have been unthinkable, provided extensive administrative support, and Niki Rodousakis, Nelson Correa and Franz Brugger provided copy-editing assistance.

The editors, Bruce Ross-Larson and Jonathan Aspin of Communications Development Incorporated, improved the language, style and structure of the report. Christopher Trott and Joe Caponio, also with Communications Development Incorporated, copy-edited and proofread the report, and Elaine Wilson was in charge of design and layout.

Technical notes and abbreviations

References to dollars ($) are to United States dollars, unless otherwise indicated.

In this report, *industry* refers to the manufacturing industry and *sectors* refers to specific manufacturing sectors.

This report defines *developed countries* or *developed economies* as the group identified as "high-income OECD countries" by the World Bank and *developing countries* or *developing economies* as all other economies. See Annex B1 for a complete list of economies by region, income level, least developed countries and largest developing economy in each region.

The annexes contain more detailed information about methodology and classifications. Annex B also contains additional tables and indicators to complement those contained in the text. Annex C is a guide to the origins of the data used for the figures and tables in the *Industrial Development Report 2016.*

Components in tables may not sum precisely to totals shown because of rounding.

CIP Competitive industrial performance
DEIE Developing and emerging industrial economies
FDI Foreign direct investment
GDP Gross domestic product
GHG Greenhouse gas
GVC Global value chain
ICT Information and communications technology
ILO International Labour Organization
IPR Intellectual property rights
ISIC International Standard Industrial Classification
ISID Inclusive and sustainable industrial development
ISO International Organization for Standardization
LDC Least developed countries
MDG Millennium Development Goal
MNE Multinational enterprises
MVA Manufacturing value added
OECD Organisation for Economic Co-operation and Development
PPP Purchasing power parity
R&D Research and development
SEZ Special economic zones
SME Small and medium-size enterprise
STI Science, technology and innovation
TFP Total factor productivity
UN United Nations
UNCTAD United Nations Conference on Trade and Development
UNDESA United Nations Department of Economic and Social Affairs
UNDP United Nations Development Programme
UNEP United Nations Environment Programme
UNIDO United Nations Industrial Development Organization

Glossary

Capital goods. Goods used in the production of other goods and services. (UNIDO 2013a)

Decoupling. Weakening or breaking the link between environmental effects and economic activity so that output increases with a less than commensurate increase (or with a decrease) in energy consumption (Von Weizsäcker 1989; Enevoldsen, Ryelund and Andersen 2007).

Deindustrialization. Long-term decline in manufacturing relative to other sectors. Typically measured in terms of a share of manufacturing employment in total employment. (UNIDO 2013a)

Elasticity. Percent change in one due to 1 percent change in another. For example, the growths of value added, employment and labour productivity as per unit increase in GDP per capita can be measured as percentage change in these variables due to 1 percentage point increase in GDP per capita. (UNIDO 2013a)

Energy. The ability to do work. In industry it commonly refers to the energy used to power manufacturing processes. This report measures energy in tonnes of oil equivalent to allow comparisons of energy from various sources. Primary energy sources include biomass-based fuels (trees, branches, crop residues), fossil fuels (coal, oil, natural gas) and renewable sources (sun, wind, water). Secondary energy sources are derived from other (usually primary) energy sources and have zero pollution at the point of use (electricity, for example). (UNIDO 2011)

Energy efficiency. The ratio of a system's energy inputs to its output. Since inputs and outputs can be measured in more than one way, energy efficiency has no single meaning. An engineer's definition will differ from an environmentalist's or an economist's—mainly reflecting differences in the level of aggregation. The energy-efficiency ratio is commonly called thermal or first-law efficiency, based on the first law of thermodynamics. In any closed energy conversion process, energy can be neither created nor destroyed; energy that goes in must come out or be accumulated in the system. But only a portion of the energy output will be in a useful form (for example, light) while the rest is waste, typically low-temperature heat. The thermal efficiency of a process is thus the ratio of useful energy outputs to total energy inputs. In engineering, energy efficiency is interpreted as conversion efficiency—the proportion of the energy input that is available as a "useful" output. For example, only 5–10 percent of the electrical energy fed to an incandescent light bulb is converted to useful light energy; the remaining 90–95 percent is lost to the environment as "waste" energy (low-temperature heat). In developed countries, the average efficiency of conversion of heat energy from fuel to electric power delivered to consumers is 33–35 percent (Ayres, Turton and Casten 2006), so if this electricity is converted to light energy using an incandescent bulb, the overall energy efficiency is just 3 percent. In economics, energy efficiency is the ratio of the value of output to the quantity or cost of energy inputs—the amount of economic activity produced from one unit of energy. (See also Energy intensity.) (UNIDO 2011)

Energy intensity. The amount of energy used to produce one unit of economic activity. It is the inverse of energy efficiency: less energy intensity means more energy efficiency. This report measures energy input in physical terms (tonnes of oil equivalent) and economic activity in monetary terms (sectoral and manufacturing value added). (UNIDO 2011)

Externalities. Costs or benefits that accrue to unrelated third parties. When it is a benefit reaped by third parties, it is called a positive externality. When it is a cost imposed on third parties, it is called a negative externality. Externality is a market failure that provides rational for industrial policy. Hausmann and Rodrik (2003, 2006) identify three main types of externalities that are particularly relevant for new activities to emerge: coordination externalities, as specific new industries or activities require simultaneous, large investments to become profitable; information externalities, as "discovery" of new activities requires an investment whose returns cannot be fully appropriated by the investor; and labour training externalities, as firms regard labour mobility as a disincentive to invest in on-the-job training, thus reducing technological spillovers. (UNIDO 2013a)

Global value chain. The value chain describes the full range of activities that firms and workers do to bring a product from its conception to its end use and beyond. This includes activities such as design, production, marketing, distribution and support to the final consumer. The activities that comprise a value chain can be contained within a single firm or divided among different firms. (Gereffi and Fernandez-Stark 2011)

Inclusive and sustainable industrial development. ISID has three elements: long-term (or sustained) industrialization as a driver for development; socially inclusive development offering equal opportunities and an equitable distribution of benefits (including all countries and all peoples, as well as the private sector, civil society organizations, multinational development institutions and all parts of the UN system); and environmental sustainability, which focuses on decoupling the prosperity generated by industrial activities from excessive natural resource use and negative environmental impacts. The Lima Declaration, adopted by UNIDO's Member States in December 2013, set the foundation for ISID. (UNIDO 2015d)

Incremental innovation. Incremental innovation concerns an existing product, service, process, organization or method whose performance has been significantly enhanced or upgraded. (OECD and World Bank n.d.)

Industrial policy. Any type of intervention or government policy that attempts to improve the business environment or to alter the structure of economic activity towards sectors, technologies or tasks that are expected to offer better prospects for economic growth or societal welfare than would occur in the absence of such intervention—that is, in the market equilibrium (Warwick 2013).

Informal economy. It is part of the economy that is operated outside the purview of government, thus not taxed and included in statistics. (UNIDO 2013a)

Innovation. An innovation is the implementation of a new or significantly improved product (good or service), or process, a new marketing method, or a new organizational method in business practices, workplace organization or external relations. (OECD and Eurostat 2005)

Intermediate goods. Goods used as inputs in the production of other goods and services. (UNIDO 2013a)

Labour intensity. Relative proportion of labour used in production. It is approximated in this report as the number of employment per value added. (UNIDO 2013a)

Manufacturing-related service/producer-related service. Service activities whose demands arise largely from manufacturing production. Wholesale, retail, transportation services for goods and business services (including, for example, renting services of machinery and equipment, research and development, and computer and related services) are considered major components of manufacturing-related services. (UNIDO 2013a)

Manufacturing value added. *See Value added.*

Marketing innovation. The implementation of new marketing methods involving significant changes in product design or packaging, product placement, product promotion or pricing. (OECD and Eurostat 2005)

Non-manufacturing industries. Industries that comprise mining and quarrying, construction and public utilities (electricity, gas and water). (UNIDO 2013a)

Organizational innovation. The implementation of new organizational methods in the firm's business practices, workplace organization or external relations. (OECD and Eurostat 2005)

Product innovation. The introduction of goods or services that are new or significantly improved with respect to their characteristics or intended uses. (OECD and Eurostat 2005)

Process innovation. The implementation of new or significantly improved production or delivery methods, including significant changes in techniques, equipment and/or software. (OECD and Eurostat 2005)

Purchasing power parity. A concept that determines the relative values of two currencies in terms of purchasing power. PPP-based GDP shows what goods and services produced in one country would cost if they were sold in the United States. Since non-tradable services of similar quality are priced lower in low-income countries than they are in the United States, their PPP-based GDPs usually become higher than their GDPs based on market exchange rates. (UNIDO 2013a)

Radical innovation. A radical or disruptive innovation is an innovation that has a significant impact on a market and on the economic activity of firms in that market. This concept focuses on the impact of innovations as opposed to their novelty. (OECD and World Bank n.d.)

Research and experimental development. R&D comprise creative work undertaken on a systematic basis in order to increase the stock of knowledge, including knowledge of man, culture and society, and the use of this stock of knowledge to devise new applications. The term R&D covers three activities: basic research, applied research and experimental development. (OECD 2002)

Resource efficiency. Resource efficiency is defined from a life cycle and value chain perspective. This means reducing the total environmental impact of the production and consumption of goods and services, from raw material extraction to final use and disposal. (UNEP n.d.)

Skill-biased technological change. Technological change that does not lead to proportional change in the demand for unskilled and skilled labour but results in greater demand for skilled labour. (UNIDO 2013a)

Structural change. Change in the long-term composition and distribution of economic activities. A normative perspective of structural change often emphasizes desirability in the direction of change. For example, Ocampo (2005), Ocampo and Vos (2008) and UNDESA (2006) define structural change as the ability of an economy to continually generate new dynamic activities characterized by higher productivity and increasing returns to scale. (UNIDO 2013a)

Sustainable Development Goal No. 9 Build resilient infrastructure, promote inclusive and sustainable industrialization and foster innovation. This goal promotes ISID via sharply raising industry's share of employment and GDP by 2030, integrating of small-scale industrial and other enterprises into value chains and markets, upgrading infrastructure and industries with greater resource-use efficiency, using clean and environmentally sound technologies and industrial processes, boosting scientific research, upgrading technological capabilities and encouraging innovation (UN 2015a).

Technological capabilities. The ability to exploit what modern technology can contribute to the economic development of the developing countries. (ILO 1991)

Technological change. Improvements in technology. Technological change involves a series of stages with multiple actors, relationships and feedback loops—from invention, as a new technology is created and prototyped, to innovation, as it becomes commercially viable (Freeman and Soete 1997; IEA 2008a). In decomposition analysis, if data on manufacturing processes were available at the lowest level of aggregation, the measure of technical change would be actual physical efficiency and the rest would be structural change (Jenne and Cattell 1983). Industrial energy intensity can be lowered by improving technology (technological change) and producing more goods that require less energy (structural change). (UNIDO 2011)

Technological levels of manufacturing industries. Manufacturing industries can be grouped into three technological categories—low tech, medium tech and high tech. They are based on research and development intensity relative to value added and production, following the technology classification of the Organisation for Economic Co-operation and Development (OECD 2005b).

Technology policy. Technology policy itself is measures implemented by the government to facilitate the development of technological capabilities and infrastructure for private and public firms. (Steinmueller 2010)

Total factor productivity. A variable that represents the amount of output not accounted for by the amount of factor inputs, such as labour and capital. (UNIDO 2013a)

Unit labour costs. Cost of labour per unit of output. It is calculated as the ratio of labour costs to real output. (UNIDO 2013a)

Value added. A measure of output net of intermediate consumption, which includes the value of materials and supplies used in production, fuels and electricity consumed, the cost of industrial services such as payments for contract and commission work and repair and maintenance, compensation of employees, operating surplus and consumption of fixed capital. Manufacturing valued added is the contribution of the entire manufacturing sector to GDP (manufacturing net output). (UNIDO 2013a)

Executive summary

The role of technology and innovation in inclusive and sustainable industrial development

Key messages

- Reaching advanced levels of inclusive and sustainable industrial development requires not only increasing incomes but also conscious efforts to sustain growth, promote social inclusiveness and move towards greener structural transformation—as well as managing the trade-offs between them.
- Industrialization, a major force in structural change, shifts resources from labour-intensive activities to more capital- and technology-intensive activities. It will remain crucial to the future growth of developing countries.
- Manufacturing's share of GDP has remained stable over the last 40 years.
- Technology and capital equipment are the main drivers of both manufacturing growth and aggregate growth in developed and developing countries, although in developing countries energy and natural resources use affects growth in medium- and low-tech industries.
- The choice of sector matters for economic growth and structural change since the technological opportunities between them vary significantly.
- Diversification into manufacturing can help to achieve rapid average growth rates, longer periods of growth and less volatility in growth—thus sustaining growth in the long run.
- Premature deindustrialization smothers economic development potential by limiting the application of technology to production and generating low productivity and informal services activities—while mature deindustrialization often leads to dynamic high-tech services.
- Technological capabilities are strengthened by investing in human capital, institutions, improving innovation systems and upgrading in industrial clusters and global value chains.
- Technological capabilities are expanded in developed countries through tinkering with the frontiers of science and technology and in developing countries by acquiring and adapting technologies created elsewhere.
- Promoting social inclusiveness in manufacturing requires matching the choice of technologies to a country's resource and skill endowment.
- Improving the environmental sustainability of industry may sometimes require adopting production technologies that are not economically viable, although the profitability of these technologies is increasing over time.
- High-tech industries produce an environmental bonus because they are less polluting than other industries.
- The recycling industry exhibits the win-win-win properties of sustaining growth, generating employment and equity and being environmentally friendly—but the trade-offs are considerable in combining these aims.
- Policy instruments for industrial development depend on the type of technology and innovation being targeted and the country's level of development, ranging from protecting property rights at one extreme to providing grants for machinery imports at the other.
- Pooling financial and research resources internationally in a global knowledge base can contribute much to building technological capabilities for inclusive and sustainable industrialization.

Under what conditions can technological change trigger structural change in developing countries and lead to long-term, socially inclusive and environmentally sustainable industrial development? That is the central question addressed in this *Industrial Development Report 2016*. The Lima Declaration, adopted by the Member States of the United Nations Industrial Development Organization in December 2013, set the foundation for a new vision of inclusive and sustainable industrial development (ISID). The ISID concept is part of the new Sustainable Development Goal 9 to build resilient infrastructure, promote inclusive and sustainable industrialization and foster innovation.

> For long-term structural change, manufacturing plays a key role

ISID has three elements, which are the framework for this report. The first is long-term, sustained industrialization as a driver for economic development. The second is socially inclusive industrial development and society, offering equal opportunities and an equitable distribution of benefits. And the third is environmental sustainability, decoupling the prosperity generated by industrial activities from excessive natural resource use and negative environmental impact. This three-dimensional structure feeds through to the policy recommendations for dealing with the many trade-offs that countries face in sustaining economic growth, promoting social inclusiveness and moving towards greener economic transformations.

For long-term structural change, manufacturing plays a key role. It creates many productive, formal jobs at an early stage of development. It also drives technological development and innovation to sustain productivity growth in manufacturing and other sectors.

And it has varying effects on employment, wages, technological upgrading and sustainability at different stages of development. The reason is that manufacturing changes economic structures, usually from labour-intensive activities to more capital- and technology-intensive activities. Each manufacturing subsector also changes products and production processes, with the increasing applications of capital and technology.

Premature deindustrialization can be a serious threat to growth in developing countries, smothering the growth potential of manufacturing when it sets in. The kind of informal service activities that emerge at this stage reduce rather than enhance growth. But when mature deindustrialization sets in at higher levels of per capita income, the kind of services that emerge—logistics, business services and information technology services—are much more dynamic and can take over and complement the growth-enhancing role of manufacturing.

Manufacturing and structural change

Manufacturing employment's share in total employment and the absolute number of manufacturing jobs

Box 1
Dimensions of innovation

Innovation may be characterized by type (product or process innovation), novelty (incremental or radical innovation), and source (technological or non-technological). There is also a social aspect (OECD and Eurostat 2005).

Non-technological innovations are generally associated with marketing and organizational innovations—although in practice, technological and non-technological innovations are highly interconnected. Technological innovations are usually associated with product and process innovation:

- Product innovation—goods or services that are new or significantly improved in their characteristics or intended uses.
- Process innovation—new or significantly improved production or delivery methods, including significant changes in techniques, equipment and software.
- Marketing innovation—new marketing methods involving significant changes in product design, packaging, placement, promotion or pricing.
- Organizational innovation—new organizational methods in a firm's business practices, workplace organization or external relations.

The distinction between radical and incremental innovation is that a radical innovation has a significant effect on a market and on the economic activity of firms in that market, whereas an incremental innovation concerns an existing product, service, process, organization or method whose performance is significantly enhanced or improved.

Innovation differs greatly by country, sector and period. The dominant form is incremental, particularly in developing countries. The rise in incremental innovation in low-wage countries has contributed to increased exports of high-quality and sophisticated manufactured goods (Puga and Trefler 2010). Some sectors are characterized by rapid change and radical innovations, others by smaller, incremental changes. In high-tech sectors, research and development (R&D) plays a central role in innovation, whereas other sectors rely to a greater degree on the adoption of existing knowledge and technology. Low- and medium-tech industries are often characterized by incremental innovation and by the adoption of foreign technology.

Developing and high-income countries display wide differences in the way manufacturing drives economic growth

are generally falling in high-income countries. On average, countries across all incomes now have a lower manufacturing share than before, and they reach their peak employment and value-added shares at a lower income than in previous decades (Rodrik 2015; Ghani and O'Connell 2014). Declining manufacturing in developed countries does not necessarily mean the same in developing countries—or a decline in the sector's importance to developed countries on value added, productivity and linkages to other sectors. In a similar way, low manufacturing shares in many developing countries (relative to past trends) might be attributed to country-specific conditions rather than to systematic and long-term reduction in manufacturing's potential contribution to the economy from a structural shift in supply and demand conditions of different sectors.

Wanted: shifting shares of low, medium and high tech

To illustrate the relationship between structural change and technological development—a key theme of this report—we look at structural change among manufacturing subsectors, grouped by technological category: low tech, medium tech, and high tech. The last 40 years have seen a relative shift in all three technological activities from developed to developing countries. In 2012, more than half of the world's value added in low- and medium-tech industries was from developing countries, and even in high-tech industries, developing countries accounted for nearly half on that measure (Figure 1).

How has the technology structure in manufacturing changed in developing countries over those 40 years? In 1972, the low-tech share in Africa was higher than in the other two regions, which had a similar technology structure by proportion of the three technology groups. In 2012, Africa increased the share of the high-tech group and reached a structure similar to that of Latin America and Asia in 1972. In the same period, Latin America went through very little change, with a slight decline in the share of the high-tech group, compensated by an increase in the medium-tech share. Asia experienced the most significant change in technology structure. Over the 40 years, its share of the high-tech group rose by 10 percentage points, at the expense of the low-tech share. Asia's economic success relative to other developing regions was thus accompanied not only by an increased manufacturing share in the economy but also by technological upgrading in manufacturing.

Figure 1
Shares of developing and developed regions in global value added of low-, medium-, and high-tech manufacturing industries, 1972 and 2012

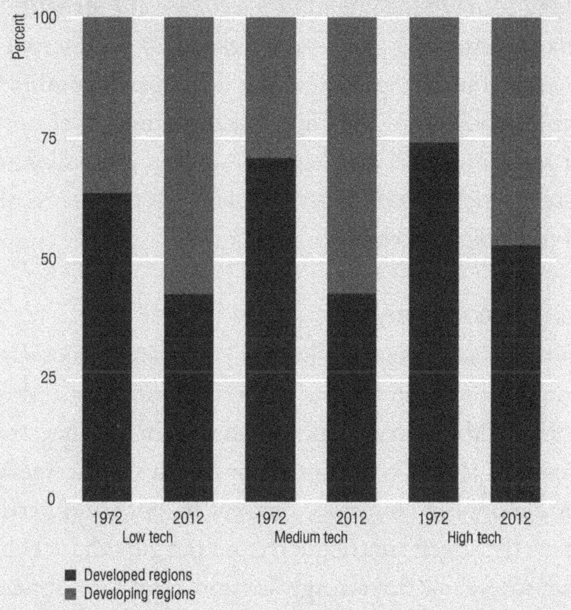

Note: Tech classifications based on Annex A2, income classification based on Annex A1, Table A1.2.
Source: UNIDO elaboration based on Lavopa and Szirmai (2015).

Big differences in the way manufacturing drives economic growth

Developing and high-income countries display wide differences in the way manufacturing drives economic growth. In developing countries, contributions to output growth derive mainly from capital investments, natural resources and energy; in high-income countries, they come from productivity. High-income countries seem to use labour- and resource-saving technology, which allows them to increase output without significantly increasing factor inputs.

> **Differences in total factor productivity growth rates between sectors are the decisive factors in structural change**

Consider the three groupings of manufacturing industries—typical low tech, medium tech, and high tech—to assess how their production characteristics affect overall growth and factor contributions along country income lines. Decomposing structural change into two parts—one related to productivity change (indicating technological change or total factor productivity—TFP) and one to changes in the use of inputs (capital and labour)—makes it possible to assess which part of structural change is a direct result of technological change.

Low-tech industries

In these industries, high-income countries had negative output growth in textiles and textile products and in leather and footwear, which is reflected in high negative shares of labour contribution or labour displacement. In developing countries, conversely, both industries grew: the largest contribution to output growth for both industries came from energy, less from capital investment and labour, whereas productivity growth made a positive contribution only to textiles. Overall, productivity made a lower contribution to growth of labour-intensive industries in developing than in high-income countries.

Medium-tech industries

These industries also show a difference between the two country income groups. Productivity was the largest source of growth for high-income countries in rubber and plastics and non-metallic mineral industries, but for developing countries in those industries—especially non-metallic minerals—the main contribution came from natural resources and energy, with productivity growth providing only a small contribution.

When countries industrialize further and move into this grouping, the pollution intensity of the manufacturing sector (here measured as carbon dioxide (CO_2) emissions per unit of value added) tends to rise. That does not, however, mean that the growth of medium-tech, resource-based industries must always be driven by heavy increases in energy and natural resource inputs, as evidenced by the relatively low contributions of energy and natural resources to the growth of these activities in high-income countries.

High-tech industries

High-income countries have an advantage in high-tech industries and clearly have the potential to achieve faster growth in those industries than in low- or medium-tech industries. That advantage drives structural change within manufacturing and shifts resources towards high-tech industries at higher income levels. Productivity is the dominant contributor to the growth of high-tech industries, and their growth does not depend significantly on an increase in the use of energy and natural resources.

In developing countries, productivity accounts for a significant share of the growth of high-tech industries. But other factors, such as energy and capital investment, made a non-trivial contribution, too. So, although the importance of productivity for the growth of high-tech industries is common to developing and high-income countries, developing countries differ in that increased use of energy and labour accompanies growth—hence, the expansion of these activities is more inclusive in job terms, but it is less sustainable.

The main reason technological change is an important determinant of structural change is that its rate differs greatly between economic sectors, providing a stimulus to economic growth that favours some sectors over others. For structural change, the differences between sectors matter most, and those differences can be substantial both within a sector (between countries) and between sectors. Differences in TFP growth rates between sectors (within a country) are the decisive factors in structural change. High values of structural change are mostly achieved by a large contribution of technological change.

Sustaining economic growth

In the long run, the ability of a country to use existing and to innovate new technology determines its economic performance through a process of structural change. However, because developing the capabilities to use and assimilate technology is very hard when they are not present, the convergence of living standards between countries has generally been very slow or even absent. Only a few countries have moved

> **Manufacturing can sustain growth by lengthening its episodes and reducing its volatility**

from relative poverty to relative development. Rich developed countries have high levels of technological sophistication and account for the large majority of investment in science and technology (primarily R&D). Poor countries have much lower technological capabilities and invest much less in R&D.

The concept—in theory, open to all

One of the three dimensions of sustainability is the ability of an economy to sustain growth over longer periods without serious interruption due perhaps to economic crises or slumps. The longer the duration of positive growth rates and the higher the rate of growth during positive growth episodes, the more likely a low- or middle-income country is to achieve sustained catch-up.

Sustained growth has three characteristics:

- *Average rates of gross domestic product (GDP) growth per capita.* Is growth rapid enough to achieve substantial increases in welfare in the foreseeable future? And is it faster than in advanced economies so that a country can catch up? Since 1950, catch-up has required growth of more than 5 percent a year, sustained over two or more decades (Szirmai 2012). Such success is rare.
- *Duration of growth episodes.* The ability to sustain growth over longer uninterrupted periods is important, but growth often is not steady, and attempting to explain differences in average growth may be misleading. More promising is finding out what initiates or halts episodes of growth, or what influences the characteristics of growth episodes (Pritchett 1998).
- *Volatility of growth.* The lower the volatility, the more sustained the growth pattern. Volatility is often much higher in low- and middle-income countries than in high-income countries, and highest in countries that remain in the "development trap."

How do countries move up the development ladder? The answer lies not in the creation of new knowledge but in the adoption and adaptation of knowledge from abroad. Poor countries tend to have high potential for rapid growth, represented by the reservoir of global technological knowledge that is available for them to tap into. The evidence suggests, however, that the tendency to realize this potential varies greatly in the group of poorer countries. In the large group of countries below, say, $15,000 GDP per capita, growth rates show a large variance. The regression line has a negative slope, indicating convergence (that poorer countries grow more rapidly), but this relationship is very weak. The regression line also divides the group of poor countries into two parts: one growing slowly and tending to fall behind or stagnate, and one showing some tendency for catch-up.

Not only is the difference in average growth rate among developing countries much higher than among developed countries, but also the volatility of a country's growth rate is higher in developing than developed countries. Thus the growth experiences of developing countries vary on the rate, duration and volatility of growth more than those of developed countries. But among developing countries, those catching up seem to have the common characteristics of higher growth rates, longer episodes of growth and lower volatility.

Interestingly, manufacturing can sustain growth by lengthening its episodes and reducing its volatility. The larger the share of the manufacturing sector at the start of a growth episode, the longer growth continues. The share of manufacturing within the modern sector yields similar results, and it has significant positive effects on duration. In line with the effects on duration, the chances of ending a growth spell are substantially reduced as the share of manufacturing at the start of the spell increases. Obviously, the longer an episode lasts, the greater the chances of it finally ending. But clearly the risk is much lower in every year in which the share of manufacturing at the start of the episode is higher.

Technological capabilities for sustained growth

One of the major sources of economic growth and catch-up in developing economies is imitating and adapting technologies streaming in from the industrially advanced economies. But that requires technological capabilities, which are mainly related to the education of the population and the allocation of human capital

> **"The scope for countries to upgrade their technological capabilities depends on national innovation systems**

and other resources to undertake R&D. The relative importance of each of these elements depends on a country's development. At early stages of development, technological gaps create the potential for fast structural change through global technological knowledge, but the extent to which such change will be realized depends on the absorptive capacities of countries, sectors and firms. Among the most important determinants of absorptive capacity are sustained investments in human capital. Strong basic and secondary education and specialized human capital are fundamental to absorb new technologies. Basic education and new skills are needed to use new technologies, and a more educated population tends to adopt new technologies faster.

But basic literacy is not enough. Certain technology-specific skills are typically needed to absorb new technologies. In some cases, skills can be provided by an improved basic education curriculum. In other cases, they have to be provided through specialized training at vocational centres. At middle ranges of development, the creation of new indigenous knowledge becomes very important. A strong tertiary education system in science and engineering and larger formal R&D efforts play a key role at this stage. In fact, the transition towards more technology-intensive manufacturing and service activities depends on a "hi-tech infrastructure," which includes—among other elements—universities and polytechnics capable of generating skilled technicians, engineers and scientists.

While learning and technological absorption take place at the firm level, the success or failure of individual firms occurs within a system. Thus, the scope for countries to upgrade their technological capabilities also depends on the functioning of national innovation systems. In this perspective, learning and innovation involve complex interactions between firms and their environment—not just the firms' network of customers and suppliers but also the technological infrastructure, institutional and organizational framework, and knowledge-creating and diffusing institutions. As innovation systems improve, countries tap into international sources of technological knowledge, which is not limited to a few modern firms but circulates rapidly among different firms and actors.

Technological upgrading needs a broad dissemination of knowledge throughout the whole economy. Such dissemination requires strong public policies to diffuse new technologies with an institutional infrastructure that includes, among other things, extension services, industrial clusters, metrology standards, productivity standards, technical information services, and quality control institutions. Upgrading technological capabilities also requires a technological commercialization infrastructure that can put into practice the new knowledge created, for example, in government research labs and universities. This infrastructure includes adequate intellectual property rights protection systems, technology-transfer offices at universities and research institutes, science and industrial parks, business incubators, and early-stage technology finance and venture capital.

Promoting social inclusiveness

During structural transformation, societies become more technologically complex and economically productive, improving incomes, wealth and subjective well-being. Demographic shifts, facilitated by rising incomes and the uptake of modern technologies, help improve outcomes in health, education and urbanization. Manufacturing is fundamental to this process. It provides productive employment in the early stages and is a catalyst for technological innovation. Over time a country's manufacturing typically evolves from being labour-intensive to being more capital- and technology-intensive, creating demand for more skilled labour. And a better skilled workforce provides incentives for technological innovation, which can enable a virtuous circle of education, innovation and productivity growth. But not everyone can access the opportunities that arise. Only with domestic capabilities and technologies better suited to match these conditions can socially inclusive industrial development distribute the fruits of economic growth more evenly.

Creating employment, distributing income

The channels for technological change to affect social inclusiveness through the transformation of the economic structure can be broadly divided into two

> **Generating direct and indirect jobs in manufacturing and manufacturing-related services brings more people into the growth process**

major areas: employment creation and income distribution (Figure 2). On the first, the relevant question is whether new technologies will lead to the creation or destruction of jobs. On the second, the interest is in whether innovations will improve or impair the distribution of incomes within society.

With the right capabilities, technology-driven structural change expands the modern, formal industrial sector and industry-related services, absorbing labour from the pool of underemployed workers in agriculture or informal services. Manufacturing plays a key role in generating and diffusing new technologies. Moreover, backward and forward linkages and spillover effects from manufacturing promote regional and country development, creating feedback loops of accumulating human capital and improving institutions. So generating direct and indirect jobs in manufacturing and manufacturing-related services not only brings more people into the growth process—it also increases average productivity, wages and family incomes. Higher family incomes, in turn, help reduce poverty.

This process can temporarily lead to income inequality. An example is the invention of the internal combustion engine, which caused substantial job losses in the horse-drawn carriage industry but eventually resulted in substantial new employment in the automobile industry. Technological innovation therefore has not only static effects in the once-off reallocation of labour but also dynamic effects, such as facilitating the growth of productivity and output in modern urban industries.

Figure 2
Conceptual framework: Technological change for inclusive structural transformation

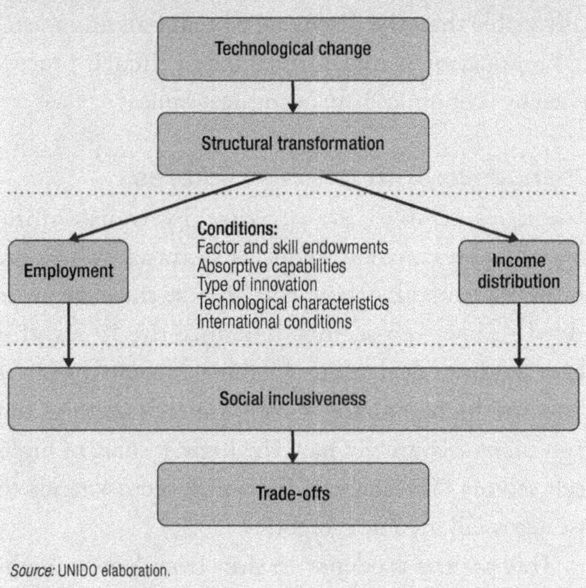

Source: UNIDO elaboration.

The expansion of the modern formal sector gives the government a tax base and more revenue in the public sector that might enable the government to improve economic, administrative and political institutions and widen social protection measures. It also helps more women participate in the labour market. With better earning opportunities, parents want their children to receive more education. And with a quantity–quality trade-off for the number of children, the expanding modern sector may reduce fertility, further allowing a shift of resources towards better education of children and enhancing human capital formation and labour productivity. Thus, a growing modern sector is also a major determinant of fertility and the demographic transition.

From this perspective, even if new technologies hurt income distribution and employment creation, it is often temporary. Persistent rising inequality ultimately reflects institutional and policy failures that perpetuate technological gaps between sectors, regions, and countries or that fail to provide adequate social buffers in times of rapid change.

What, then, are the conditions for getting technology to drive social inclusiveness? Regulations and incentives help steer the direction of technological change, and more can be done to guide innovation to complement rather than replace humans. It may also be necessary to support technological innovations with organizational change, helping to flatten hierarchies and decentralize management responsibilities. Countries should try to use technologies that are better suited for their characteristics, reflecting their factors, skills and endowments. Innovation and industrial policies are therefore fundamental in shifting the innovation path towards a more inclusive trajectory —determining the structure of prices, factor costs,

> **Countries tend to industrialize by transitioning towards more emissions-reducing sectors**

infrastructure and the availability of alternative technologies (and the knowledge that firms have about these technologies).

Moving towards greener structural transformation

Technological change for environmental sustainability operates mainly through two channels—the production process and the production structure—involving environmental, economic and social trade-offs (Figure 3).

Changes in production processes

The changes in production processes happen through more efficient use of natural resources, such as non-renewable energy and materials, helping firms to be more cost competitive. Under ideal conditions, costs of renewable inputs are comparable to fossil fuel energy. Pollution abatement technologies that reduce any incurring pollution are cheap, and production processes are re-engineered to minimize resource use. Waste, normally considered a bad outcome of the production process, becomes a key input to be reused directly as inputs through materials recovery or waste-to-energy technologies. But such transformations are possible only if the environmental technologies exist and the conditions, including the relative prices faced by producers, enable environmentally positive change in production. Some transformations such as a global transition to the use of renewable energy or a drastic reduction of costs for pollution abatement technologies are still far from materializing, but evidence shows that firms tend to use more efficient energy inputs even if not necessarily driven by policies.

An increase in energy prices is an important vehicle for environmentally friendly innovation in the medium to long term, as rising energy costs stimulate firms to invest in energy-efficient technologies. Firms tend to maximize output by minimizing input costs. The more innovative sectors, such as manufacturing, are more exposed to profit-driven measures. But a short-term increase of energy prices generates a reduction of real GDP, especially for energy-importing countries.

Efficiency also pushes firms to invest in technologies that recycle waste or materials. There is increasing awareness of technologies that reuse materials as inputs in the production process. Sharp price increases in primary materials in the last decade indicate that resources are scarce and need to be managed more sustainably. So, recycling becomes more economically viable than the discharge of materials and waste, and production is transformed into a circular process whereby economic "bads" acquire a value.

Figure 3
Conceptual framework: Technological change for environmental sustainability

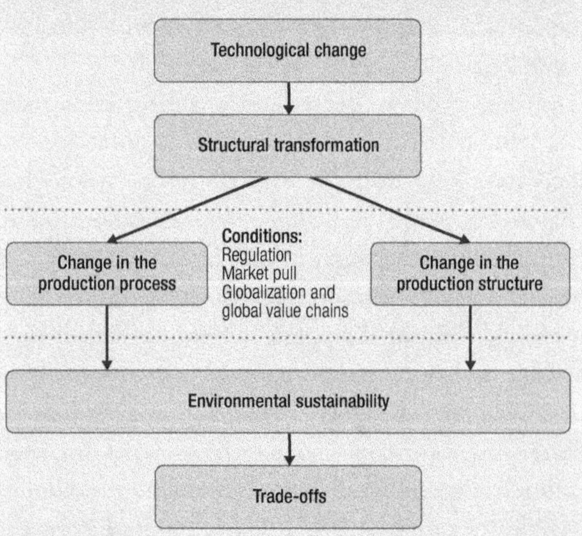

Source: UNIDO elaboration.

Changes in production structures

Countries tend to industrialize by transitioning towards more emissions-reducing sectors. Low-income countries generally show the highest share of value added in low-tech sectors, but since the 1970s, this share has been decreasing. Medium-income countries show the highest share of medium-tech sectors, and high-income countries have the highest share of high-tech sectors. The share of high-tech sectors tends to rise across all income categories.

This natural tendency to shift from low- to high-tech sectors comes with a natural tendency to pollute.

> **The greening of global value chains can create opportunities for collaborative approaches to eco-innovation**

The lowest environmental productivity (expressed as the value added/pollution ratio) is associated with medium-tech sectors. The medium-tech sector also shows the highest pollution intensity for other pollutants beyond carbon dioxide emissions, such as particulates, sulphur dioxide (SO_2) and nitrogen dioxide (NO_2) though with lower abatement costs than other sectors. Low- and high-tech sectors have higher environmental productivity—in other words, they generate fewer emissions when producing $1 of value added. Sectoral specialization towards high-tech sectors reduces emissions intensity. In short, a natural economic tendency contributes to ISID.

But environmental protection improvements deriving from the low- to high-tech transition may not be sufficient to decouple economic growth from pollution. Countries need to enforce actions to curtail environmental harm, even if they are not strictly related to the production process (like using environment-friendly pollution abatement technologies). Yet this non-profit-driven technological change is often expensive.

Facilitating the adoption of environmentally friendly technologies

Market conditions and the way markets are organized play a role in driving—or deterring—eco-innovation. The demand for new products and the progressive incorporation of environmental features in existing products have driven the adoption and diffusion of eco-innovations. Market demand has also been shaped by developments in the policy agenda that define what consumers expect from the environmental impact of products and services. Rising prices of, for example, metal products have created incentives to reuse metal elements in buildings. Firms may be interested in polluting reduction actions simply because they are profitable, but market externalities may prevent them from exploiting market opportunities. In those cases, policy-makers need to correct biases to create the right market environment.

Different types of regulatory approaches may trigger different types of innovations. While regulatory standards may trigger pollution abatement solutions, environmental management systems or integrated regulatory systems can incentivize cleaner and more resource-efficient technologies. And for resource-efficient eco-innovations and cleaner technologies, both regulatory pressure and cost savings seem to be pivotal.

The innovation effects of regulations can also vary according to the environmental area targeted. Whereas standards may set minimums for recycled or recyclable content in products, packaging and other eco-design considerations, economic instruments tackle market failures such as externalities of environmental impacts linked to resource use.

The greening of global value chains can create opportunities for collaborative approaches to eco-innovation that permeate and benefit all actors involved. More companies are committed to stricter and more stringent ways to identify material sources and to certification schemes that ensure the sustainable supply of different materials. Regional and national support systems that provide access to specific knowledge and that help companies (especially smaller ones) in introducing, adopting, or even developing new technologies may be particularly important.

For global pollutants in a post-Kyoto world, the main problem is to reach coordinated agreement for cutting emissions globally. Even mild emissions-reduction agreements such as the Kyoto Protocol represent a cost for all signatory countries. For Europe, the cost is estimated to be 0.31–1.50 percent of GDP, and for the United States 0.42–1.96 percent. Even flexible efficient mechanisms, such as the emission permit market, do not completely eliminate costs. That market allows sellers of carbon credits to gain money from the sale of those permits and purchasing countries to abate emissions by purchasing emission allowances that minimize total expenditures, reducing costs for Europe to 0.13–0.81 percent and for the United States to 0.24–0.91 percent.

Self-enforcing international environmental agreements can sustain a large number of signatories but only when the difference in net benefits between the non-cooperative and fully cooperative outcomes is

> **“ If productivity growth goes hand in hand with accelerated growth of output, the net effects on employment can be positive**

very small. That happens when the benefit from additional expanded participation is marginal. Fairness is another issue that counts overwhelmingly in negotiations. The reluctance of poor countries to join international agreements is prompted by the historical responsibility of rich regions in generating atmospheric carbon concentration. Likewise, rich countries claim that emissions-stabilizing policies will be effective only when developing countries contribute fully to reducing emissions. Rich countries are reluctant to sign heavy agreements, aware that developing countries will generate future emissions.

Designing and implementing ISID policies

To support a country's competitiveness, technology and industrial policies for innovation need to be complemented by infrastructure policies, industry representation, and business-enabling trade and investment. These policies are prerequisites for integrating into GVCs, but they should be complemented with a more radical macroeconomic approach and strategic investment policies. Complementary policies should address possible trade-offs and ensure a balance between environmental and social objectives.

Managing trade-offs and seeking complementarities

There are possible complementarities and possible trade-offs between sustained growth and inclusive development. One important trade-off is that the kind of productivity growth associated with rapid upgrading tends to reduce the demand for labour (Massa 2015). But this trade-off is not inevitable because, at lower levels of per capita income, manufacturing tends to be more labour intensive. And if productivity growth goes hand in hand with accelerated growth of output, the net effects on employment can be positive. So, if structural change and industrialization promote rapid growth in the whole economy due to linkages and spillovers, this can increase total employment and labour absorption. In poverty reduction, synergies between sustained growth and inclusive development are most prominent.

Trade-offs between sustained growth and income inequality can be very pronounced. In almost all countries experiencing sustained growth and catch-up, there have been increases in inequality as measured by the Gini index. This has to do with the balance between the supply and demand for skilled labour. Where technological change is skill-biased and the labour supply fails to keep up with the demand for skilled labour, inequality will tend to increase. This is not an inevitable outcome, but it does seem to characterize growth experiences in the past decades.

The final trade-off is between sustained growth and environmental sustainability. Here the record so far has been disappointing, and the negative environmental impacts of growth on CO_2 emissions and global warming have been larger than the positive impacts of technological advance.

Clusters of policies

Policy-makers thus have to weigh economic pros and environmental cons, social pros and environmental cons, and environmental pros and economic cons. In order to support a country's competitiveness, technology policies need to be complemented by policies focusing on the macroeconomic environment, business-enabling trade and investment, and industry institutionalization as well as infrastructure. These policies are prerequisites for integrating into GVCs but should be complemented with a more radical macroeconomic approach and strategic investment policies. Complementary policies are also needed to address possible trade-offs and ensure an environmental and social equilibrium.

Technology policies

Technology policies vary by an economy's development stage: early, middle and late. Each stage is characterized by some regularity in factors, such as the complexity of market structures, technological content, productivity and degrees of specialization and qualification of the labour force. In each stage, there is a choice between general horizontal measures available to all firms and selective vertical ones applied

"A sound policy mix of innovation and competitiveness policies is crucial

selectively to priority targets, whether subsectors or specific firms. In addition, there are market-based interventions and public inputs. The former affect prices and taxes and thus operate through pricing links. The latter reflect the provision of goods or services, which firms themselves would not supply adequately either because they cannot be marketed or because significant external benefits are involved.

Industrial policies

Industrial policies for innovation are a broad concept for combining technological and non-technological policies for different kinds of innovations at different stages of development. One crucial element determining the emergence, development and expansion of innovation activities is government intervention. Governments in developed and developing countries are increasingly making innovation a key issue, recognizing its potential to promote economic growth and address social and environmental challenges.

The main argument for government support is that a market economy cannot generate the optimal levels of investment in innovation by itself because of market failures and information asymmetries that lead to serious funding gaps. These market failures inhibit private firms from investing the optimal amount of resource (in fact they do not invest enough) in innovation activities, thus depriving the economy of one of the key levers of sustained growth. To counter this, governments aim to restore optimality by providing different forms of support to firms' investment in innovation, often through (sometimes overlapping) policy instruments.

Competitiveness policies

The innovation toolbox has to be extended to competitiveness policies in order to achieve structural transformation. A sound policy mix of innovation and competitiveness policies is crucial; the orthodox competitiveness approach is too timid.

GVC lead firms might require their local suppliers to adopt international standards, if they are skilled and fully competent or when the product is a commodity. Lead firms can also require them to adjust to specific technical and quality standards and to take full responsibility for the process technology. As lead firms do not become directly involved in the learning process but impose pressure on their suppliers for innovating and keeping abreast of technological advancements, they can be seen rather as a crucial stimulus for inducing learning and innovation but not as participants in the process. Nor do lead firms always enrich local firms with knowledge transfer and support upgrading processes. So, it is crucial to understand the structure of the value chains, the processes of structural change and the power asymmetries between firms that determine how entry barriers are created and how gains and risks are distributed.

Complementary policies

Technological change can lead to enormous advantages for economy and society, but it can also result in awkward trade-offs, often in manufacturing and in three main dimensions: economic vs. social, social vs. environmental and environmental vs. economic. Understanding these trade-offs is a precondition for developing the right complementary policies. To achieve gains on all three dimensions, integrative policy approaches are needed, which consider the full range of positive and negative consequences of innovation and promote interactions between all actors and sectors of the economy.

Another important key is to provide incentives to innovate and diffuse technologies. National policies have failed to achieve this objective so far because governments have been unable to develop integrative approaches to the full range of consequences of technological change, partly because of knowledge and implementation gaps.

There is no single, correct recipe; nor can all governments privatize, stabilize and liberalize in similar ways. Industrial policy-makers, especially in developing countries, might gradually shift their attention from investigating and imitating international best practices to identifying and reproducing *national* success stories. This approach underlines the need for sound measuring, monitoring and evaluation, especially in the context of serious budget constraints,

> **Industrial policy-makers might gradually shift to identifying and reproducing national success stories**

since it is essential to know whether a policy intervention is effective (or not) and whether the benefits outweigh the associated public costs.

International cooperation can help in all this. Technology and innovation policy-making is usually conducted nationally. As suggested by the subsidiarity principle, interventions should be accomplished where results are expected to be best. International collaboration is needed with trans-border and global problems driving collaboration in this area. Globalized technology (and innovation in general), the rise of emerging and developing countries as champions of globalization, and the growing role of individuals, small firms and open modes are further reasons for the need of international technology and innovation policy cooperation. The OECD emphasizes the need for effective international cooperation and sharing of burdens and benefits to protect the global commons and the world's public goods (including technology and innovation). This implies not only pooling financial resources and sharing a large research infrastructure but also improving the global knowledge base.

Trends in manufacturing value added, manufactured exports and industrial competitiveness

Key messages

- Global manufacturing value added (MVA) reached an all-time high of $9,228 billion in 2014. By 2014, the MVA of developing and emerging industrial economies (DEIEs) increased 2.4 times from 2000, while their GDP doubled.
- World export growth rates averaged 7.7 percent over 2005–2013, and in 2013 world trade reached a peak of more than $18 trillion, with 84.0 percent comprising manufacturing products.
- Manufacturing exports by industrialized countries expanded by an annual average of 4.3 percent over 2005–2013, reaching $11,998 billion in 2013. In the same period, DEIEs expanded their manufactured exports by an average 11.5 percent, to peak at $6,327 billion, 2.4 times more than in 2005.
- Around 58 percent of the world's manufactured exports consists of medium- and high-tech products, such as chemical machinery and equipment, communication equipment, and motor vehicles.
- On UNIDO's industrial competitiveness index, most industrialized countries lost ground in the last three years. Among the five most competitive are four high-income countries (Germany, Japan, the Republic of Korea and the United States), along with China ranking fifth. The four are among the world's most industrialized countries and, with China, account for 59 percent of world MVA.

Over the last few decades, global manufacturing has shifted from West to East and from North to South. Since the beginning of the century, rapid growth in MVA has been a major source of poverty reduction in many DEIEs through employment creation and income generation. Those countries still have considerable capacity for manufacturing growth and technological progress in the coming decades.

Manufacturing value added—recovering, but not to precrisis levels

World MVA climbed strongly until the 2008–2009 global financial crisis. Industrialized countries contributed the highest share of world MVA but, along with DEIEs, experienced a slowdown in MVA growth. Since 2010, MVA has recovered in both groups but has so far not reached the pre-crisis level within the industrialized country group (Figure 4).

Global MVA reached an all-time high of $9,228 billion (at 2005 constant prices) in 2014. The MVA share of industrialized countries in gross domestic product (GDP) fell from 15.4 percent in 1990 to 14.5 percent in 2014; in DEIEs it increased from 16.2 percent in 1990 to 20.5 percent in 2014. The share of MVA in world GDP increased from 15.6 percent to 16.2 percent over the period. Since 1990, MVA

"Global manufacturing value added reached an all-time high of $9,228 billion in 2014

Figure 4

World manufacturing value added, by country group and worldwide, 1990–2014

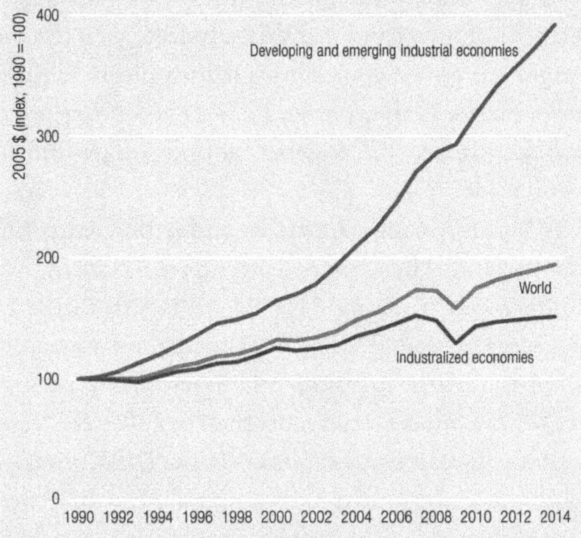

Note: Development level classification based on Annex B1, Table B1.2.
Source: UNIDO elaboration based on Manufacturing Value Added Database (UNIDO 2015e).

growth has remained consistently higher in DEIEs. By 2014, the MVA of DEIEs had expanded almost four times compared with 1990. Higher MVA growth has led to sustained economic growth in many developing countries.

Manufacturing remains a key driving force of overall economic growth of DEIEs. From 1990 to 2014, global MVA doubled from $4,753 billion to $9,228 billion at 2005 constant prices (Table 1). In DEIEs since 1992, MVA growth has stayed consistently higher than GDP growth (aggregate economic output). By 2014, the MVA of DEIEs had increased 2.4 times from 2000 at constant 2005 prices, while GDP doubled; industrialized countries saw their MVA increase overall by only 51.3 percent.

DEIEs as a whole improved their share in total MVA, but performance varied widely. Among the top five, China's share in world MVA increased by 6.5 times over 1990–2014. China's manufacturing industry has become the largest sector in the country and accounted in 2012 for more than 30 percent of GDP and more than 18 percent of global MVA,

Table 1

Manufacturing value added in developing and emerging industrial economies by development group and region, 1990, 2000 and 2014

	Manufacturing value added (billions, constant $ 2005)			Percentage of manufacturing value added		
	1990	2000	2014	1990	2000	2014
World	4,753	6,295	9,228	100	100	100
Industrialized countries	3,907	4,902	5,914	82	78	64
Developing and emerging industrial economies	846	1,393	3,314	18	22	36
By development group						
Emerging industrial countries	708	1,222	2,994	84	88	90
Least developed countries	20	22	54	2	2	2
Other developing countries	118	148	266	14	11	8
By region						
Africa	79	92	144	9	7	4
Asia and Pacific	315	746	2,362	37	54	71
Europe	151	164	300	18	12	9
Latin America	301	391	508	36	28	15

Note: Regional and development level classification based on Annex B1, Tables B1.1 and B1.2.
Source: UNIDO elaboration based on Manufacturing Value Added Database (UNIDO 2015e).

> **In 2013 world trade reached a peak of more than $18 trillion, with 84.0 percent comprising manufacturing products**

second only to the United States. Although China—and India—improved their group share, the other three of the five faltered, particularly Brazil.

Manufactured exports—84 percent of world trade

An increasingly export-oriented manufacturing sector, accompanied by a rising share of manufacturing in total exports, is part of a normal pattern of structural change in the growth process of DEIEs. Following this pattern, developing countries today have increased their presence in the export of manufactured goods. More developing countries are now benefiting from integration into the global economy through manufactured export growth and diversification. In most of these instances, export promotion has played a critical role in long-term growth by supporting a virtuous circle of investment, innovation and poverty reduction.

It is widely recognized that the benefits of exporting manufactured goods are greater than those from exporting primary commodities, largely due to the higher value added. Successful DEIEs have pursued export-led economic growth policies, diversifying from primary commodities to manufactured goods. As with their industrialized peers, the success of these countries stems from concentrating on manufactured exports.

World export growth rates averaged 7.7 percent over 2005–2013, and in 2013 world trade reached a peak of more than $18 trillion, with 84.0 percent comprising manufacturing products (Table 2). Over the period, world output expanded at an average 2.3 percent a year, though many countries saw a decline during the crisis. Global manufacturing trade recovered fully after a sharp decline during 2007–2009, largely due to the fast-expanding DEIEs. Indeed, their relative weight has grown enormously, mainly due to China's meteoric rise as an exporter. Exports of primary products surged but still account for only 1.6 percent of world trade.

Manufacturing exports by industrialized countries expanded by an annual average 4.3 percent over 2005–2013, reaching $11,998 billion in 2013. In the same period, DEIEs expanded their manufactured exports by an average 11.5 percent, to peak at $6,327 billion, 2.4 times more than in 2005. The three largest manufacturing exporters in the DEIE group—China, Mexico and India—accounted for 62.1 percent of the total of the country group in 2013, up from 55.3 percent in 2000, indicating the rapid growth of larger economies and the increasing gap with smaller economies.

The fast-growing share of DEIEs in world manufacturing exports reflects their dynamism. The group accounted for 6.1 percent of world manufacturing trade in 1990, 17.6 percent in 2000 and 34.5 percent in 2013 (Figure 5). The emerging industrial economies contributed most to the DEIE growth path by increasing their share in global manufactured exports to 15.2 percent and 31.7 percent in 2000 and 2013, respectively, from 5.6 percent in 1990. It is expected that the role of DEIEs as exporters will increase

Table 2
World exports by product category, 2005–2013

Category	Exports (billions, current $)									Average growth rate 2005–2013 (percent)
	2005	2006	2007	2008	2009	2010	2011	2012	2013	
Manufacturing	8,130	9,367	10,772	12,050	9,421	11,409	13,422	13,363	13,866	6.9
Primary	1,146	1,411	1,543	2,197	1,422	1,939	2,511	2,442	2,620	10.9
Other	102	137	163	193	141	185	224	214	196	8.5
Total trade	9,378	10,915	12,478	14,440	10,984	13,533	16,157	16,018	16,682	7.5

Note: Product category classification based on ISIC Rev. 3, ITC (2015).
Source: UNIDO elaboration based on United Nations Comtrade database (UNSD 2015a).

"Around 58 percent of the world's manufactured exports consists of medium- and high-tech products

Figure 5
Share in world manufactured exports by country group, 1990–2013

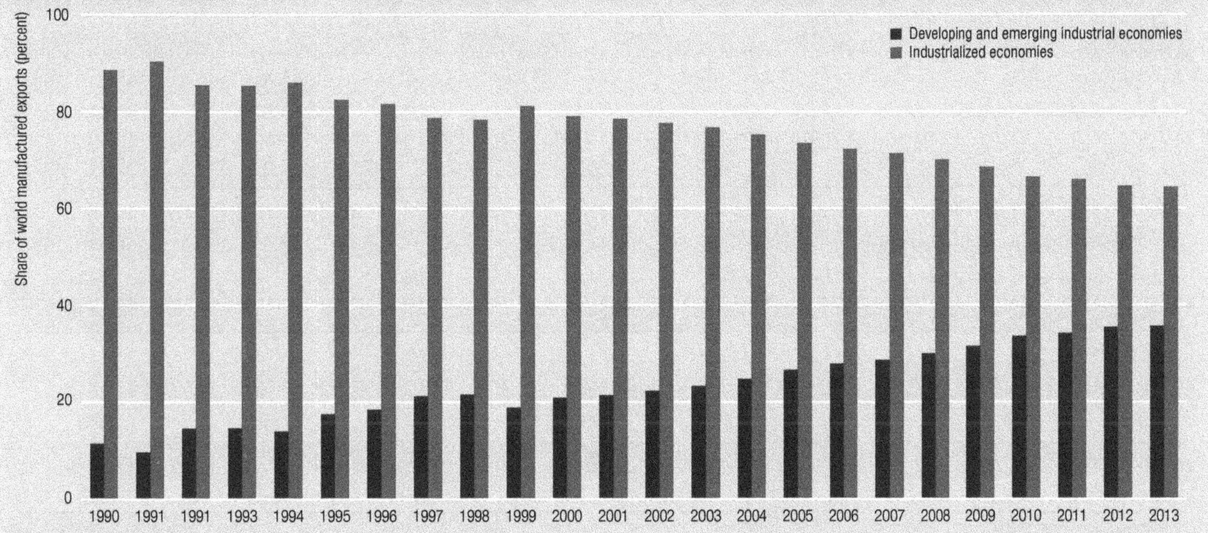

Note: Development level classification based on Annex B1, Table B1.2.
Source: UNIDO elaboration based on United Nations Comtrade database (UNSD 2015a).

significantly over the next years, reflecting their high growth rate and the development of the middle class. In addition, their dependence on developed-country markets is expected to decline as they move towards more advanced manufacturing sectors.

Led by China, Asia and the Pacific recorded a new peak of $7,145 billion in manufacturing exports in 2013, with an average growth of 11.6 percent a year over 2009–2013 (Table 3). Lower prices with the high competitiveness of China's market caused many manufacturing firms to move production there from more expensive, industrialized countries.

Europe as a whole contributed to a higher share in global manufacturing exports, though its pace of recovery was more moderate, with average growth of 7.0 percent a year over 2009–2013. Manufacturing exports in Latin America grew at a high 11.1 percent a year during the period, but the region failed to maintain its share of world manufacturing exports, contributing a low of 5.0 percent in 2013.

Africa followed a similar pattern to Latin America but with less strong growth of 10.4 percent, taking its share to a low of 1.4 percent in 2013. The region concentrates on resource-based manufacturing exports, which are the key factor in the overall growth as product prices and demand from industrializing countries have increased. High-tech products account for only 3.8 percent of manufacturing exports.

Despite some signs of progress, the least developed countries (LDCs) remain highly vulnerable to geopolitical tensions and political instabilities. Lack of proper infrastructure to support manufacturing adds to the problem. In 2013, LDCs accounted for 0.2 percent of world manufacturing exports. The group traditionally concentrated on low-tech manufactured products, but in the past few years that share has dropped dramatically due to lack of support in industry and the struggle of some countries with war. LDCs' manufacturing exports slumped by an average 19.3 percent a year.

Around 58 percent of the world's manufactured exports consists of medium- and high-tech products, such as chemical machinery and equipment, communication equipment and motor vehicles. The high-tech sector reached its peak, 25 percent, in 2000, and fell to 20 percent in 2013. This could be due to the high investment risk in the sector, which can hold markets back. While the export share of low- and medium-tech

> **Countries can become more industrially competitive if they develop their technological capabilities, expand their production capacity and invest in their infrastructure**

Table 3
World manufacturing exports by development group, region and income, selected years, 1995–2013 (billions, current $)

	1995	2000	2005	2010	2013
World	3,901	5,079	8,130	11,409	13,866
Industrialized countries	3,218	4,015	5,967	7,579	8,929
Developing and emerging industrial economies	683	1,064	2,163	3,831	4,937
By development group					
Emerging industrial countries	653	938	1,944	3,451	4,526
Least developed countries	7	14	24	49	39
Other developing countries	24	113	195	330	372
By region					
Asia and Pacific	346	566	1,291	2,509	3,371
Europe	83	127	302	483	620
Latin America	213	309	460	632	733
Africa	41	62	110	207	212
By income (world)					
High income	3,407	4,221	6,225	7,914	9,269
Upper middle income	417	669	1,570	2,872	3,771
Lower middle income	72	178	313	578	794
Low income	6	12	22	45	33

Note: Regional, development level and income classification based on Annex B1.
Source: UNIDO elaboration based on United Nations Comtrade database (UNSD 2015a).

products fell during 2000–2013, the share of resource-based manufacturing increased from 17.8 percent to 23.7 percent. The increasing size of the middle classes in industrialized and developing countries has generated higher demand for processed food.

Industrial competitiveness—most industrialized countries losing ground

UNIDO assesses and benchmarks industrial competitiveness through its Competitive Industrial Performance (CIP) index, building on a concept of competitiveness that emphasizes countries' manufacturing development, implying that industrial competitiveness is multidimensional. Industrial competitiveness is defined as the capacity of countries to increase their presence in international and domestic markets while developing industrial sectors and activities with higher value added and technological content.

Countries can learn in international markets and become more industrially competitive if they develop their technological capabilities, expand their production capacity and invest in their infrastructure. Hence, increasing industrial competitiveness requires selective policy interventions, through which comparative advantages are exploited while new competitive advantages are created.

The CIP index is a performance (or "outcome") indicator rather than a potential (or "process") indicator. It consists of output indicators only. Given its focus on industrial competitiveness and structural economic variables, it provides country rankings that tend to remain relatively stable over short periods of time. The reason is that processes of technological learning are cumulative and take time. The effects of learning are reflected in industrial statistics and structural economic variables only in the medium to long term, and those effects can be captured through detailed longitudinal studies, in particular by tracking changes of key dimensions over time. The CIP index allows us to observe not only the absolute level of key

Countries in the top Competitive Industrial Performance quintile account for nearly 83 percent of world manufacturing value added and more than 85 percent of global manufactured trade

indicators at any particular time but also their rate of change.

Based on their CIP values, countries are divided into five, colour-highlighted quintiles: top, upper middle, middle, lower middle and bottom. Countries in the top quintile account for nearly 83 percent of world MVA and more than 85 percent of global manufactured trade. Among the five most competitive are four high-income countries (Germany, Japan, the Republic of Korea and the United States), along with China ranking fifth. The four are among the world's most industrialized countries and, with China, account for 59 percent of world MVA.

Germany's manufacturing sector is a key factor in its macroeconomic performance, with a strong industrial core and an ability to control complex industrial value creation chains. Its medium- and high-tech exports account for 73 percent of its total manufactured exports, and it has maintained its technological lead against newcomers in the global economy. Germany thus has strong technological upgrading and deepening, on both the production and trade sides.

Japan's industrial competitiveness is supported by its large manufacturing base, high-tech exports and high manufacturing per capita. United States industrial competitiveness arises from its large manufacturing base, although it is more aimed at the domestic market than Japan or any other developed country. The United States alone accounts for nearly 20 percent of world MVA. The Republic of Korea has a competitive manufacturing sector based on a high share of medium- and high-tech industries.

In the top quintile, given the population size and stage of development, China has the lowest per capita values on both trade and production sides. China's position in the ranking is attributable to its high share in global trade (though low per capita values indicate that manufacturing still has the potential to grow further). China has increased its share of manufacturing exports to 17 percent of global manufacturing trade in 2013 and is the largest exporter in the world today. It has also started positioning itself as a high-tech manufacturing exporter: the export share of medium- and high-tech products almost doubled over 1995–2013. China's manufacturing industry has become the largest sector in the economy and accounted for more than one third of GDP and 18 percent of global MVA in 2013, second only to the United States.

Others in the top quintile include Switzerland, Singapore and the Netherlands, thanks to their very high exports per capita in general and high-tech exports in particular. Other top-quintile members include major EU transition economies, such as the Czech Republic, Poland, Slovakia and Hungary—due to their export orientation, more focused on the European market. Completing the list are Mexico, Malaysia and Thailand, whose competitiveness arises from their participation in global value chains.

The upper-middle quintile includes some of the most populous countries in the world, such as Turkey, the Russian Federation, Brazil, Indonesia, South Africa, India and the Philippines. The production and export performance of high-tech products in the Philippines and Indonesia is strong, while the Russian Federation and South Africa have higher MVAs per capita but low manufacturing exports due to their dependence on foreign sales of natural resources. India and Brazil accounted for 2.2 percent and 1.7 percent, respectively, of global MVA in 2013.

The middle quintile has populous countries, such as Iran, Egypt and Bangladesh, and some less populous nations, such as Costa Rica, Iceland, Oman and Uruguay. Countries in the lower-middle and bottom quintiles include less developed countries by income, accounting for roughly 0.8 percent of world MVA in 2013. Their level of industrialization is on average less than one third that in countries in the middle quintile.

The CIP ranking for 2013 shows that most industrialized countries have lost ground from the 2010 ranking. Denmark and Finland have been replaced by Mexico and Poland during the past three years. Germany, Japan, the Republic of Korea and the United States, although not among the winners, show very stable and enduring industrial competitiveness that relies on long-term advantages, such as high technology, good education and advanced infrastructure.

Part A

The role of technology and innovation in inclusive and sustainable industrial development

Chapter 1

Moving towards inclusive and sustainable industrial development

Developing countries, especially at an early stage of industrialization, have more opportunities to pursue inclusive industrial development with a potential for rapid growth and limited environmental damage. The take-off of labour-intensive industries exporting to major world markets could boost both output and employment, thus promoting sustained and inclusive growth. And the limited output volume and concentration on less polluting activities tend to make manufacturing less damaging for the environment than they will become at a later stage.

As countries acquire skills and expand their infrastructure, the opportunities for growth and employment generation rise in other industries but usually proceed by drawing in increasing amounts of production factors, as well as natural resources and energy. Most industries at the middle-income stage are resource intensive and have relatively poor emission performance. So countries emerging from the low-income stage have good prospects for continuing the path of fast and inclusive development, but they start facing sustainability challenges.

Entry into the high-income group at a mature level of industrialization comes with structural and technological changes in manufacturing. High-income countries tend to have slower growth in manufacturing, except for high-tech industries, and experience a reduction in employment. At this stage, productivity is the main driver for growth across manufacturing industries, leading to output growth without much increase in inputs of capital, labour and materials. People employed in the manufacturing sector might receive a fairly high wage, but the sector is not expanding or often is shedding employment. So, the sector has limited opportunities for inclusive development in employment absorption, but it is more environmentally friendly.

Although employment prospects in manufacturing diminish as incomes grow beyond a certain level, high-tech industries could create a large number of manufacturing-related service jobs—with a wage often comparable to that in manufacturing—which could fully offset the reduction in manufacturing employment. Germany, for example, gradually reduced manufacturing employment for 10 years before the global financial crisis, but manufacturing-related service jobs increased over the same period, fully making up for the decline in manufacturing jobs (Figure 1.1).[1]

Unless countries make conscious efforts on all three fronts—sustained economic growth, social inclusiveness and environmental sustainability—and on managing the trade-offs among them, regardless of development stage they are not likely to make much progress towards inclusive and sustainable industrial development (ISID). The foremost challenge for low-income countries is sustaining the process of industrialization. For middle-income countries, it is environmental sustainability. And for deindustrializing, high-income countries, it is continued employment generation and inclusive industrial development. However, in different ways at different stages, skill development, technological change and innovation remain crucial for successful industrialization.

Pursuing rapid, long-run and stable growth

One dimension of sustainability is an economy's ability to sustain growth over long periods without serious interruption. The longer the episodes of positive growth and the higher the rate of growth during positive growth episodes, the more likely a low- or middle-income country is to achieve sustained catch-up.[2]

The concept—in theory, open to all

Sustained growth has three characteristics: it is rapid, its episodes of growth are long and its volatility is low.

Average rates of gross domestic product (GDP) growth per capita. Is growth rapid enough to achieve substantial increases in welfare in the foreseeable future? And is it faster than in advanced economies so that a country can catch up? Since 1950, catch-up has required growth of more than 5 percent a year,

> **Sustained growth is rapid, its episodes of growth are long and its volatility is low**

Figure 1.1
Manufacturing-related employment in Germany, 1995–2008

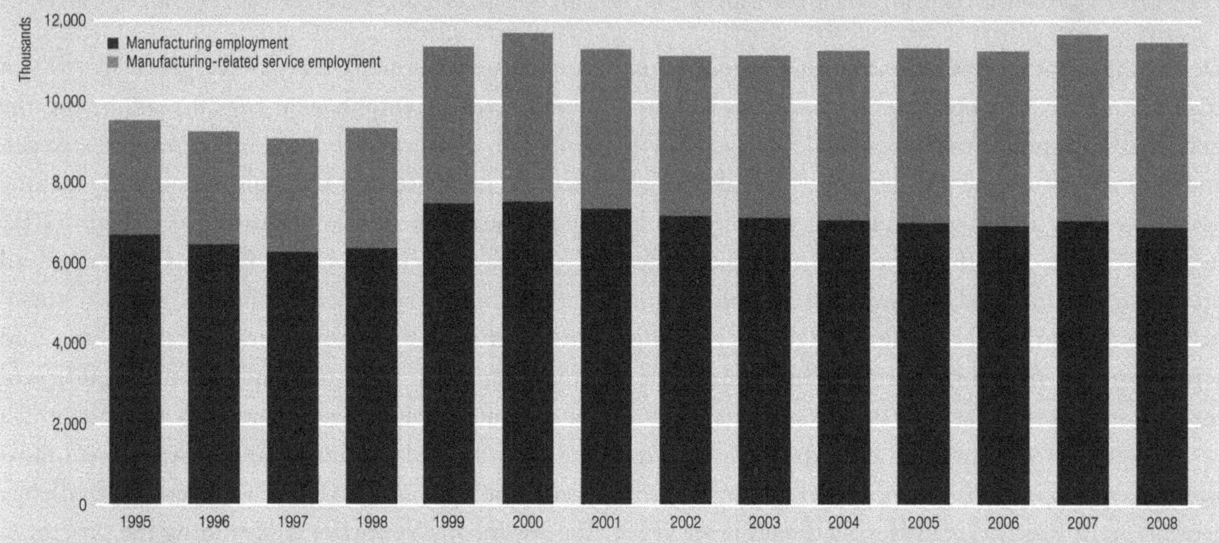

Source: UNIDO elaboration based on World Input-Output Database (Timmer and others 2015).

sustained over two or more decades (Szirmai 2012a). Such success is rare (see Chapter 2).

Duration of growth episodes. The ability to sustain growth over long, uninterrupted periods is important, but often growth is not steady. And as Pritchett (1998) has argued, attempts to explain differences in average growth may be misleading—so it is more promising to find out what initiates or halts growth episodes or what influences their characteristics. The various types of growth episodes (slumps, recoveries, accelerations, plateaus) are the building blocks of the long-run growth process.[3]

Volatility of growth.[4] The lower the volatility, the more sustainable the growth pattern. Volatility often is much higher in low- and middle-income countries than in high-income economies, and it is highest in countries caught in the "development trap."[5]

The global reality—in practice, very few succeed

How do countries move up the development ladder? The answer lies not in the creation of new knowledge but rather in the adoption and adaptation of knowledge from abroad. Poor countries tend to have high potential for rapid growth, residing in the repository of available global technological knowledge. But the evidence suggests that the ability to realize that potential varies greatly within the group of poorer countries (Figure 1.2). Within the large group of countries with a per capita GDP of less than, say, $15,000, growth rates show a large variance. The regression line shows a negative slope, indicating convergence (that is, poorer countries growing more rapidly), but it is a very weak relationship. The regression line also divides the group of poor countries into two subgroups: one, below the regression line, growing slowly and tending to fall behind or stagnate and one, above the line, showing some tendency for catch-up with richer countries.

Growth to last

Average rates of GDP per capita growth: Catching up and falling behind

Figure 1.3 shows the quintile distribution of GDP per capita for 154 countries, based on GDP per capita in

Poor countries tend to have high potential for rapid growth

Figure 1.2
GDP per capita and growth rate, 1998–2013

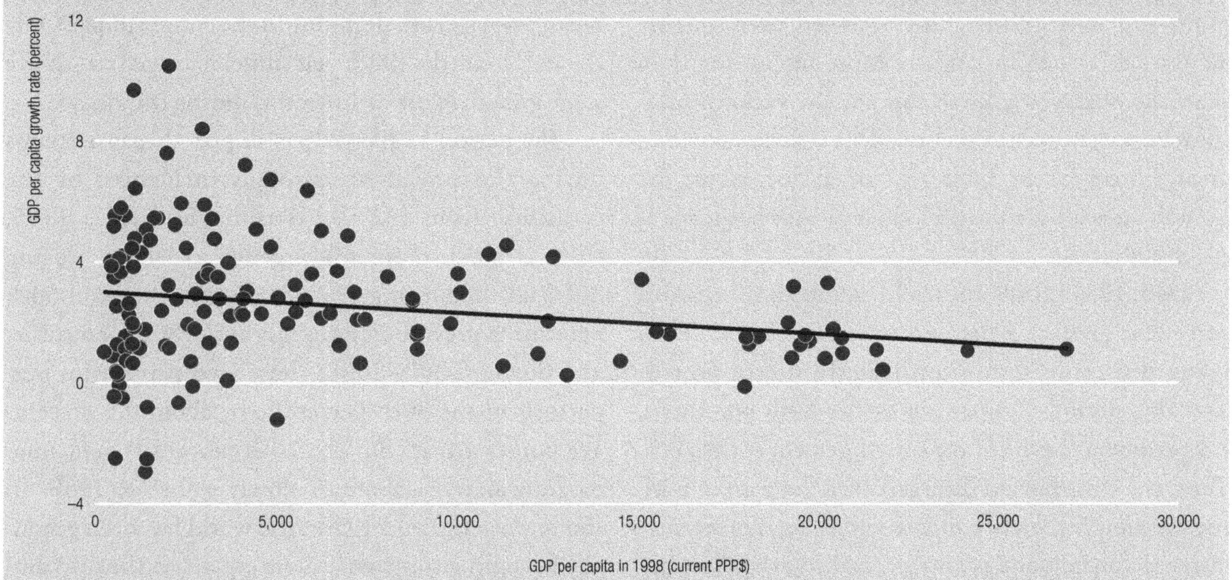

Note: GDP is gross domestic product.
Source: Kaltenberg and Verspagen (2015).

Figure 1.3
Quintile distribution of GDP per capita, 1950–2008

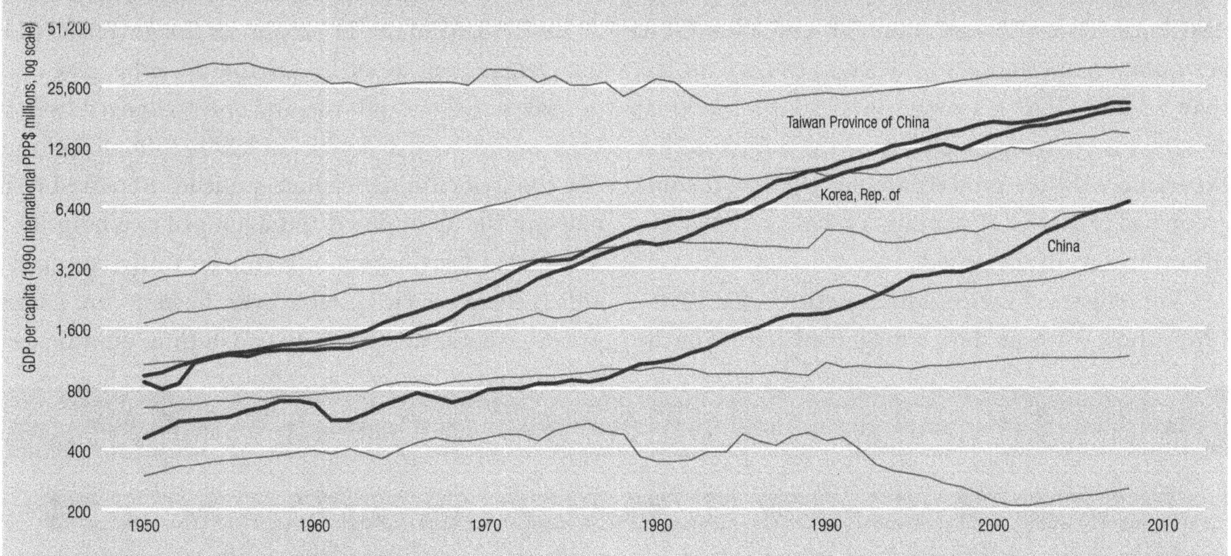

Note: GDP is gross domestic product; PPP is purchasing power parity.
Source: Kaltenberg and Verspagen (2015).

1990 international dollars, in logscale on the vertical axis. The blue lines indicate borders between quintiles of the distribution at each year. Hence, the bottom blue line is the minimum value, and the top blue line is the maximum value. The second blue line from the bottom marks the border between the lowest 20 percent values and the next group of 20 percent observations, and similarly for the other lines.

"The world has been gradually becoming somewhat more equal, though the poorest countries remain an exception

The blue lines are interesting indications of trends at parts of the distribution. For example, the minimum and maximum values observed over roughly 50 years do not change much. More interesting than that, the observed growth rates at the various quintiles show significant variation. This is documented in greater precision in Table 1.1, which documents the growth rates for various periods at various positions of the distribution.

Over 1950–2008, the (rich) countries at the 80th percentile grew at 3 percent a year (Table 1.1). Each lower percentile shows consistently slower growth over this period: 2.4 percent at the 60th percentile, 2.1 percent at the 40th, and only 1 percent at the 20th. Thus, the distribution diverged over the period, with the top-ranking (rich) countries growing significantly faster than the poor countries, making the distribution wider (in relative terms). As a result, although the difference in GDP per capita among the countries at the 80th percentile steadily narrowed over the nearly six decades, it widened considerably among the countries in the bottom 40 percent of the GDP per capita distribution. So, although countries with a low GDP per capita in absolute terms (such as those with less than $3,000 GDP per capita in 1990 international dollars) generally had higher growth, countries that experienced faster growth among the low-income group had higher GDP per capita within each percentile at the start of the period.

This long trend is especially associated with 1950–1980. Those 30 years show a magnified picture of the overall divergence, with the growth rate declining strongly at each percentile. The 1980s show a major break, with growth declining in the entire sample. Only the rich countries (80th percentile) managed to achieve some growth (close to 1 percent) during this decade.

The period 1990–1995 is skipped because trends during those years are strongly influenced by the transition from 132 to 151 countries.[6] After 1995, there is much more convergence, especially during 2000–2008. During 1995–2008, the 40th and 60th percentiles grew roughly as fast as the 80th percentile, and during 2000–2008, those two percentiles outperformed the 80th percentile by about 0.5 percentage points a year. But the 20th percentile continues to grow slowly—about as slowly as before 1980. In short, since the mid-1990s, the world has been gradually becoming somewhat more equal (on the distribution of living standards among countries), though the poorest countries remain an exception.

How about the growth experiences of individual countries? To summarize them, look at countries that move upward in the distribution over the long run. For 1950–1989, 65 of the 132 countries (roughly half) did not change their quintile. Another 51 either rose one quintile or fell one quintile, and only 9 changed two or three quintiles up or down. For 1990–2008, 109 of the 151 countries did not change quintile, 40 moved just one quintile up or down and 2 jumped two quintiles. The global distribution of countries in living standards is therefore fairly stable over the long run. Large jumps—countries moving up two or three quintiles, as

Table 1.1
Annual growth rates at selected percentiles of the world income distribution, 1950-2008 (percent)

	1950–2008	1950–1980	1980–1989	1995–2008	2000–2008
Minimum	–0.3	0.5	3.7	–1.3	1.7
20 percentile	1.0	1.6	–1.0	1.2	0.8
40 percentile	2.1	2.5	0.2	3.1	3.6
60 percentile	2.4	3.1	–0.5	3.5	3.6
80 percentile	3.0	4.2	0.9	3.4	3.0
Maximum	0.2	–0.1	–2.0	2.0	1.4

Source: Kaltenberg and Verspagen (2015).

“Countries stuck in the bottom quintile have the shortest episodes of growth

Taiwan Province of China, the Republic of Korea and China have (Figure 1.3)—are very rare. Household surveys add some nuance to this picture (Box 1.1).

Duration of growth episodes

Table 1.2 gives a first indication of the importance of the duration of positive growth episodes (see endnote 3). The table arranges 10 groups of countries according to their incomes relative to their income positions (by quintile) in 1960 and 2008, distinguishing countries that remained in the same quintile over the period and countries that improved their relative positions.[7]

Two patterns can be discerned. First, countries that remain stuck in the bottom quintile have the shortest episodes of growth (7 years on average). Countries that have maintained their position in the top quintile have much longer growth episodes (17 years on average). But growth rates do not differ much. Countries trapped in the bottom quintile grew faster during their growth episodes than did countries in the top quintile. That fact is in line with the observation that achieving growth is easy for poor countries—they can grow at least as fast as rich countries—but they are more vulnerable to interruptions in the growth process and find sustaining growth over long periods to be harder (Hausmann, Pritchett and Rodrik 2005).

Second, developing countries that improved their relative position over the period tend to have much longer growth episodes than countries that remained in the same quintile or even dropped: the three countries that moved from the fourth to fifth quintile have an average duration of no less than 26 years.[8] So, not only is the duration of growth episodes longer in catch-up countries but they also tend on average to have much higher growth rates during those episodes.

In plotting the duration of each episode against its growth rate, there is no relationship between the length of an episode and the rate of growth within that episode. That means that duration itself has an independent influence on the average growth rate between 1960 and 2008. The longer the growth episodes for a country, the greater the average rate of growth over the period. So, duration matters. It captures the notion of sustained growth. The variation in within-episode growth rates is much higher in shorter episodes than in longer ones. That pattern remains visible even after removing the shortest episodes and keeping only the episodes of three years and longer.

Box 1.1

Real income growth based on household surveys

Milanovic (2012) uses a compilation of household surveys from more than 120 countries between 1988 and 2008 to examine the changes in real income, in purchasing power parity (PPP) dollars, for various shares of the global income distribution during a period of globalization.

He finds corroborating evidence that the lowest income group has not improved much, while the global middle class has improved, with a compound annual growth rate of around 2.7 percent between the 20th and 60th percentiles. The median income rose the most (almost 80 percent), representing the corresponding growth in income, predominantly in China and India but also in Indonesia, Brazil and Egypt. Further, incomes of the richest 5 percent increased considerably, accounting for around 44 percent of the total increase in income. In contrast, incomes of the upper-middle class (the 70th to 95th percentiles of the global income distribution) stagnated. This group includes many people from the former Eastern European bloc and Latin America, as well as some of the rich countries.

Global growth incidence curve, 1988–2008

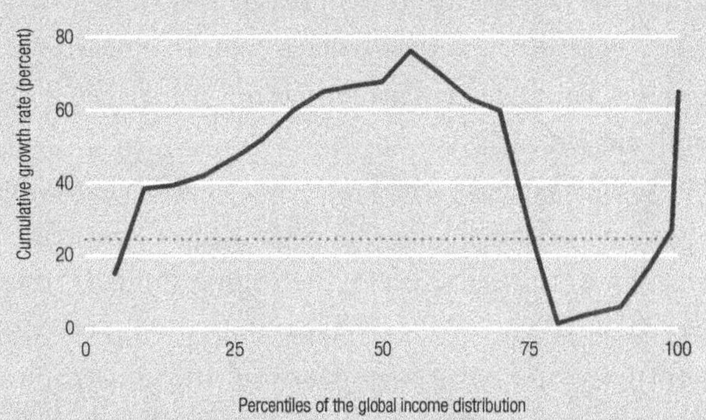

Note: Y-axis displays the growth rate in average income of the fractile group (in 2005 PPP$). Population-weighted. Growth incidence evaluated at ventile groups (such as the bottom 5 percent); top ventile is split into top 1 percent and 4 percent between percentiles 95 and 99. The horizontal line shows the growth rate in the mean of 24.3 percent (1.1 percent a year).
Source: Lakner and Milanovic (2013).

“The lower the volatility, the more sustained the growth

Table 1.2
Duration and volatility of growth episodes, by income group, 1960–2008

	Duration (years)	Volatility	Average growth during episode, 1999 PPP$ (percent)	Number of countries
1. Bottom quintile in 1960 and 2008	6.6	3.7	3.4	13
2. Relative improvement: bottom quintile in 1960, second or third quintile in 2008	11.4	1.5	5.0	7
3. Second quintile in 1960 and 2008	7.9	10.6	3.0	12
4. Relative improvement: second quintile in 1960, third quintile or higher in 2008	13.6	1.3	5.0	7
5. Third quintile in 1960 and 2008	8.8	2.9	3.0	17
6. Relative improvement: third quintile in 1960, fourth or fifth quintile in 2008	8.8	1.1	6.0	2
7. Fourth quintile in 1960 and 2008	9.3	1.8	4.0	16
8. Relative improvement: fourth quintile in 1960, fifth quintile in 2008	26.2	0.8	4.0	3
9. Falling behind: Fifth quintile in 1960, fourth quintile in 2008	7.4	3.9	4.0	6
10. Fifth quintile in 1960 and 2008	16.8	0.9	3.0	14

Note: PPP is purchasing power parity. GDP per capita in 1960 and 2008 is ranked in quintiles for 97 countries. Countries formerly part of the Union of Soviet Socialist Republics (USSR) are excluded because no data on growth rates are available before 1989.
Source: UNIDO elaboration based on The Maddison Project (2013).

Volatility of growth

Another characteristic of the ability to sustain growth is volatility—or, rather, the lack of it. The lower the volatility, the more sustained the growth. Volatility can be calculated by measures such as the standard deviation of growth or the coefficient of variation. Typically, growth volatility is higher in low- and middle-income countries than in advanced economies.

Table 1.2 shows the coefficient of variation in growth for the 97 countries between 1960 and 2008, in the volatility column. Again we distinguish 10 groups of countries, based on their quintile ranking in 1960 and 2008.

Two clear messages emerge. First, volatility is much higher in low-income than in high-income countries. Thus, for countries stuck in the second quintile, the coefficient of variation is no less than 10.6, and for countries in the bottom quintile, 3.7. In comparison, countries in the top quintile in both 1960 and 2008 have a coefficient of variation of 0.9. Second, for each quintile in 1960, the volatility of growth of countries that improved their income rankings is much lower than for countries stuck in the same quintile. The clear implication is that in the long run, lower growth volatility is a key ingredient in successful economic development.

Those findings complement the results in Figure 1.2. Not only is the difference in average growth rate among developing countries much higher than among developed countries, but the volatility of a country's growth rate is higher in developing than in developed countries. Thus, the growth experiences of developing countries vary more on the rate, duration and volatility of growth than do those of developed countries. But among developing countries, those successful in catching up seem to have the common characteristics of higher growth rates, longer growth episodes and lower volatility.

Traveling together—structural change and growth

Structural change correlates with growth—a leitmotif throughout this report. We start by looking at structural change in a basic way, simply as the rate at which

“Countries that experience faster structural change tend to experience faster growth

sectoral shares in the economy change. We adopt an indicator similar to the Finger-Kreinin Index[9] that is defined as the sum of absolute changes in sectoral shares over 1995–2013, divided by 2. A value of 0 corresponds to the case in which all shares remain the same, and a value of 1 would correspond to the (extreme) case in which in 1995, one sector has a share of 1, and in 2013, another sector had a share of 1.

Figure 1.4 shows the relationship between structural change and the rate of growth between 1995 and 2011 for a sample of 108 countries with a breakdown of GDP into 10 sectors.[10] We calculate the structural change index for GDP (value added) in constant 2005 prices. There is a clear and positive association between structural change and growth, although it explains only a limited fraction of the total growth variance. Countries that experience faster structural change tend to experience faster growth. But a variation in growth rates can be observed at any level of structural change.

Among different types of structural change, industrialization—an increase in the share of manufacturing value added (MVA) in GDP—has been the engine of growth for developing countries. So, the following section takes a closer look at structural change from that angle.

Manufacturing development and structural change

During the first three quarters of the 20th century, in the context of structural change, the literature mainly discussed manufacturing as the engine of growth because of the high productivity and the increasing or high manufacturing shares in rapidly growing or high-income countries. In recent years, however, the literature has paid increasing attention to deindustrialization (or premature deindustrialization)—the decline of the relative importance of manufacturing in the economy.

The assertion that manufacturing has declined in importance comes from both empirical evidence and a general overview of manufacturing's current state in the world. Manufacturing employment's share in total employment and the absolute number

Figure 1.4
Structural change and GDP per capita growth, 1995–2011

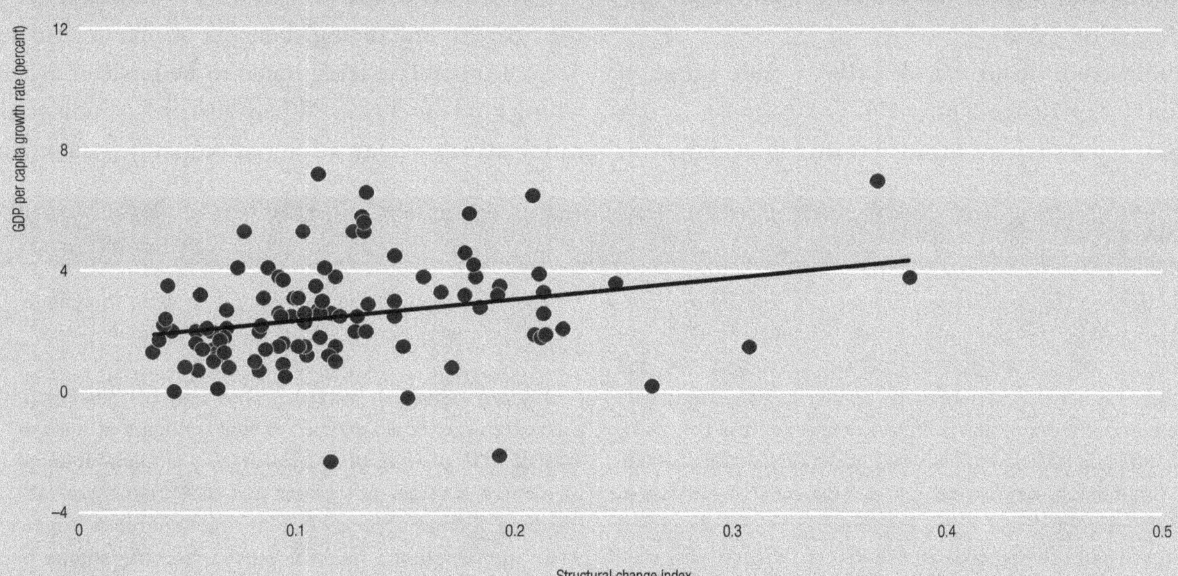

Note: The index of structural change is defined as the sum of absolute changes in sectoral shares over the period 1995–2013, divided by 2. A value of 0 corresponds to the case where all shares remain the same, and a value of 1 would correspond to the (extreme) case where in 1995 one sector has a share of 1, and in 2013, another sector has share 1.
Source: UNIDO elaboration based on The Maddison Project (2013), UN National Accounts Statistics (UN 2014b), Groningen Growth and Development Centre 10-Sector Database (Timmer, de Vries and de Vries 2014) and World Input-Output Database (Timmer and others 2015).

> **If countries start deindustrializing prematurely, they are prone to growth-reducing structural change**

of manufacturing jobs are generally falling in high-income countries. Studies also show that, on average, countries across all incomes now have a lower manufacturing share than before—and that they reach their peak employment and value-added shares at a lower income level than in previous decades (Ghani and O'Connell 2014; Rodrik 2015).

This mix of empirical evidence and general observations has perhaps given many the idea that manufacturing's contribution to economic development is declining. But the declining role of manufacturing in developed economies does not necessarily mean the same trend is at work in developing countries. The low manufacturing shares in many developing countries (relative to past trends) might be attributed to country-specific conditions rather than to a systematic and long-term reduction in manufacturing's potential contribution to the economy—that is, as a result of a structural shift in supply and demand conditions of different sectors.

So we dig a bit deeper, first looking at deindustrialization before determining whether manufacturing still has a key role in economic growth.

Are developing countries deindustrializing? In the main, No

Deindustrialization can describe a wide range of country experiences (Box 1.2). For example, in one country, the share of manufacturing in employment may fall because very rapid technological progress in manufacturing leads to its productivity rising more than productivity in other sectors. So, employment is growing, but more slowly than it was previously. This can go hand in hand with healthy growth in manufacturing output, exports and sometimes even employment itself. In another country, the share of employment may be increasing, but due to slow productivity growth, the share of manufacturing in GDP is in decline. In a third country, manufacturing could be collapsing when a country experiences productivity declines, stagnant output growth and shrinking jobs in manufacturing.

If countries start deindustrializing prematurely (when their per capita income and degree of industrialization are too low), they are prone to growth-reducing structural change, involving the wrong kind of low-productivity informal services, which in many countries in Africa and Latin America are currently expanding their shares in value added and employment. They offer little potential for growth. Such premature deindustrialization is a threat to sustained economic growth in low- and middle-income countries for two reasons.

First, such countries will have obtained fewer of the "growth-enhancing" benefits of manufacturing. Second, manufacturing tends to be replaced by the wrong kind of services. When "mature" deindustrialization sets in—in an advanced economy—subsectors

Box 1.2
Types of deindustrialization

Mature (or normal) deindustrialization. The normal pattern (unlike the premature variant) occurs as GDP per capita increases beyond a certain level and services become more important in the economy. Associated with advanced economies at higher incomes, it is driven by increasing demand for services, rapid productivity growth in manufacturing (resulting in a reduced share of manufacturing jobs) and rising outsourcing of manufacturing activities in global value chains (GVCs). It is not seen as a threat to economic development. Technologically dynamic service sectors—logistics, trade, information and communications technology, and financial services—emerge as alternative drivers of the economy. Nor is it problematic that growth rates slow as countries reach high incomes.

Premature deindustrialization. One way of thinking of premature deindustrialization is that it begins at a lower level of GDP per capita or a lower level of manufacturing as a share of total employment and GDP than is typically the case. If that happens—or if deindustrialization begins when manufacturing has not yet reached the shares of employment and GDP typically associated with the normal turning point of industrialization—it could be considered premature.

“The growth-pulling role of manufacturing is especially important in the earlier stages of development

of the expanding service sector have the dynamic characteristics attributed to manufacturing in the past: strong linkages, productivity increases and technological innovations. That kind of service sector can act as an engine of growth. In an economy characterized by premature deindustrialization, the service activities that emerge often are informal services lacking dynamism and growth potential.

In all types of deindustrialization, very austere macroeconomic policies—especially high interest rates and overvalued exchange rates—are likely to have more pronounced negative effects on industry (and on the rest of the real economy) than on the financial sector. Such policies are likely to financialize and deindustrialize the economy. Similarly, trade liberalization affects tradables more than non-tradables and has uneven effects among tradables, depending on their competitive position at the time the domestic market opens. Liberalizing tariffs too quickly without giving manufacturing time to restructure is a major contributor to deindustrialization.

The perils of premature deindustrialization —you have to have something to lose it

The statistical evidence for the effects of deindustrialization on growth is inconclusive, in part because of difficulties in capturing the different types of deindustrialization. Special country characteristics may also be germane to whether deindustrialization affects growth. Still, a few general conclusions can be made.

The lower the GDP per capita at which a country begins to deindustrialize, the more the process is likely to affect growth and growth prospects. Similarly, the lower the share of manufacturing in value added when deindustrialization sets in, the more deindustrialization is likely to affect growth.

The extent to which deindustrialization is triggered or accelerated by a policy change, as opposed to just gradually taking place over time with economic development, also influences the likely effects of deindustrialization on growth. Policy-induced deindustrialization—as associated with, for example, trade liberalization—is more likely to kick in before the full benefits of industrialization have been obtained, before manufacturing has matured and before a dynamic and advanced service sector has developed.

The aggregate effects of deindustrialization depend on the characteristics both of the manufacturing activities in decline and of the service activities that are increasing their shares in employment and GDP. For instance, if the manufacturing activities have little scope for increasing returns to scale and limited scope for cumulative productivity increases while the service activities are growing, negative effects on growth need not occur.

Still, the growth-pulling role of manufacturing is especially important in the earlier stages of development—and it is more important for developing than for developed countries. When a country begins deindustrializing after manufacturing's share has reached 30 percent of GDP, the benefits of manufacturing likely have diffused through the economy over an extended period. Those benefits include skill development through learning by doing, technological benefits to other sectors, foreign exchange relieving balance-of-payments constraints to growth, and stimulating other sectors through forward and backward linkages. But when a country fails to industrialize or when it prematurely deindustrializes before manufacturing accounts for even 5 percent of total jobs, those pro-growth economic benefits will not be realized.

Furthermore, premature deindustrialization can jeopardize the potential of the services sector to act as an alternative growth engine. With mature deindustrialization, certain advanced and dynamic services activities may have the kinds of growth-enhancing properties attributed to manufacturing. But the types of services activities likely to replace manufacturing at premature deindustrialization are more likely to be low-skilled, low-productivity, non-tradable activities, such as retail or personal services, which do not have strong increasing returns or the potential for cumulative productivity increases. Although those activities may be important for job creation, they are not likely to drive growth. Nor are they likely to allow countries

"Manufacturing's share in global GDP has been stable for more than 50 years

to leapfrog to dynamic growth-pulling services activities before they have industrialized. A partial developing-country exception may be India, the "office of the world," which has enclaves of dynamic service activities but whose employment numbers are tiny relative to the country's population.

Overall, however, with deindustrialization at low incomes per capita, a country is unlikely to have enough effective demand to support the sustainable development of dynamic services that can act as an alternative engine of growth. The non-tradable nature of many services makes domestic demand more of a constraint than it is for manufacturing. To the extent that services can be such an engine of growth, the situation is more likely to be feasible in advanced than in developing countries.

Premature deindustrialization generally is likely to be more sudden than mature deindustrialization in advanced economies, partly because it is more likely to be brought about by policy changes such as trade liberalization. Advanced countries can suddenly liberalize their economies, thereby accelerating deindustrialization, but those patterns tend to be more common in developing countries, which are more likely to be dependent on international financial institutions and to have liberalization programmes as part of loan conditions. The relatively low diversification in most developing countries also leaves their economies more susceptible to shocks, so the impact of sudden liberalization in triggering or accelerating deindustrialization is likely to be more pronounced.

Prospects for manufacturing expansion remain strong for most of the developing world

Given the early deindustrialization in some developing countries, have the opportunities for manufacturing production and employment generation diminished in recent years? This section shows how the share of manufacturing in value added and jobs has differed among developing and developed countries over the past 40 years, based on a raft of valuation measures. It sheds light on (de-)industrialization trends between developing and developed countries and shows whether they differ from long-term historical trends. It then highlights regional variations from global patterns.

Value added

First, we look into structural change, focusing on the relative position of manufacturing in the economy, using value-added data with different pricing methods to elucidate manufacturing's contributions to developing and developed economies and how those contributions might have changed over time. The left panel of Figure 1.5 includes the Union of Soviet Socialist Republics (USSR) before 1990 and former USSR countries after 1990, but the right panel excludes those countries because their economic collapse and subsequent consolidation of manufacturing industries are unlikely to reflect the normal trend of structural change. In both figures, manufacturing shares of developed and all countries have been steadily declining since 1970, while that of developing countries has been stable since at least 1993, hovering around 18–20 percent of GDP. This stability extends to the entire period if the USSR and the countries of the former USSR are excluded.

The same analysis is made in constant prices in Figure 1.6. The share of developing countries (including former USSR countries) increased after 1970 except for a short period of decline at the end of the 1980s and beginning of the 1990s. Even the MVA share of developed countries has been stable since 1993, at least until the global economic crisis of 2008–2009.

Another valuation approach based on sector-specific PPPs confirms that in constant prices, manufacturing's share in global GDP has been stable for more than 50 years (Box 1.3). But the share is lower by 2 or 3 percentage points than that based on manufacturing value-added deflators, because values of non-tradable sectors (typically services) are higher in PPP terms than those of tradable sectors (typically manufacturing).

Among developing countries, all regions except Asia have reduced their manufacturing shares in GDP

Figure 1.5
Manufacturing value added shares, worldwide, current prices, 1970–2013

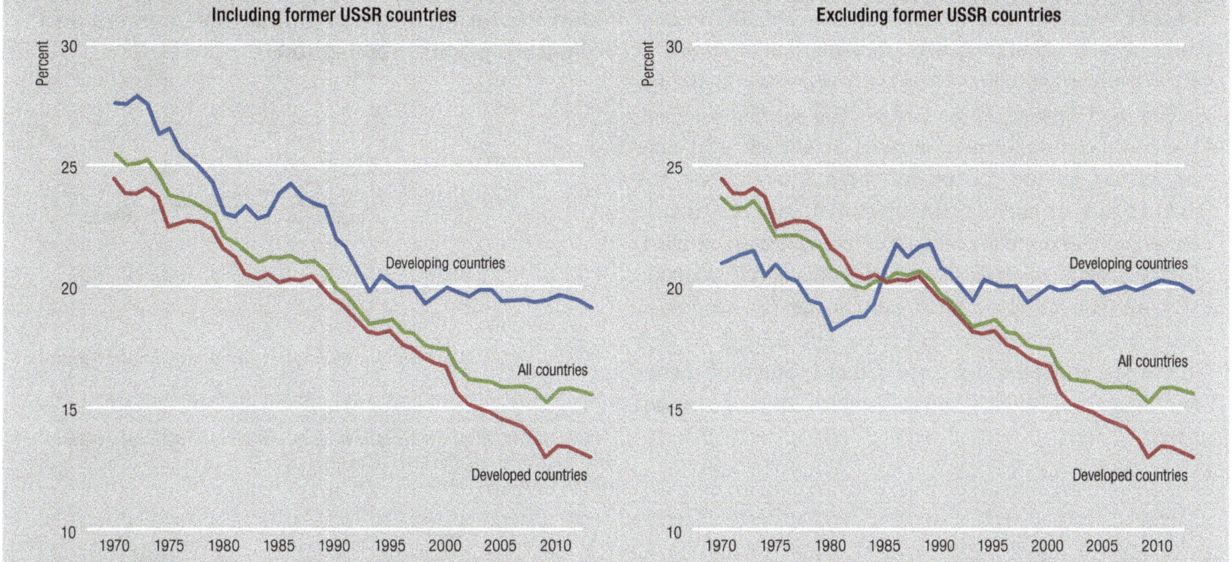

Note: Shares are calculated as world manufacturing value added (current $) over world GDP (current $). Number of country observations (n) varies according to period. Figure a: n=184 (1970–1990), n=191 (1991), n=191 (1992–1993), n=192 (1994–2010) and n=193 (2011–2013). Figure b: n=185 (1970–1990), n=204 (1991), n=205 (1992–1993), n=206 (1994–2010) and n=207 (2011–2013). Income classification based on Annex A1, Table A1.2.
Source: UNIDO elaboration based on UN National Accounts Statistics (UN 2014b).

Figure 1.6
Manufacturing value added shares, worldwide, constant prices, 1970–2013

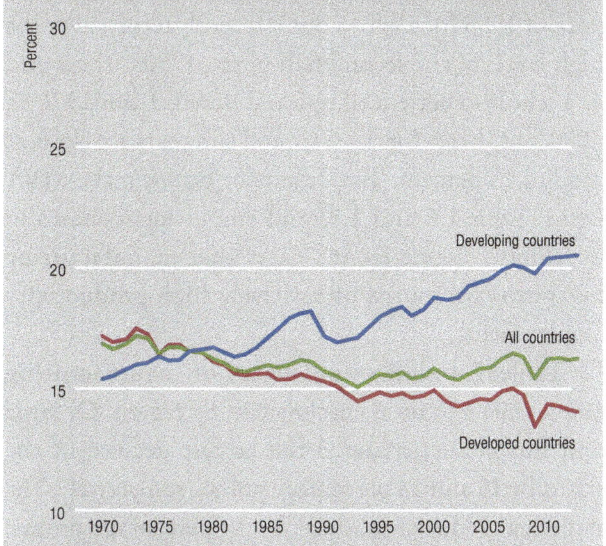

Note: Shares are calculated as world manufacturing value added (constant 2005 $) over world GDP (constant 2005 $). Number of country observations (n) varies according to period: n=185 (1970–1990), n=204 (1991), n=205 (1992–1993), n=206 (1994–2010) and n=207 (2011–2013). Income classification based on Annex A1, Table A1.2.
Source: UNIDO elaboration based on UN National Accounts Statistics (UN 2014b).

In constant prices, manufacturing shares in global GDP have therefore been stable for a long time. Even in developed countries over the past 20 years, a declining trend is hard to find. In developing countries, no evidence of a declining trend emerges in their manufacturing shares no matter which valuation approach is used. Comparisons in constant and current prices indicate that manufacturing prices have been falling much faster relative to the prices of other sectors. But the faster growth of labour productivity in manufacturing than in other sectors has prevented developing countries' manufacturing share from declining in current prices.[11]

Table 1.3 shows the regional trends of manufacturing shares in GDP in current prices. As seen also in Figure 1.5, the manufacturing share in developing countries as a whole has been quite stable over the past 40 years, with regional differences. Among developing countries, all regions except Asia have reduced their manufacturing shares in GDP—the Americas and Europe have reduced theirs by more than 10 percentage points.

> **“Manufacturing employment shares of developing countries increased from 11.6 percent in 1970 to 14.2 percent in 2010**

Box 1.3
Sector-specific purchasing power parity

Lavopa and Szirmai (2015) put forward an original approach to estimating manufacturing value added at current sector-specific PPPs for the countries covered in Inklaar and Timmer (2012). In short, the approach *adjusts* the observed sectoral structures at current domestic prices to take into consideration the differences in the relative size of sectors (manufacturing and non-manufacturing) that appear when a PPP conversion is applied. In developing countries the relative price of non-tradable sectors (relative to tradable sectors) tends to increase after a PPP conversion. So, their share in GDP also increases, reducing the corresponding share of manufacturing. That effect tends to diminish as countries get richer.

Manufacturing share in global GDP, current and constant prices, 1960–2009

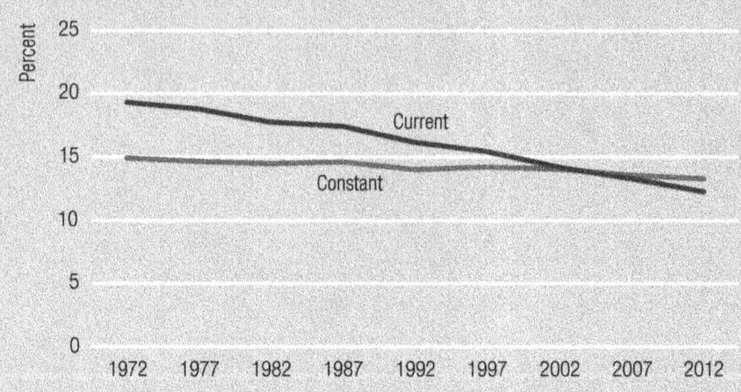

Note: Calculations are five-year averages. Each series calculates the manufacturing value added–to-GDP ratio, using different valuations for the sectoral and aggregate value added. PPP (PPPsh) is the value added at current PPP using sector-specific converters; PPPk05 (PPPsh) is the value added at 2005 PPP using sector-specific converters.
Source: Lavopa and Szirmai (2015).

Although Asia’s increased share has not been as much as the declines of the shares of the Americas and Europe, the larger economic size of developing Asia has made its increase enough to compensate for the declines of other developing regions. South-East Asia had the largest increase in manufacturing share—more than 7 percentage points. East Asia experienced a decline over the period but still keeps a remarkably high share of more than 30 percent, mainly due to increases in China and South Korea. Africa’s share dropped sharply after 2000, and both North Africa and the Middle East and Sub-Saharan Africa are at their lowest in the past four decades, at around 10 percent. Oceania steadily decreased its manufacturing share from an already low point in the 1970s and now has the lowest share of all regions.

In brief, developing countries’ stable manufacturing share masks variations among regions, but it essentially reflects increases in Asia and declines in other developing regions. Premature deindustrialization seems evident in African countries, many of which started reducing their manufacturing share from an already low base.

Employment

For developed and developing countries, manufacturing employment shares in total employment show opposite trends since 1970 (Figure 1.7). The share of developed countries fell by more than 10 percentage points over the past 40 years to the lowest point of the period, 13.3 percent. In contrast, that of developing countries increased from 11.6 percent in 1970 to 14.2 percent in 2010. It increased from 1970 to 1988 and moved to a few percentage points lower than the peak of 1988 for a while, but it recently returned to the high levels of the second half of the 1980s. The world as a whole experienced ups and downs from 1970 to 2010, but shares for the start and end years are close, at around 15 percent. The difference between the MVA (see Figures 1.6 and 1.7) and employment shares in developing countries indicates that manufacturing has been the source of relatively high-productivity employment.

Table 1.4 shows the shares of manufacturing employment in total employment by region. Oceania and Europe experienced the largest decline in the share, by 15 and 13 percentage points, respectively. The Americas reduced its share by 10 percentage points. The decline of the manufacturing employment share in developing countries in Asia has been modest compared with that of Oceania, Europe and the Americas, with a drop of 7 percentage points over the 40 years: Asia still has a more than 20 percent manufacturing

"The role of manufacturing in economic development continues as important as ever

Table 1.3
Manufacturing value added share in GDP by region, current prices, 1970–2010 (percent)

	1970–1974	1975–1979	1980–1984	1985–1989	1990–1994	1995–1999	2000–2004	2005–2009	2010–2013
World	23.4	22.0	20.3	20.5	19.0	18.0	16.5	15.7	15.8
Developed countries	24.0	22.6	20.9	20.3	18.7	17.5	15.5	14.0	13.3
Americas	22.2	21.2	19.4	18.2	16.6	16.1	14.0	12.5	12.0
Asia	31.3	26.9	26.1	25.3	23.5	21.0	19.3	18.8	18.0
Europe	24.7	23.3	21.3	20.5	18.5	17.4	16.0	14.4	13.6
Africa	12.3	8.5	9.9	13.1	12.7	13.3	12.2	10.0	9.6
Oceania	20.8	18.8	17.7	15.5	13.6	12.8	11.3	9.5	7.5
Developing countries	21.1	20.1	18.7	21.5	20.2	19.8	20.1	19.9	20.1
Americas	23.6	24.3	22.9	24.6	20.6	16.6	16.7	15.4	13.5
Central America	19.8	19.7	19.9	19.1	18.7	18.7	17.9	16.4	15.0
South America	25.1	26.4	24.9	26.0	21.2	15.4	15.8	14.9	12.5
Asia	23.3	23.0	20.4	22.6	23.0	25.2	26.0	26.8	26.8
East Asia	33.5	31.8	25.6	28.8	26.6	29.0	29.9	30.7	30.8
South-East Asia	16.4	16.9	18.5	21.6	24.4	25.8	26.9	26.4	24.1
South Asia	13.0	14.4	14.3	14.3	14.2	14.5	13.9	14.8	13.5
Europe	27.3	26.8	25.8	25.4	18.8	16.6	15.5	15.5	15.8
Western Europe	17.7	16.9	16.6	16.0	14.1	13.0	11.6	10.0	9.9
Eastern Europe	31.2	31.8	30.7	31.1	23.5	19.3	18.2	18.6	18.3
Africa	12.2	10.8	10.5	13.8	15.2	15.1	12.5	11.2	10.7
North Africa and Middle East	12.9	10.9	10.0	14.4	15.7	16.0	12.8	11.8	11.4
Sub-Saharan Africa	11.7	10.7	11.2	12.9	14.2	13.2	11.8	9.9	9.1
Oceania	12.6	12.6	11.1	11.2	10.3	9.3	9.2	9.4	8.3

Note: Calculations are five-year averages. Number of country observations (n) varies according to year: n = 136 (1989) and n = 157 (2013). Former Union of Soviet Socialist Republics (USSR) countries are not included, but including them would not significantly change the trend since 1995. Income and regional classification based on Annex A1, Table A1.2.
Source: UNIDO elaboration based on UN National Accounts Statistics (UN 2014b).

employment share, the highest among all regions (including developed countries).

Although developing countries as a whole have increased their manufacturing employment share for the past four decades, the increase has been concentrated in Asia, where all three regions increased the manufacturing share in total regional employment. East Asia increased its share by 10 percentage points over the 40 years. Given that it includes China, the world's most populous country, this increased manufacturing employment—by 130 million jobs since 1970.

All other regions, except Central America and the Middle East and North Africa (where the manufacturing share barely changed) have reduced the share. The largest decline is in the countries of the former USSR, especially after 1990, when manufacturing industries were consolidated after the demise of the USSR. Even the regions that have industrialized little, such as Sub-Saharan Africa, have further reduced the share of manufacturing employment from an already low level.

This empirical evidence for value added and employment together with the earlier discussion on growth suggests that the role of manufacturing in economic development continues as important as ever and that the prospect of manufacturing expansion for developing countries has not diminished in recent years. At least, no indication has emerged of premature deindustrialization in developing countries as a whole, even if Africa stands out. But industrialization remains highly concentrated in a few regions,

> **Industrialization remains highly concentrated in a few regions, especially in Asia**

Figure 1.7
Manufacturing employment share of total employment, worldwide, 1970–2010

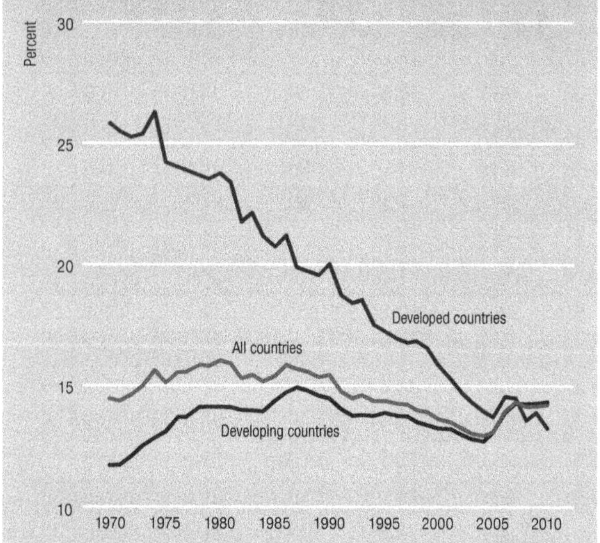

Note: Shares are calculated as world manufacturing employment over total employment. Income classification based on Annex A1, Table A1.2.
Source: UNIDO elaboration based on Groningen Growth and Development Centre 10-Sector Database (Timmer, de Vries and de Vries 2014), ILOSTAT (ILO 2015a) and KILM Datbase (ILO 2015b).

especially in Asia, which is a very large, diverse and populous region.

Technological change within manufacturing

Subsectors with different technologies

To illustrate the relationship between structural change and technological development—a key theme of this report—we now look at structural change in manufacturing subsectors, grouped by low, medium and high technological level (Annex A2).

Figure 1.8 shows how low-, medium- and high-tech shares of global MVA have changed for developing and developed regions from 1972 to 2012 in current PPP. In 1972, the developed region generated 64 percent of the world's value added in low-tech production. Its shares for medium- and high-tech production were higher, at 71 percent and 74 percent, respectively.

The past 40 years have seen a relative shift in all three technological activities from developed to developing countries. In 2012, developing countries accounted for more than half the world's value added in low- and medium-tech industries and for nearly half in high-tech industries.

Figure 1.9 shows how the technology structure in manufacturing (based on Annex A2) has changed in developing countries over 40 years. In 1972, the low-tech shares in Latin America and Asia—regions with similar technology structures—were lower than in Africa. In 2012, Africa increased its share of the high-tech group and reached a structure similar to that of Latin America and Asia in 1972. In the same period, Latin America had very little change, with a slight decline in the share of the high-tech group, compensated by an increase in the medium-tech share.

Among the three developing regions, Asia experienced the most significant change in technology structure. Over the 40 years, its share of the high-tech group rose by 10 percentage points—at the expense of the low-tech share. Asia's economic success relative to other developing regions was thus accompanied not only by an increased manufacturing share in the economy but also by technological upgrading in manufacturing.

Productivity and employment growth: six industries, two country-size groups, many different incomes[12]

The previous section looked into the technological composition of industry, linking shifts into high technology to higher manufacturing shares and faster economic growth, as exemplified by the case of Asia. Technological upgrading however, while being instrumental to growth isn't that straightforward in terms of employment. This section takes a look at how six manufacturing industries present opportunities for productivity gains, while facing trade-offs between productivity and employment.

An industry's growth in value added per capita is roughly equal to its employment per capita growth plus labour productivity growth:

$$\frac{\textit{Value added}}{\textit{Population}} = \frac{\textit{Employment}}{\textit{Population}} \times \frac{\textit{Value added}}{\textit{Employment}}$$

“ Asia's economic success was accompanied by technological upgrading in manufacturing

Table 1.4
Manufacturing employment share in total employment by region, 1970–2010 (percent)

	1970–1974	1975–1979	1980–1984	1985–1989	1990–1994	1995–1999	2000–2004	2005–2009	2010–2013
World	14.8	15.6	15.6	15.6	14.7	14.2	13.3	14.0	14.2
Developed countries	25.6	23.8	22.4	20.2	18.7	16.9	14.9	14.1	13.3
Americas	21.9	20.3	18.5	16.5	14.7	13.8	12.1	11.0	10.2
Asia	28.1	26.3	25.7	25.4	24.4	21.1	18.9	18.6	21.1
Europe	27.2	26.0	23.5	21.1	19.3	18.0	16.3	15.0	13.2
Africa	20.3	17.9	16.1	15.4	14.5	13.7	12.5	11.4	10.7
Oceania	24.6	21.6	19.8	17.0	15.0	13.7	12.4	10.7	9.7
Developing countries	12.2	13.7	14.0	14.7	14.0	13.7	13.0	13.9	14.3
Americas	15.5	15.6	15.4	16.0	15.4	14.3	13.5	13.7	13.1
Central America	13.1	13.2	14.3	15.1	16.2	16.3	15.3	14.2	13.2
South America	15.1	15.0	14.5	15.3	14.4	13.0	12.2	12.9	12.4
Asia	9.1	11.2	12.0	13.4	13.4	14.0	13.4	14.8	15.7
East Asia	9.5	12.6	13.9	15.9	15.7	15.8	14.2	17.6	19.4
South-East Asia	7.9	8.9	9.6	10.4	11.6	12.8	13.6	12.8	12.8
South Asia	9.0	9.4	9.6	10.1	10.3	11.4	12.1	11.9	12.2
Europe	27.2	27.5	27.4	27.0	22.9	18.2	17.1	16.1	14.7
Western Europe	22.4	22.4	22.4	22.1	20.8	18.8	17.4	15.3	13.7
Former USSR	27.5	27.6	27.6	27.3	22.3	16.9	15.9	14.7	13.5
Eastern Europe	27.2	28.3	28.1	27.0	25.8	23.1	22.5	22.5	20.5
Africa	9.3	10.5	10.3	9.2	8.9	8.5	8.4	8.6	8.7
North Africa and Middle East	13.6	14.7	14.7	14.0	14.2	14.2	13.8	14.0	13.8
Sub-Saharan Africa	7.2	8.6	8.2	6.8	6.2	5.5	5.6	5.8	6.0

Note: USSR is Union of Soviet Socialist Republics. Calculations are five-year averages. Number of country observations (n) varies according to year: n = 92 (1989) and n = 109 (2010). Income and regional classification based on Annex A1, Table A1.2.
Source: UNIDO elaboration based on Groningen Growth and Development Centre 10-Sector Database (Timmer, de Vries and de Vries 2014), ILOSTAT (ILO 2015a) and KILM (2015b).

Expressing logs of the growth rate in relation to GDP per capita:

Value added per capita growth ≈ employment per capita growth + labour productivity growth

Figures 1.10 and 1.11 use the estimated growth of employment per capita and of labour productivity in six industries to show their changes and contributions to value-added per capita growth at different incomes.[13] The growth rate (Y axis) is an elasticity defined as the percentage change in each variable associated with a 1 percent change in GDP per capita. Thus, an elasticity of 1 indicates that an industry grows at the same rate as GDP per capita (PPP). The differences in the patterns across six manufacturing industries reveal the development characteristics of each industry: each is unique in its production process, factor intensity and technological development potential.

For large countries (Figure 1.10),[14] three low-tech industries (as defined in Annex A2)—food and beverages, textiles, and wearing apparel—show marked differences in employment generation, productivity growth and sustainability of growth. Food and beverages could sustain value-added growth (more than zero) over a wide range of incomes due to the combination of continued growth of labour productivity at a rate similar to GDP per capita growth and a very slow decline of employment. Textiles and wearing apparel show a similar development pattern in value-added growth, but textiles have less potential for increasing employment.[15]

> **“Countries need to lay the foundation for the growth of other industries**

Figure 1.8
Shares of developing and developed regions in global value added of low-, medium-, and high-tech manufacturing industries, 1972 and 2012

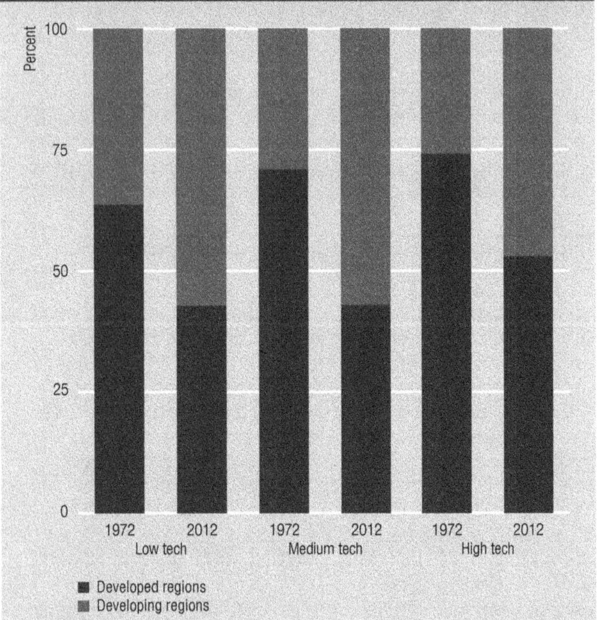

Note: Tech classifications based on Annex A2, income classification based on Annex A1, Table A1.2.
Source: UNIDO elaboration based on Lavopa and Szirmai (2015).

Figure 1.9
Changes in technology structure in manufacturing by developing region, 1972 and 2012 (percent of total manufacturing value added)

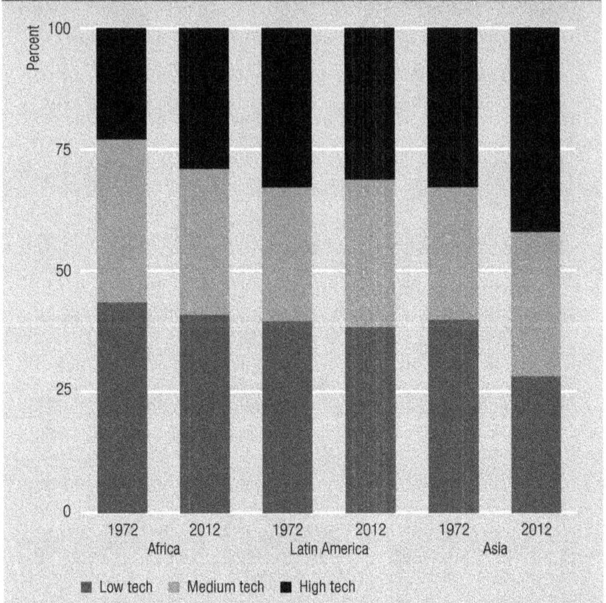

Note: See tech classifications based on Annex A2. Regional classification based on Annex A1, Table A1.2., except Africa here includes also Israel, Qatar, Oman, Kuwait, Bahrain and Cyprus.
Source: UNIDO elaboration based on Lavopa and Szirmai (2015).

That industry's growth relies on labour productivity growth, but since that is stable or decreases very slowly at high incomes, the decline of value added per capita is mainly caused by the contraction in employment.

Wearing apparel has huge employment-creation potential at low and lower middle incomes. Because of the limited room for labour productivity growth (especially at low and middle incomes), a rapid decline in employment growth leads to a fast decline in value-added growth as country income rises. That implies that technological change in wearing apparel is limited. So, before the employment potential of the wearing apparel industry is fully exhausted, countries need to lay the foundation for the growth of other industries.

The rubber and plastic industry is unique in its ability to sustain growth over a long period from low to high incomes. At low incomes, employment growth contributes to the value-added growth of the industry, whereas at high incomes, a gradual increase in labour productivity growth and a very slow decline in employment growth help the industry sustain its value-added expansion. The high-tech industry of electrical machinery and apparatus also sustains its value-added growth across incomes but differently from the rubber and plastic industry.

Electrical machinery and apparatus has high potential for labour productivity growth, even at a low income. That, combined with decent employment growth, allows the industry to develop rapidly at low and lower middle incomes, where its productivity growth can be two or three times higher than the growth in GDP per capita. Its employment growth rate declines faster than that of rubber and plastics as countries increase their income, but very fast growth in labour productivity more than offsets the decline in employment growth to sustain the fast growth of the industry's value added.

In large countries, the high-tech motor vehicle industry could experience rapid growth at low and middle incomes. Although it maintains a high growth rate of labour productivity well into high income, the

“Developing and high-income countries display wide differences in the way manufacturing drives economic growth

Figure 1.10
Patterns of employment, labour productivity and value added growth (large countries), 1963–2010

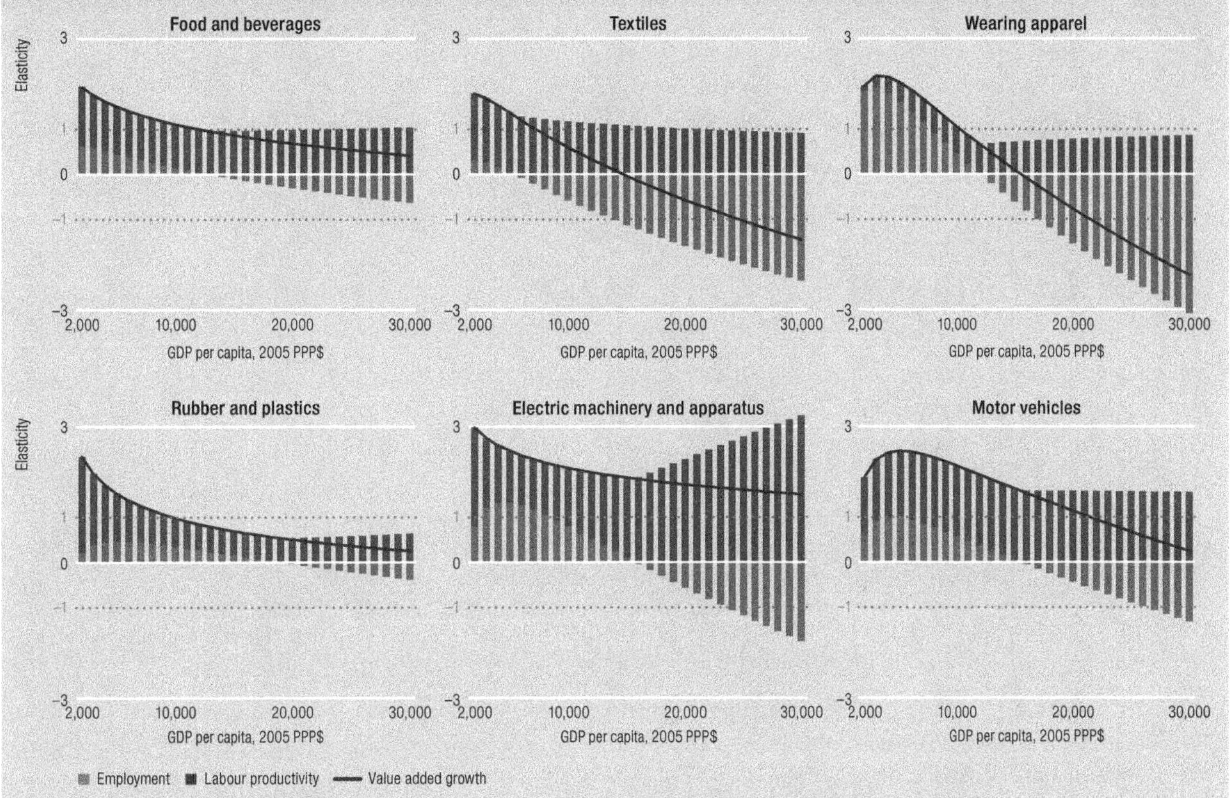

Note: Elasticity is measured as the percentage change in each of the three variables over the percentage change in gross domestic product per capita.
Source: UNIDO elaboration based on INDSTAT2 (UNIDO 2014c).

steady decline in its employment growth causes its value-added growth to slow.

Small countries (see Figure 1.11) generally exhibit change in value-added growth similar to those of large countries (Figure 1.10). The main differences are the lower growth of labour productivity and slower decline in employment in small countries. But the motor vehicle industry follows a very different development pattern: small countries are unlikely to see much development in that industry until they reach upper-middle-income stage, then the industry increases its value-added growth, mainly resulting from the growth of labour productivity.

How manufacturing sustains growth: High-income versus developing countries

Developing and high-income countries display wide differences in the way manufacturing drives economic growth (Figure 1.12). Average output growth rates are shown at the top of each figure.[16] In developing countries, contributions to output growth come mainly from capital investments, natural resources and energy. In high-income countries, they come from productivity—using labour-saving and resource-saving technology to increase output without significantly increasing factor inputs.

The following text examines three groupings of manufacturing industries—typical low tech, medium tech, and high tech—to assess how their production characteristics affect overall growth and factor contributions along country income lines.[17]

Low-tech industries. In these industries, high-income countries had output growth of –1.1 percent in textiles and textile products and of –3 percent in leather and

> **Productivity contributed less to the growth of labour-intensive industries in developing than in high-income countries**

Figure 1.11
Patterns of employment, labour productivity and value added growth (small countries), 1963–2010

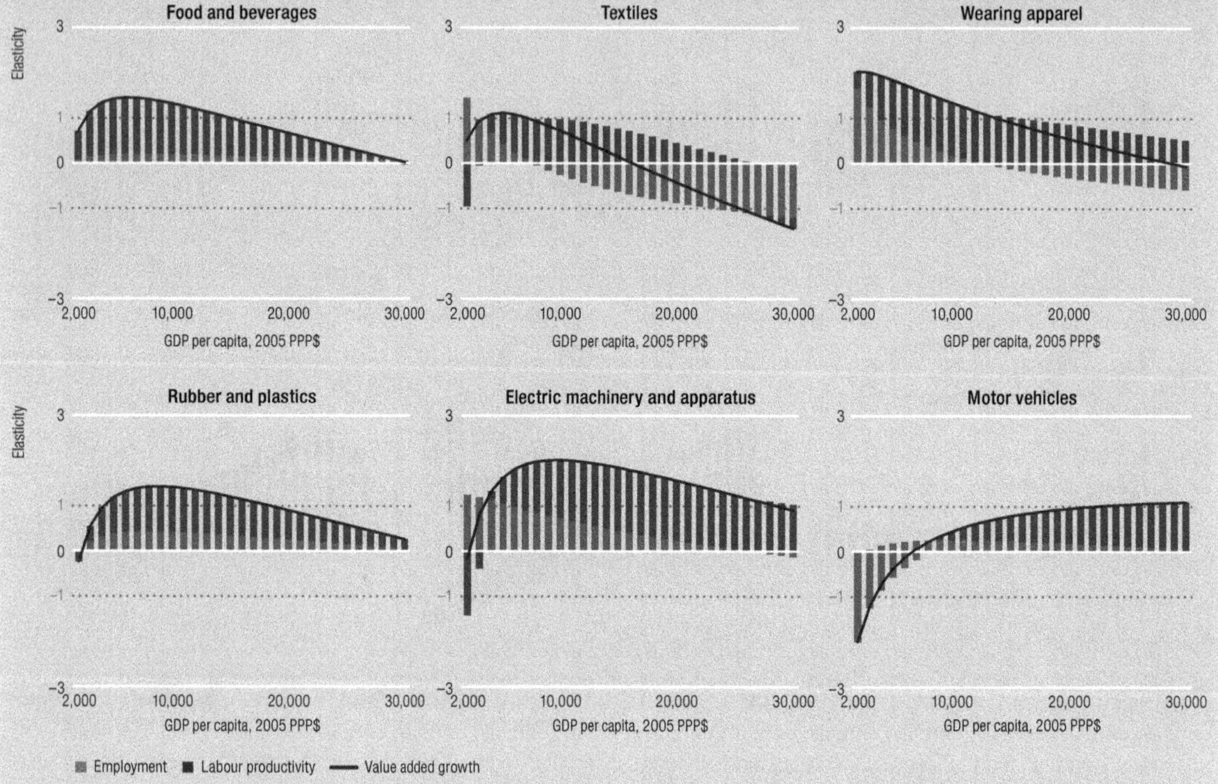

Note: Elasticity is measured as the percentage change in each of the three variables over the percentage change in gross domestic product per capita.
Source: UNIDO elaboration based on INDSTAT2 (UNIDO 2014c).

footwear (Figure 1.13), especially due to high negative shares of labour contribution (labour displacement). Conversely, in developing countries, both industries grew. The largest contribution to output growth for both industries came from energy, less from capital investment and labour,[18] whereas productivity growth made a positive contribution only to textiles. Overall, productivity contributed less to the growth of labour-intensive industries in developing than in high-income countries.

Medium-tech industries. Productivity was the largest source of growth for high-income countries in rubber and plastics and non-metallic mineral industries (Figure 1.14). But for developing countries in those industries—especially non-metallic minerals—the main contribution came from natural resources and energy, with productivity growth providing only a small contribution.

When countries industrialize further and move into the medium-tech group, the pollution intensity of manufacturing tends to rise (carbon dioxide emissions per unit of value added). That does not mean, however, that the growth of medium-tech, resource-based industries always must be driven by heavy increases in energy and natural resource inputs, as evidenced by the relatively low contributions of energy and natural resources to the growth of these activities in high-income countries.

High-tech industries. High-income countries have an advantage in high-tech industries and clearly have the potential to achieve faster growth in those industries than in low- or medium-tech industries (Figure 1.15). That advantage drives structural change

> **In developing countries, productivity accounts for a significant share of the growth of high-tech industries**

Figure 1.12
Annual average manufacturing growth and factor contributions, high-income and developing countries, 1995–2007

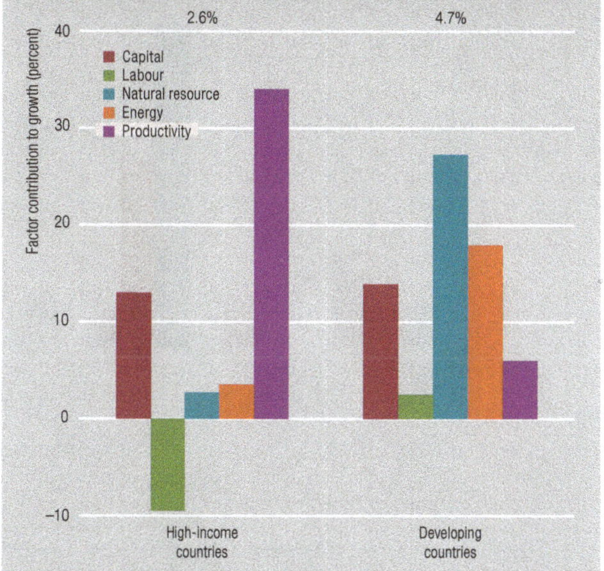

Note: The analysis covers 40 countries. Based on income, eight are developing countries and the rest are high income. Income classification based on Annex A1, Table A1.2.
Source: UNIDO elaboration based on World Input-Output Database (Timmer and others 2015).

in manufacturing and shifts resources towards high-tech industries at higher incomes. Productivity is the dominant contributor to the growth of high-tech industries, and their growth does not depend significantly on an increase in the use of energy and natural resources.

In developing countries, productivity accounts for a significant share of the growth of high-tech industries. But other factors—energy and capital investment—make a nontrivial contribution, too. So, although the importance of productivity for the growth of high-tech industries is common to developing and high-income countries, developing countries differ in that the increased use of energy and labour accompanies growth—hence, expanding these activities is more inclusive in job terms, but it is less sustainable.

Structural change and inclusive and sustainable industrial development

ISID has the three complementary dimensions of sustained economic growth, social inclusiveness and environmental sustainability. The objective is thus industrial development that maximizes the synergies and

Figure 1.13
Selected low-tech, labour-intensive industries, 1995–2007

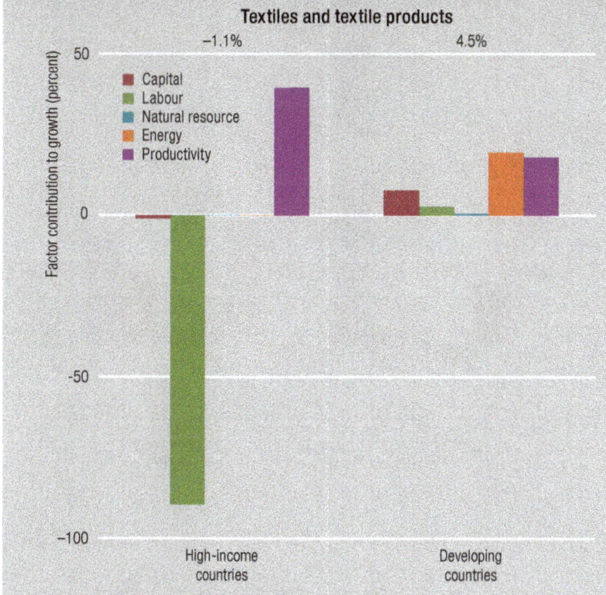

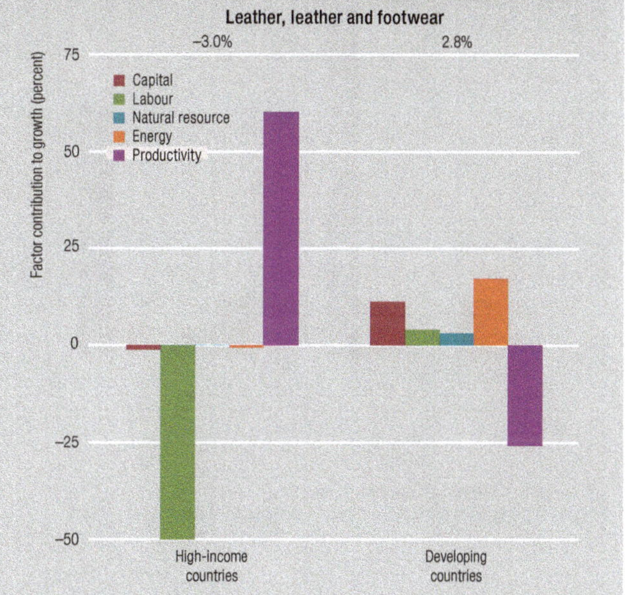

Note: Income classification based on Annex A1, Table A1.2
Source: UNIDO elaboration based on World Input-Output Database (Timmer and others 2015).

The importance of productivity for the growth of high-tech industries is common to developing and high-income countries

Figure 1.14
Selected medium-tech, resource-based industries, 1995–2007

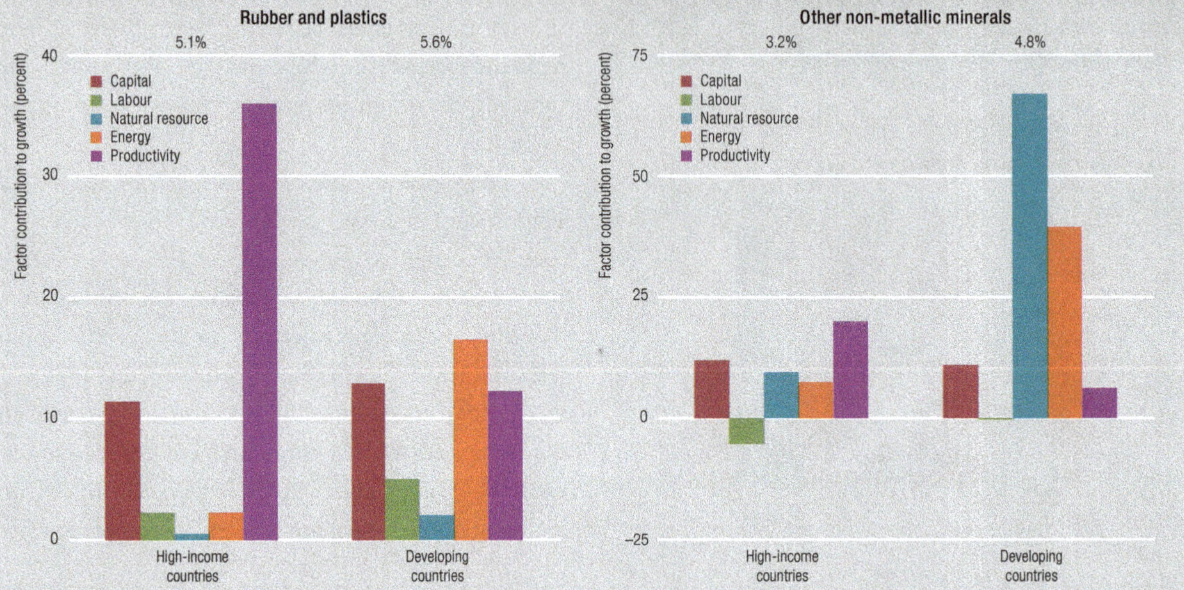

Note: Income classification based on Annex A1, Table A1.2.
Source: UNIDO elaboration based on World Input-Output Database (Timmer and others 2015).

Figure 1.15
Selected high-tech, technology-intensive industries, 1995–2007

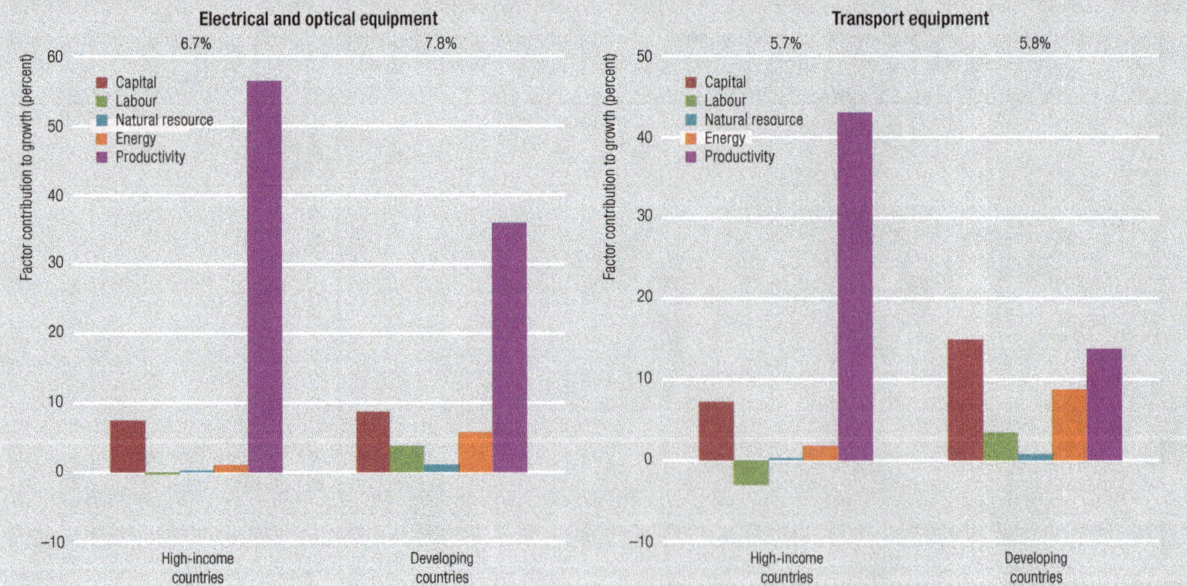

Note: Income classification based on Annex A1, Table A1.2
Source: UNIDO elaboration based on World Input-Output Database (Timmer and others 2015).

> **ISID has the three complementary dimensions of sustained economic growth, social inclusiveness and environmental sustainability**

minimizes trade-offs among the three dimensions for greater economic, social and environmental welfare for countries and the world.

This section first analyses the two additional aspects of inclusive and sustainable manufacturing development separately by looking at the trends of their individual components over the long term. After their individual patterns are observed, the two aspects will be combined to see how the two change together as countries develop.

Manufacturing's inclusiveness

Equation *E1* indicates how value creation by the manufacturing sector translates into inclusive industrial development. Inclusive industrial development is defined here in two parts. The first is manufacturing wage level adjusted by the wage distribution in the manufacturing sector, which is (*Mwage_equality*)[19] times the manufacturing average manufacturing wage (*Mwage*), and the second is total manufacturing employment (*Memp*). A higher value of each term in the numerator—the adjusted wage or employment—contributes to a higher level of industrial inclusiveness.

$$\frac{M\ Inclusiveness}{M\ Output} = \frac{(Mwage\_equality)^{*}\ Mwage \times Memp}{MVA} \qquad \text{E1}$$

Note: M is manufacturing; MVA is manufacturing value added.

Offsetting trends in wages and employment generate the largely flat trend observed in inclusiveness (Figure 1.16). As expected, the adjusted wage increases along a country's development (Figure 1.17), a positive relationship that becomes clearer after income reaches around \$2,000–3,000 GDP (PPP) per capita. But employment intensity (manufacturing employment per unit of value added) first increases and then steadily declines as countries move towards higher incomes (Figure 1.18).

The decline in employment intensity in manufacturing (see Figure 1.18) stems from structural changes in manufacturing, reflecting a combination of higher concentration in capital-intensive industries and an

Figure 1.16
Manufacturing inclusiveness

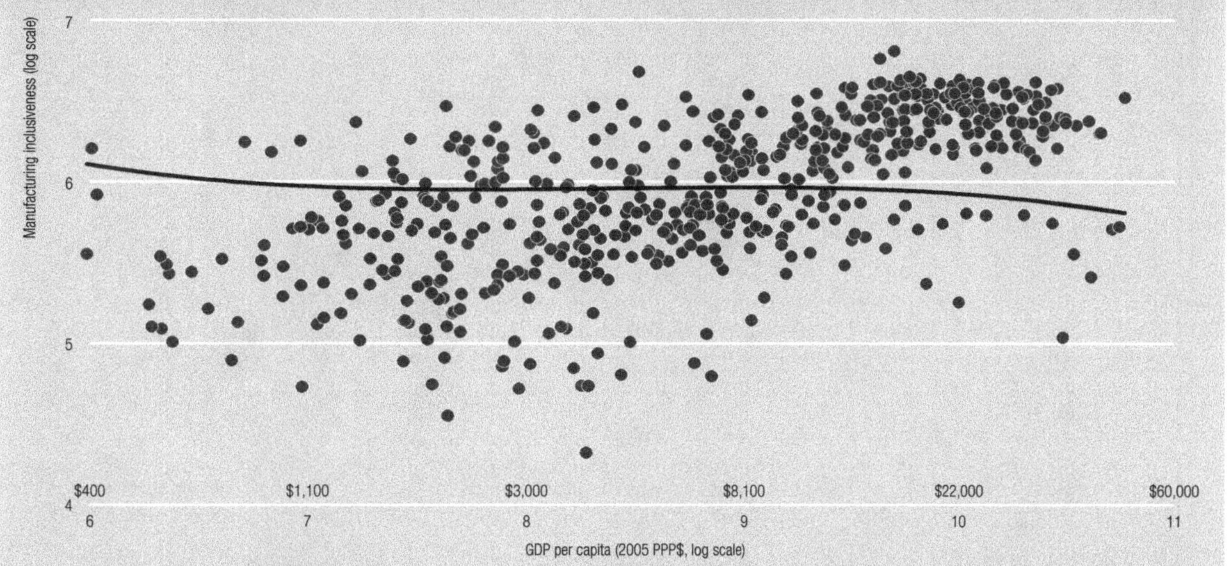

Note: PPP is purchasing power parity. Calculations are five-year averages and cover 98 countries between 1970 and 2013. Manufacturing inclusiveness is defined in equation E1.
Source: UNIDO elaboration based on INDSTAT2 (UNIDO 2014c), Penn World Tables (Feenstra and others 2015), UN National Accounts Statistics (UN 2014b), World Input-Output Database (Timmer and others 2015), Groningen Growth and Development Centre 10-Sector Database (Timmer, de Vries and de Vries 2014), ILOSTAT (ILO 2015a), KILM Database (ILO 2015b), EU KLEMS Database (O'Mahony and Timmer 2015), CAIT Climate Data (WRI 2015) and UTIP-UNIDO Industrial Pay Inequality Database (University of Texas and UNIDO 2015).

> **"The equity-adjusted wage increases along a country's development**

Figure 1.17
Equity-adjusted wage

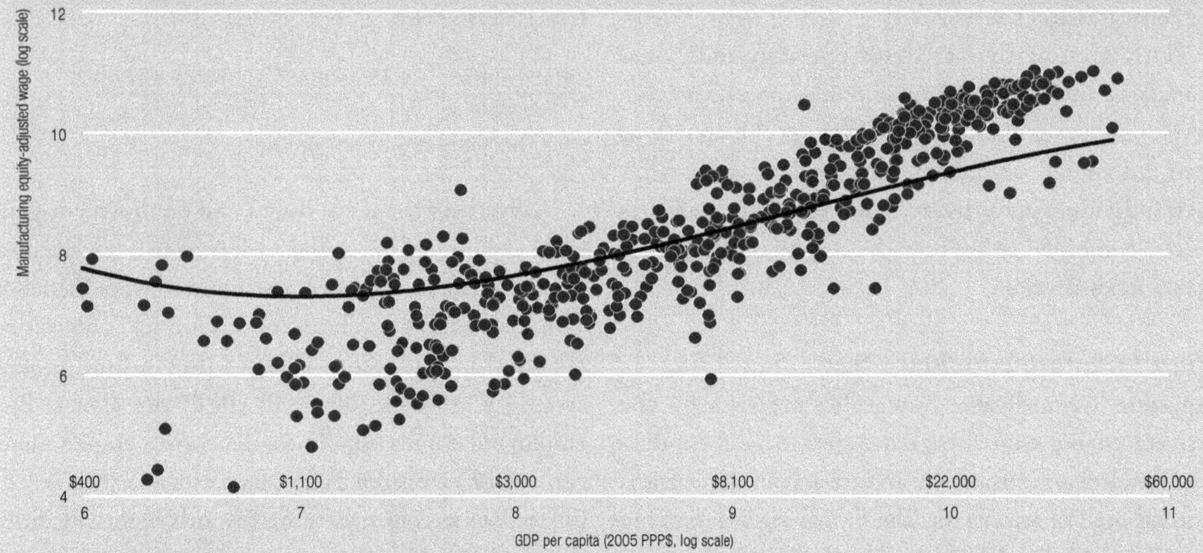

Note: PPP is purchasing power parity. Calculations are five-year averages and cover 98 countries between 1970 and 2013. Manufacturing equity-adjusted wage is calculated as the average manufacturing wage level adjusted by the wage distribution within the manufacturing sector. Wages are defined as the yearly average compensation to employees in constant 2005 $ and are "weighted" by one minus the Theil index of wage inequality; see Industrial Pay Inequality Database (UTIP UNIDO 2015).
Source: UNIDO elaboration based on INDSTAT2 (UNIDO 2014), Penn World Tables (Feenstra and others 2015), UN National Accounts Statistics (UNSD 2015b), World Input-Output Database (Timmer and others 2015), EU KLEMS (O'Mahony and Timmer 2015) and Industrial Pay Inequality Database (UTIP UNIDO 2015).

Figure 1.18
Employment intensity

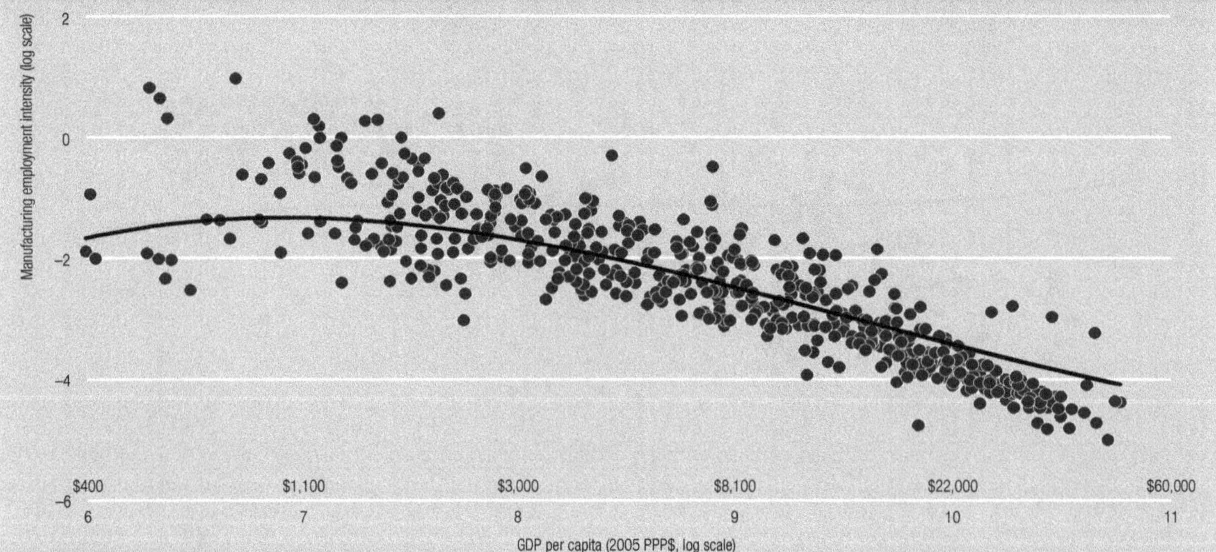

Note: PPP is purchasing power parity. Calculations are five-year averages and cover 98 countries between 1970 and 2013. Manufacturing employment intensity is defined as the number of manufacturing workers per unit of manufacturing value added in constant 2005 $.
Source: UNIDO elaboration based on Penn World Tables (Feenstra and others 2015), UN National Accounts Statistics (UN 2014b), Groningen Growth and Development Centre 10-Sector Databas (Timmer, de Vries and de Vries 2014), ILOSTAT (ILO 2015a), KILM Database (ILO 2015b), EU KLEMS Database (O'Mahony and Timmer 2015) and UTIP-UNIDO Industrial Pay Inequality Database (University of Texas and UNIDO 2015).

> **The decline in employment intensity in manufacturing stems from structural changes in manufacturing**

Figure 1.19
Changes in the level of employment in manufacturing industries, 1963–2010

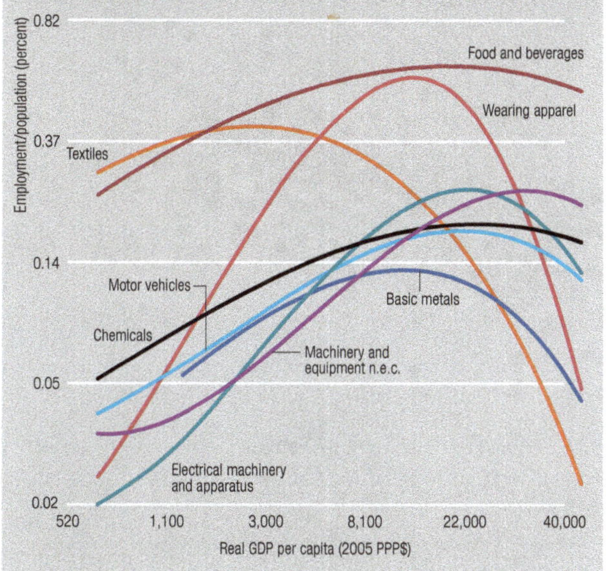

Note: GDP is gross domestic product; PPP is purchasing power parity; n.e.c. is not elsewhere classified.
Source: UNIDO elaboration based on Penn World Tables (Feenstra and others 2015) and INDSTAT2 (UNIDO 2014c).

Figure 1.20
Changes in the value added of employment in manufacturing industries, 1963–2010

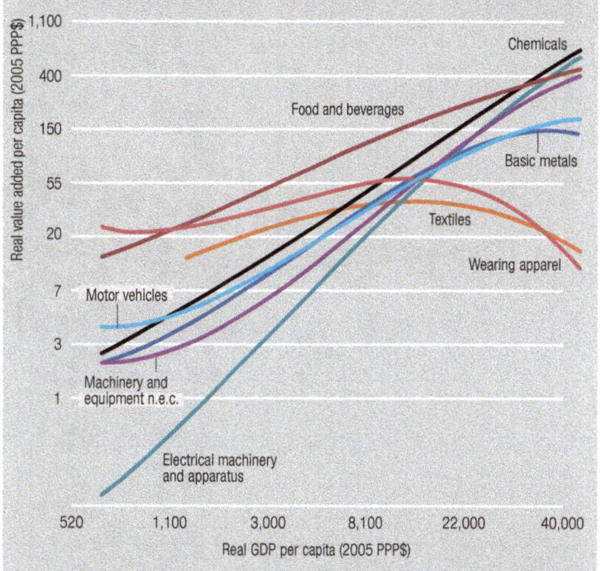

Note: GDP is gross domestic product; PPP is purchasing power parity; n.e.c. is not elsewhere classified.
Source: UNIDO elaboration based on Penn World Tables (Feenstra and others 2015) and INDSTAT2 (UNIDO 2014c).

overall rise in capital intensity in manufacturing industries. The three major sources of manufacturing employment—food and beverages, textiles and wearing apparel—are more labour intensive than other industries,[20] but textiles and wearing apparel generally cease to generate employment by the time countries graduate from upper-middle-income status (Figure 1.19).

After those labour-intensive industries start reducing employment, employment may still increase in emerging capital-intensive industries, such as chemicals and electrical machinery and apparatus. But although those industries contribute to MVA (Figure 1.20), they do not generate as much employment as do the labour-intensive industries, which are shedding labour (see Figure 1.19). Further, the capital and technology intensity of many manufacturing industries increases as countries move to higher incomes (as the U-shaped employment curves in Figure 1.19 and the rising value-added curves of many manufacturing industries in Figure 1.20 imply).

Manufacturing's environmental sustainability

One way to express manufacturing's environmental sustainability is MVA per unit of manufacturing carbon dioxide (MCO_2) emission ($E2$), which shows the emission efficiency of value creation by the manufacturing sector.[21] Value added per unit of manufacturing CO_2 does not fully capture manufacturing sustainability but is a measurable proxy of its trends.

$$\frac{M\,Output}{M\,Environmental\,Impact} = \frac{MVA}{MCO_2\,emission} \quad \text{E2}$$

Note: M is manufacturing; MVA is manufacturing value added.

In Figure 1.21, we see an improvement of $E2$ at higher incomes after an initial worsening. Subsector results reveal some underlying changes (Figure 1.22). Many manufacturing industries improve their emission performance as countries move to higher incomes. Industries that do not improve their performance—or that see a modest worsening, such as food and beverages, textiles and wearing apparel, and non-metallic mineral industries—are at

> **The U-shape of ISID is the result of an improvement in manufacturing's environmental sustainability**

Figure 1.21
Manufacturing environmental sustainability

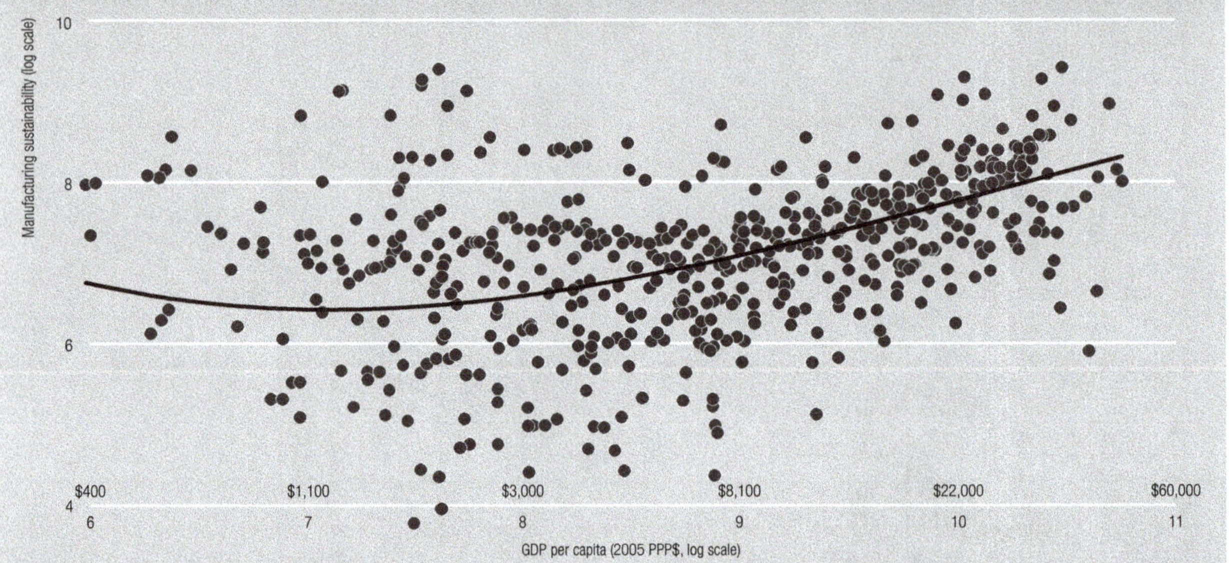

Note: PPP is purchasing power parity. Calculations are five-year averages and cover 98 countries between 1970 and 2013. Manufacturing sustainability is defined in equation E2 as the manufacturing value added in constant 2005 \$ per megaton of manufacturing CO_2 emissions.
Source: UNIDO elaboration based on Penn World Tables (Feenstra and others 2015), UN National Accounts Statistics (UN 2014b) and CAIT Climate Data (WRI 2015).

Figure 1.22
Real value added per unit of CO_2 emission in various manufacturing industries, 1963–2010

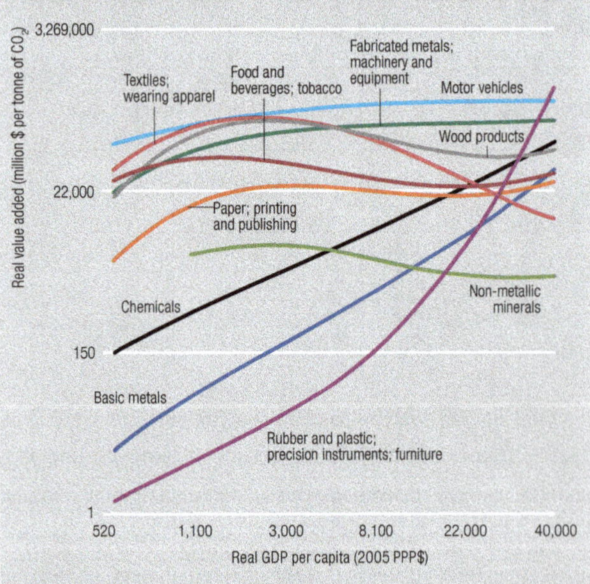

Note: CO_2 is carbon dioxide; PPP is purchasing power parity.
Source: UNIDO elaboration based on Fuel Combustion Statistics (IEA 2015b) and INDSTAT2 (UNIDO 2014c).

a relatively early stage of development. Except non-metallic minerals, they are not heavy emitters. As countries develop, industries such as chemicals, basic metals and rubber and plastic greatly improve their emission performance, transforming from dirty to relatively clean industries. Their value added per unit of emissions approaches the levels of the less polluting sectors, such as machinery and equipment and motor vehicles.

Linking inclusiveness and environmental sustainability—the ISID index

As the following equation illustrates, putting the separate analyses on manufacturing inclusiveness and sustainability together, we can observe how structural change in manufacturing along income levels shifts ISID as a whole. Equation *E3* is a proxy for the impact of manufacturing output per unit of manufacturing CO_2 emission on people's well-being in the manufacturing sector or inclusiveness. It addresses the extent of inclusive industrial development achieved per unit of environmental impact.

Countries at low and lower middle incomes have opportunities to create a large number of formal manufacturing jobs

$$\underbrace{\frac{(Mwage\_equality) * Mwage \times Memp}{MVA}}_{\text{Inclusive industrial development}} \times \underbrace{\frac{MVA}{MCO_2 emission}}_{\text{Sustainable industrial development}} = \underbrace{\frac{(Mwage\_equality) * Mwage \times Memp}{MCO_2 emission}}_{\text{Inclusive and sustainable industrial development}} \quad \text{E3}$$

Note: M is manufacturing; MVA is manufacturing value added.

Figure 1.23 shows the trend of ISID expressed in *E3* as economies develop. The ISID index tends to deteriorate slightly in the early stage of industrialization up to around $3,000 GDP per capita (2005 PPP) and then to improve as income increases.

The U-shape of ISID is the result of an improvement in manufacturing's sustainability (rising value added per unit of emissions), with a largely steady level of manufacturing inclusiveness. The inclusiveness component does *not* have a clear trend across incomes, although that does not mean that countries have similar inclusiveness: the scattered plots show significant differences across countries for the same income (see Figure 1.16).

The sustainability component exhibits an upward trend after deterioration at low incomes, as the carbon dioxide efficiency of manufacturing output first decreases and then starts improving (see Figure 1.21). Even though the figure shows that carbon dioxide inefficiency bottoms out at a fairly low income, variances among countries are very high up to around $8,000 GDP per capita. Only then does the upward pattern become clearer.

Trade-offs over time, emanating from structural change

The three components—equity-adjusted wage, employment intensity and environmental sustainability of ISID (the first two belong to inclusiveness)—take either U or inverted U shapes, which means that a change can occur in all three, typically between the lower and upper middle incomes, and there is usually a trade-off between components.

Once industrialization takes off, countries at low and lower middle incomes have opportunities to create a large number of formal manufacturing jobs because their cheaper wages provide them with a comparative advantage in labour-intensive industries, such as textiles and wearing apparel (see Figure 1.19). Their

Figure 1.23
ISID index and GDP per capita—a shallow U shape

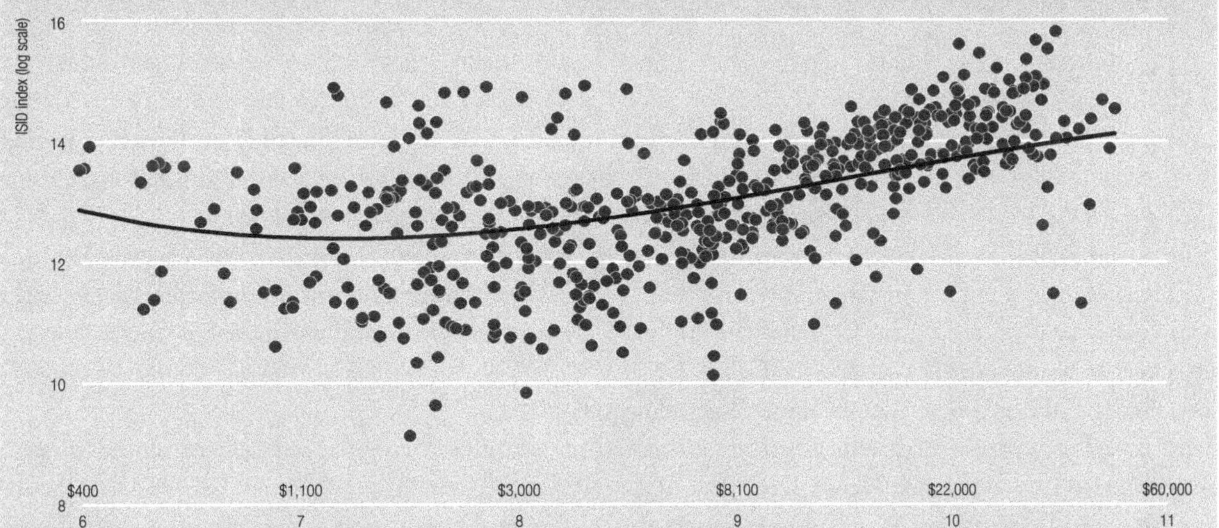

Note: ISID is inclusive and sustainable development; PPP is purchasing power parity. Calculations are five-year averages and cover 98 countries between 1970 and 2013. The ISID index is defined in equation E3.

Source: UNIDO elaboration based on INDSTAT2 (UNIDO 2014c), Penn World Tables (Feenstra and others 2015), UN National Accounts Statistics (UN 2014b), World Input-Output Database (Timmer and others 2015), Groningen Growth and Development Centre 10-Sector Database (Timmer, de Vries and de Vries 2014), ILOSTAT (ILO 2015a), KILM Database (ILO 2015b), EU KLEMS Database (O'Mahony and Timmer 2015), CAIT Climate Data (WRI 2015) and UTIP-UNIDO Industrial Pay Inequality Database (University of Texas and UNIDO 2015).

> **At low incomes, countries' low-tech manufacturing industries are relatively clean**

manufacturing wages in these activities might be much lower than those in capital-intensive industries, so the wage inequality across manufacturing industries can be high, which lowers the manufacturing wage equity part of the inclusiveness term in the ISID index. However, what matters for countries in transition from an agrarian to a modern economy is generating many formal manufacturing jobs that pay more than jobs in the agricultural and subsistence sectors. For that to happen, rapid growth of export-oriented, labour-intensive industries is important.

At low incomes, countries' low-techmanufacturing industries are relatively clean because the labour-intensive industries—such as textiles, wearing apparel and food and beverages—have high value-added performance per unit of CO_2 emission (see Figure 1.22). So, from a structural change perspective, industrialization for low-income countries can be conducive to inclusive and sustainable growth, which is often more difficult at higher stages of development. Even though labour-intensive industries are less emission-intensive than are heavy industries, emissions for the economy as a whole might increase as countries shift from an agricultural to a more industrial economy. So, mitigation measures will still be important for low-income countries.

As countries move to middle incomes, their rising skills and capital accumulation often open the door to more capital-intensive resource-processing industries, such as basic metals and chemicals industries. Inclusiveness is likely to improve further due to the continued expansion of labour-intensive industries, the increase in employment in capital-intensive industries as they emerge (see Figure 1.19) and the gradual increases in manufacturing wages (see Figure 1.17). The share of labour compensation in manufacturing value added could remain constant because the value added is also increasing (see Figure 1.16). Increases in the equity-adjusted wage and employment are important since they contribute to the inclusiveness component of the ISID index (the right-hand side of the ISID equation, *E3*). As capital-intensive, resource-based industries emerge, however, sustainability can deteriorate since those industries tend to be less emission-efficient relative to labour-intensive industries —at least at an earlier stage of their development (see Figure 1.22). This is the stage indicated by the bottom of the U-shaped curve of sustainability (see Figure 1.21), which countries would reach before they move to upper middle income.

Finally, as countries develop further and move to upper middle and high incomes, they tend to experience a decline of labour-intensive industries but an increase in opportunities to develop capital- and technology-intensive industries (see Figure 1.20). At higher incomes, these industries usually have high output-to-emission performance (see Figure 1.22), so manufacturing's sustainability usually improves (see Figure 1.21). But those industries employ much less labour to produce one unit of MVA than labour-intensive industries. And manufacturing as a whole intensifies the use of capital and technology relative to labour in production. So, even though manufacturing wages increase as GDP per capita rises, employment intensity steadily falls at higher incomes (see Figure 1.18).

As the trend continues, countries eventually reach the mature stage of industrialization (or deindustrialization). As a result of higher wages and better wage equity across manufacturing industries (see Figure 1.17), inclusiveness within manufacturing may not deteriorate, but its contribution to the inclusiveness of the whole economy certainly declines at very high incomes because of the sector's limited capacity to absorb the country's labour force.

Chapter 2 looks deeper into the centrality of technological change and innovation for long-term, inclusive and environmentally sustainable growth.

Notes

1. Manufacturing-related service employment is defined as the employment generated in the service sector due to the demand from the manufacturing sector based on the existence of an input–output relationship.
2. Positive episodes are defined as follows: For each country, a year is considered to be part of a

positive growth episode if its GDP per capita is higher than that of the previous year for two successive years.

3. An important strand of literature argues that developing countries are more vulnerable to interruptions of growth, in part because of their institutional characteristics, and that this very vulnerability determines long-run differences in growth performance (Bluhm and Szirmai 2012; North, Wallis and Weingast 2009).
4. Volatility is defined as the degree of variation of the rate of growth in any given period. Volatility can be measured by the standard deviation of growth, the coefficient of variation, the Gini index or the Theil index. In the tables we primarily use the Theil index.
5. A country is defined as being in a development trap if it remains stuck in a given income category (low income, lower middle income, etc.) significantly longer than the average period it takes for a country to graduate out of that category on average (Lavopa and Szirmai 2014).
6. After 1990, the database drops Czechoslovakia, the USSR and Yugoslavia. At the same time, these new countries appear: Armenia, Azerbaijan, Belarus, Bosnia, Croatia, Czech Republic, Estonia, Georgia, Kazakhstan, Kyrgyzstan, Latvia, Lithuania, Macedonia, Moldova, Russian Federation, Serbia/Montenegro/Kosovo economy, Slovakia, Slovenia, Tajikistan, Turkmenistan, Ukraine and Uzbekistan.
7. Improving relative position is based on relative income rankings. It is not the same as catch-up, which means reducing the percentage gap in GDP per capita relative to the lead economy. A country could improve its ranking even though its GDP gap increases.
8. The only exception to this finding is the two countries that shift from the fourth to the fifth quintile, which have the same average duration as countries that remain in the third quintile.
9. This index sums the absolute value of changes in sectoral shares of value added. Obviously, these changes in shares occur as a result of differences in growth rates of value added between the sectors. These growth rates, in turn, are the result of two sources: productivity changes and changes in the use of inputs (labour and capital).
10. Those 10 sectors are agriculture; mining; manufacturing; utilities; construction; trade; restaurants and hotels; transport, storage and communication; finance, insurance, real estate and business services; public services and other. These data are taken from a database that was created for this project and that combines data from the GGDC 10-sector database (Timmer, de Vries, and de Vries 2014), World Input-Output Database (Timmer and others 2015), and national accounts data reported to the UN (UN 2014b).
11. Relative to the share of manufacturing employment (Figure 1.7), the share of MVA of developing countries (in constant price) increased more.
12. A population of 12.5 million is the threshold to distinguish large and small countries.
13. The analysis includes 60 to 100 countries, depending on the industries, that have the necessary data.
14. See endnote 12.
15. That is partly because the textile industry tends to have a high share of employment (as does the food and beverage industry) even at a very low income.
16. The analysis is based on the World Input-Output Database (Timmer and others 2015), which covers 40 countries. Based on income, eight are developing countries and the rest are high income. To focus on inclusiveness, sustainability and productivity aspects, the analysis assesses for intermediate inputs, for example, only the contributions of energy and mining (from domestic as well as foreign sources) to output growth; other intermediate inputs are excluded. Thus, the shares of each factor contribution do not add up to 100 percent. The natural resource data come from "Mining and Quarrying" in the World Input-Output Database.
17. For the classification, refer to Annex A2.
18. None of the eight developing countries is from the low-income group, and only one country, India, is

from the lower middle income group. The rest of the developing countries belong to the upper middle income category. So the results may not reflect the conditions of countries at the early stage of development. That might be why the labour contribution to the growth of labour-intensive industries in the developing countries group is relatively low.

19. We use the Theil index for manufacturing income inequality, so 1–Theil is used.
20. To determine the level of labour intensity, employment per value added was estimated for 18 manufacturing industries at two income levels of $5,000 and $20,000 GDP per capita because labour intensity changes along income level. If an industry's labour intensity is higher than the median of 18 manufacturing industries at both incomes, it is considered to be a labour-intensive industry. But if an industry's labour intensity is lower than the median at both income levels, it is considered to be relatively capital intensive. The five most labour-intensive industries are wearing apparel, textiles, wood products, fabricated metals and food and beverages.
21. Environmental impacts include factors in addition to CO_2 emissions, such as material waste and water pollution. But due to the lack of data, we illustrate the ISID concept using only CO_2 emission data.

Chapter 2

Technological change, structural transformation and economic growth

Four stylized facts can describe how technology and innovation drive economic development and growth. First, the abilities of a country to use existing technology and to innovate determine its economic performance in the long run through structural change. But because developing new capabilities to use and assimilate technology is very hard, the convergence of living standards between countries has generally been very slow or even absent. Only a few countries have moved from relative poverty to relative development. The rich developed countries have high levels of technological sophistication and account for the large majority of investment in science and technology, primarily research and development (R&D). Poor countries have much lower technological capabilities and invest much less in R&D.

Second, in the past 15 years or so, capabilities revealed in export markets have been a good way to distinguish poor countries likely to grow slowly or rapidly. In other words, the kinds of capabilities that enable a country to catch up with the global frontier are closely related to performance in global markets. Globalization offers opportunities for catch-up through knowledge diffusion, but opening to global markets does not automatically lead to growth.

Third, using international export markets as a vehicle for economic growth requires firms to move into new product categories with higher complexity and technology. Structural change is driven mainly by technological change that occurs at very different rates in different sectors in the economy. Moving into new export markets means that technological change affects some areas of the economy more than others and those parts can stimulate rapid structural change.

Fourth, global markets, growing in recent decades, have been the locus of a high degree of structural change. The growth in developing countries coincides with the rising importance of internationally tradable goods. Especially since 2001, the share of production for foreign final demand has been climbing. Manufacturing and market services, both internationally tradable, lead in the globalization of production, with resource-based growth important in fewer countries.

Wanted: Technology and innovation to drive productivity and economic growth

The direct effect of technological change—fundamental and structural—on economic production and employment is hard to measure, so economists usually resort to measures of productivity. This section explores two aspects of productivity-increasing technological change: the relationship of technological change to structural change, and the different roles of productivity in developing and emerging economies, and in developed economies.

We look at two forms of productivity. The simplest and most directly relevant is labour productivity—that is, the value added per worker. The other more comprehensive form is total factor productivity (TFP), which takes into account the role of capital goods (for example, machinery and equipment, buildings).[1]

Total factor productivity—for structural change

Technological change is an important determinant of structural change because its rate differs greatly between economic sectors, thus stimulating economic growth that favours some sectors over others.

Structural change can be measured by the Finger-Kreinin index (see endnote 9 in Chapter 1). By decomposing that index into two parts—one related to productivity change (indicating technological change—TFP) and one to changes in the use of inputs (capital and labour)—one can assess which part of structural change is a direct result of technological change. The technological part of the index looks at the weighted (by value added shares) differences between sectoral rates of productivity change and the economy's rate of productivity change.

At low levels of development, the potential for rapid technological change and growth is highest

Figures 2.1 and 2.2 show the positive relationship (marked by the black regression lines) between the value of the structural change index and the contribution of the TFP part of the index. The relationship is very strong within manufacturing but weaker in the total economy.

Since most values are in the upper-right or lower-left corners of the figures, differences in TFP growth rates between sectors (within a country) are the decisive factors in structural change. High values of structural change, shown on the vertical axis, are mostly achieved by a large contribution of technological change, shown on the horizontal axis. (High values of structural change achieved by high factor contributions would lie in the lower-right part of the figures, which are unpopulated.)

The countries with high values of structural change and a high contribution of technological progress are mostly Eastern European countries, such as Slovakia, Hungary and the Czech Republic. Sweden, Finland and the Republic of Korea also have high scores on both dimensions. Figure 2.1 contains some exceptions in the upper-left quadrant. They are Eastern European countries where factor movements led to relatively fast structural change. The figure also shows a far larger group of countries with a low contribution of technological change and low structural change.

In the figures, the black line is the regression line and the red line is the 45-degree line. Points that lie below the 45-degree line have a higher contribution of technological progress than the value of the total index, which means that the contribution of changes in the production factors is negative. That happens mostly as a result of production factors moving out of sectors with rapid productivity change—say, because demand does not match productivity growth. This happens most often when overall structural change and the contribution of technological change are high —in the upper-right quadrant of the figures.

Labour productivity and technology

At low levels of development, the potential for rapid technological change and growth is highest, but limited absorptive capability often checks their realization (Chapter 3), so overall growth remains low. At

Figure 2.1

Total factor productivity growth as a source of structural change, nine-sector breakdown of total economy, 1995–2007

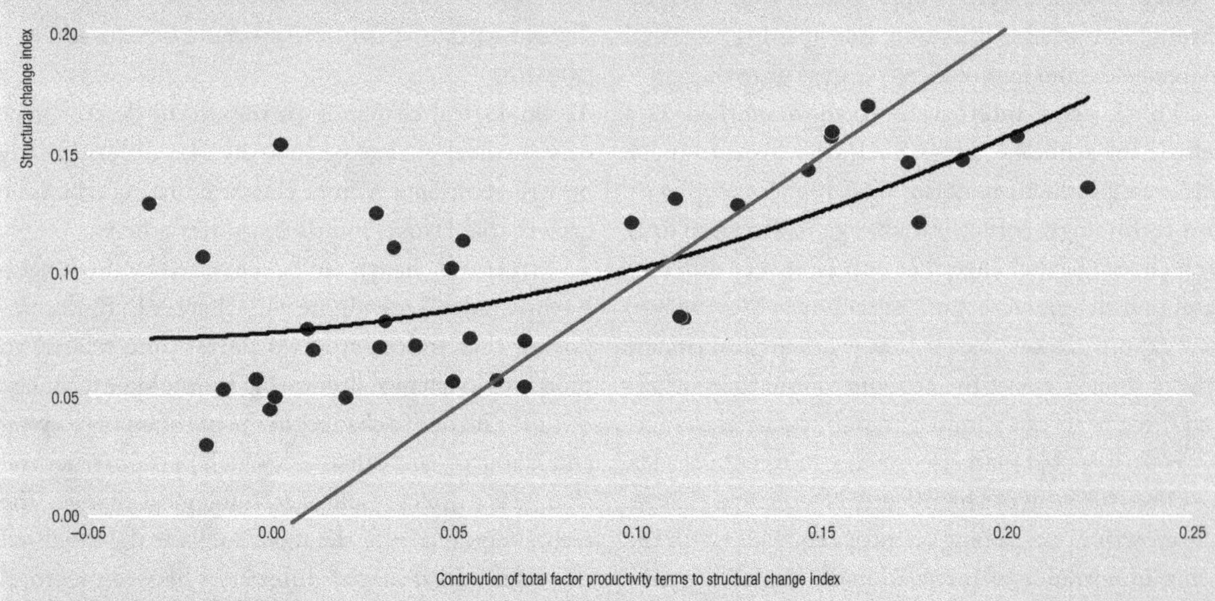

Note: The figure excludes one outlier, Slovenia.
Source: UNIDO elaboration based on Kaltenberg and Verspagen (2015).

At higher levels of development, such potential declines

Figure 2.2
Total factor productivity growth as a source of structural change, 14-sector breakdown of manufacturing, 1995–2007

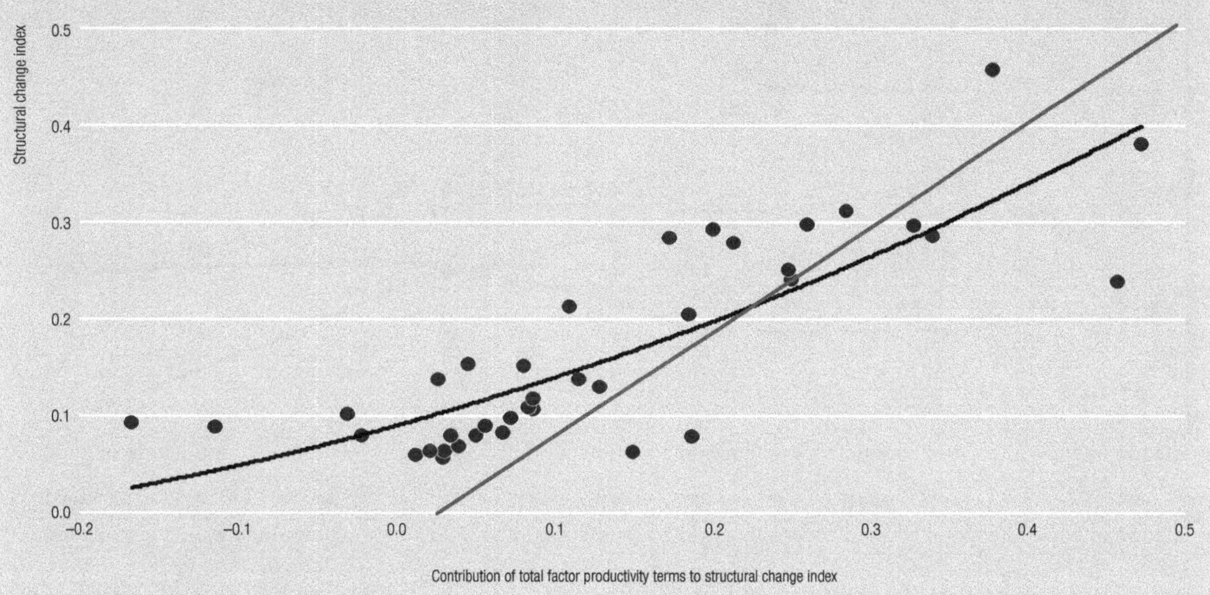

Source: UNIDO elaboration based on Kaltenberg and Verspagen (2015).

higher levels of development, such potential declines, but the probability of higher absorptive capability (see the following discussion) increases.

How have countries at different levels of development used technological progress for economic growth at the sectoral level? Kaltenberg and Verspagen (2015) apply a method of locally weighted regression-smoothing, which performs regressions on moving windows of subsets of an entire dataset to construct a smooth, but non-parametric, plot of one variable against another. They look at the evolution of growth rates of value added and labour productivity over 1995–2009 against the level of development indicated by the log of gross domestic product (GDP) per capita in 1995. They consider the same 14 manufacturing sectors as before and 73 countries at all levels of development. The 14 sectors are split into low, medium, and high tech, according to the definition in Annex A2.

Low-tech

In low-tech industries (Figure 2.3), labour productivity growth rates are small at low levels of development, but highly variable across countries, whereas growth rates of value added are high. Combined, the findings indicate that those countries are growing in a labour-intensive manner. The variability of value added growth rates is also high at low development levels, though it is lower than for labour productivity growth.

At higher levels of development, the labour productivity growth curve follows an S-shaped pattern, with a strong rise starting just before China's development level, and flattening again at the Czech Republic's development level. The variability of labour productivity growth rates gradually falls with development. While labour productivity growth rises with development, the growth rate of value added falls, monotonically. This implies that with rising levels of development, growth becomes more productivity (technology)-intensive, on average, with the richest countries showing declining employment, as productivity rises faster than value added. The variability of value added growth also falls with development, but only slowly.

> **In low-tech industries, labour productivity growth rates are small at low levels of development; the level of development appears as a strong determinant of the opportunities for growth**

Figure 2.3

Rates of growth of labour productivity and value added against level of development, low-tech manufacturing sectors, 1995–2009

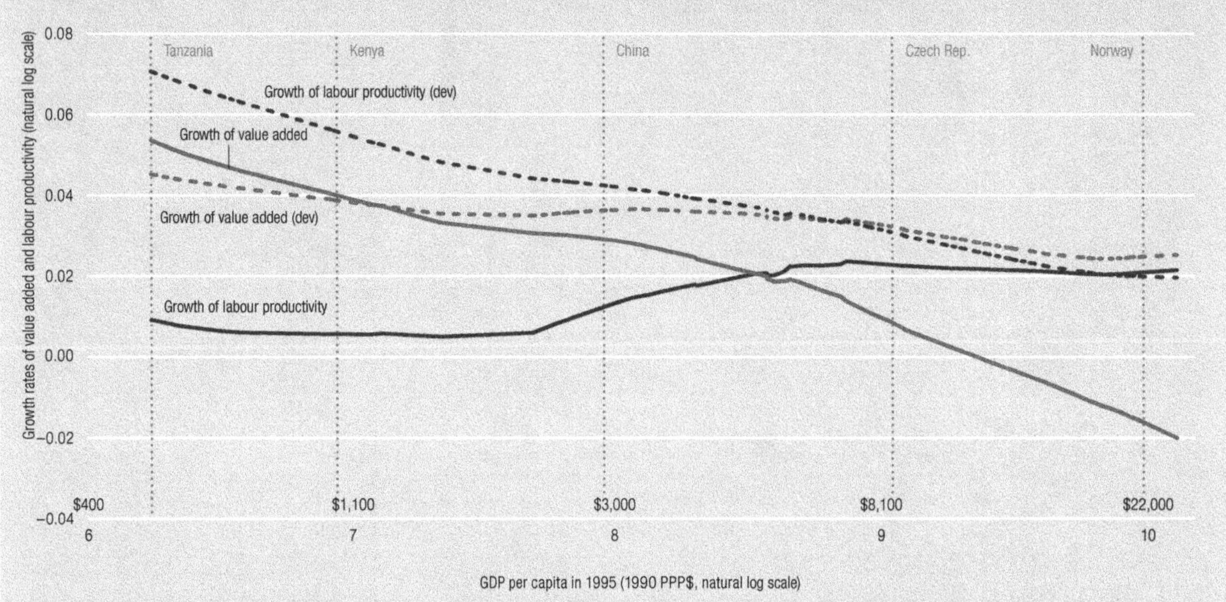

Note: GDP is gross domestic product; PPP is purchasing power parity. The solid lines show the smoothed growth rates, the dashed lines show (average) deviations from these smoothed rates, at the indicated development level. The difference between the valued added growth line and the labour productivity growth line is the growth rate of employment (factor use). For reference, the vertical grey lines indicate development levels of selected countries.

Source: UNIDO elaboration based on Kaltenberg and Verspagen (2015).

Medium-tech

In medium-tech industries, the pattern is similar to that in the low-tech sectors, apart from the low and high extremes of the income range (Figure 2.4). For low levels of development, both growth rates are lower, and labour productivity growth is negative. That means that for the first part of the paths, the growth of value added and labour productivity are rising with development. The value added growth curve peaks just before China's development level, whereas the labour productivity growth curve flattens just after that level. The variability of both growth rates is still very high at low levels of development and falls gradually with development. At the high end of the development range, both growth rates, as well as their variability, increase sharply. This pattern reflects the results of only a few (rich) countries. As with the low-tech sectors, value added growth is combined with employment growth at low levels of development, but with employment shrinking at high levels.

High-tech

The pattern for the high-tech sectors looks very similar to that for the medium-tech sectors (Figure 2.5). But at low levels of development, labour productivity growth is negative for only a very short income range, and it peaks at a much higher rate than for the other two technology sectors. Employment shrinkage in high-income countries is much smaller than in the medium- or low-tech sectors.

These three figures show that the level of development (as approximated here by income per capita) appears as a strong determinant of the opportunities for growth, a tendency associated with development traps (see later in this chapter). The results therefore show the importance of low-tech industries as avenues by which the absorption of foreign knowledge is easy, relative to other industries. But development is not destiny: the variability across countries is high, and even at low levels of development, high productivity growth is possible—again reflecting the importance of capabilities to absorb technological knowledge from abroad.

The variability across countries is high, reflecting the importance of capabilities to absorb technological knowledge from abroad

Figure 2.4

Rates of growth of labour productivity and value added against level of development, medium-tech manufacturing sectors, 1995–2009

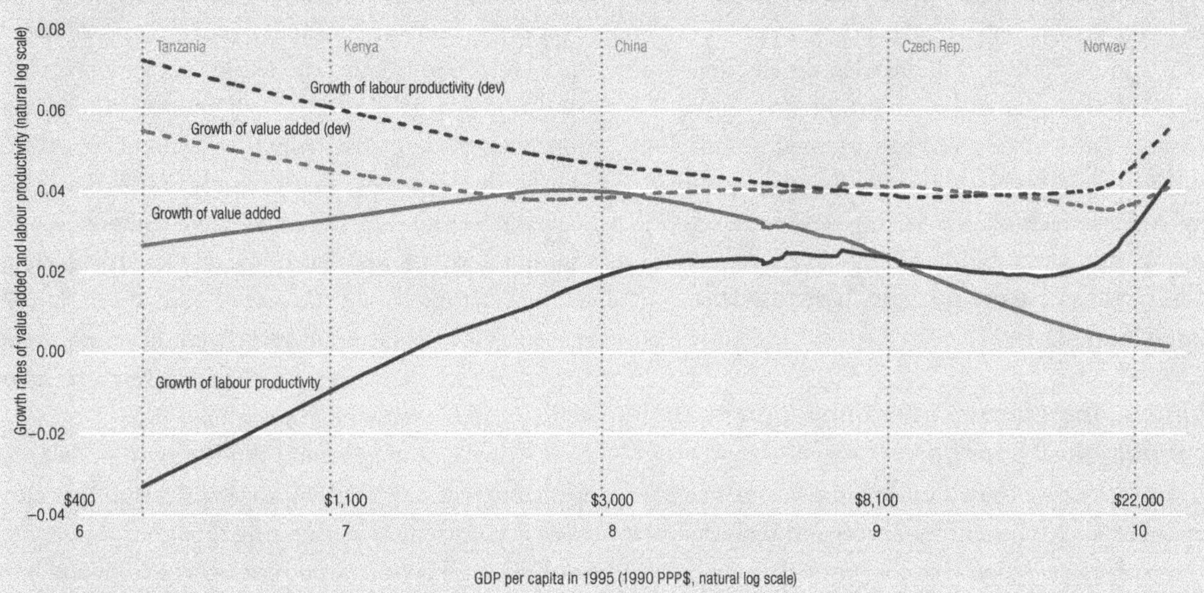

Note: GDP is gross domestic product; PPP is purchasing power parity. See Figure 2.3.
Source: UNIDO elaboration based on Kaltenberg and Verspagen (2015).

Figure 2.5

Rates of growth of labour productivity and value added against level of development, high-tech manufacturing sectors, 1995–2009

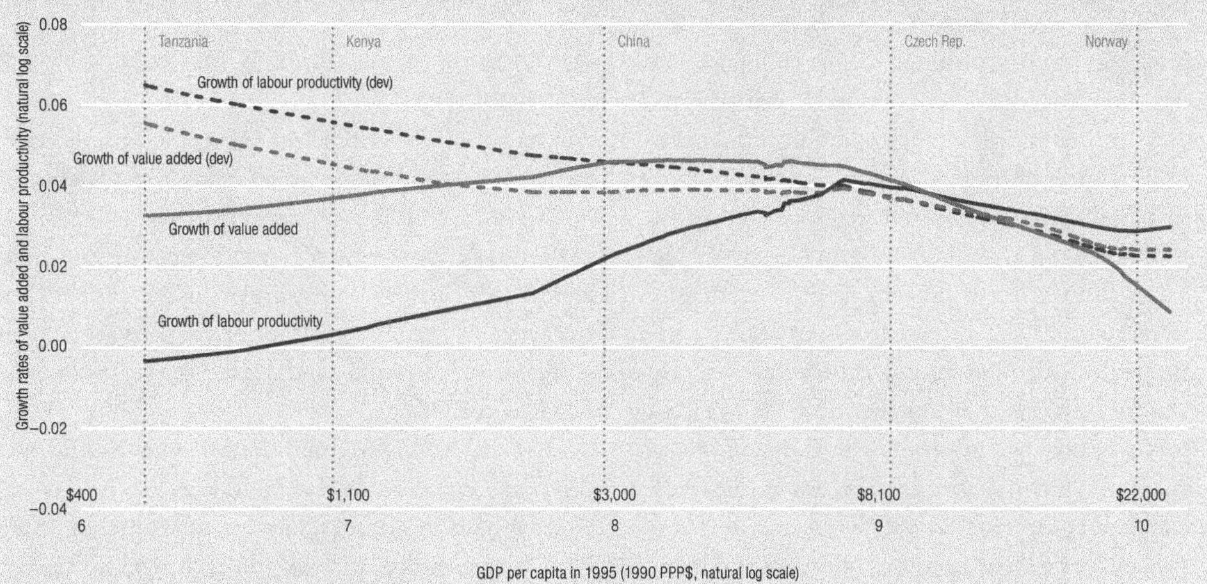

Note: GDP is gross domestic product; PPP is purchasing power parity. See Figure 2.3.
Source: UNIDO elaboration based on Kaltenberg and Verspagen (2015).

> **The imitation and adaptation of technologies streaming in from industrially advanced economies demands technological capabilities**

Value added growth rates for all three groups of industries are, on average, higher in developing countries. Those countries have advantages other than productivity growth—particularly low wages. Even in high- and medium-tech industries, developing countries find segments where labour-intensive growth is possible but does not lead to sustained productivity increases that enable higher levels of development, which is why technological change is so important.

Making technology and innovation work together

Radical and incremental innovation for technological change

New technologies based on broad areas of scientific research—such as energy, material and biological sciences and information technologies—that are rapidly diffused are examples of technological breakthroughs. Such new technologies are probably fuelling the next wave of global economic growth. A dozen new economically disrupting technologies that might have a huge effect in years to come include mobile Internet, cloud technology, advanced robotics, autonomous vehicles, energy storage, 3-D printing, advanced materials and renewable energy (Manyika and others 2013). These technologies have the potential to affect billions of consumers, hundreds of millions of workers and trillions of dollars of economic activity across different industries.

The types of radical technological advances described in Box 2.1, however, represent only a fraction of what the economic literature typically identifies as innovation and technological change. At the extreme, radical innovations can lead to what Schumpeter called "technological revolutions," clusters of innovations that together have a far-reaching effect in a range of industries or in the economy as a whole. Such technologies are also sometimes called "general purpose technologies"—that is, technologies that affect the entire economy, transforming both household life and the way firms conduct business.

Incremental innovations also drive economic growth. Their cumulative effect on long-run economic and social change may be even greater than that of radical innovations (Fagerberg 2006). In fact, the realization of economic benefits from radical innovations typically requires many incremental improvements. Such innovations enter the world in a very primitive condition and go through a long process of technical improvement and cost reduction (Rosenberg 2006). Some of today's most extended electronic devices—such as televisions, mobile phones and computers—are examples. When first introduced, their commercial use was restricted, and the cost of production so high that very few members of society could afford them. Enabling their later massive diffusion was a series of widespread incremental innovations.

One type of incremental innovation that deserves special attention, particularly for developing countries, is absorbing and imitating foreign technologies. Introducing a product or process in a new context is by definition an innovation and often requires considerable effort and capability to adapt it to the local context. In fact, the imitation and adaptation of technologies streaming in from industrially advanced economies is one of the major sources of economic growth and catch-up in developing economies. Such an adaptation demands technological capabilities.

Technological capabilities for imitation and adaptation

Abramovitz (1986) noted that technological advancement has preconditions, which he termed social capabilities and technological congruence.[2] But while social capabilities are factors within a country, technological congruence is a measure of relatedness between countries in matching resources, markets, consumer preferences, scales and capital intensities (Abramovitz and David 1996).

Lall (1992) highlighted the role of capabilities at the firm or national level.[3] National technological effort is hard to measure, but as a proxy, he suggested R&D expenditure, patents and technical personnel. He also noted that importing technology is necessary but must be done in such a way that countries can also gain the learning benefits from the innovative process.

“ A radical innovation benefits producers by increasing profits from drastic cost reductions

Box 2.1

Radical innovations drive disruptive change

A radical innovation drives disruptive changes in the production process and in living standards. It benefits producers by increasing profits from drastic cost reductions. It benefits consumers if it improves their quality of life and is easily accessible at cheap prices. When technology-intensive goods are supplied massively, they can be produced by reaping enormous economies of scale. Steam power exemplifies how a technology can change economies and societies.

Manyika and others (2013) identify and analyse the potential effect of radical technologies that can revolutionise the production process and the life of future societies. The box table mentions the technologies, provides a definition and indicates their possible economic impact, especially in developing countries. The technologies with the greatest economic potential for developing countries by 2025 are mobile Internet, Internet of things and cloud computing.

The potential of radical innovations for developing countries

Technology	Definition	Economic effect
Mobile internet	Combination of mobile computing devices, high speed wireless connectivity, and applications.	$3.7 trillion to $10.8 trillion per year by 2025. $1.85 trillion to $5.4 trillion per year in developing countries.
Knowledge work automation	The use of computers to perform tasks that rely on complex analyses, subtle judgments and creative problem solving.	$5.2 trillion to $6.7 trillion per year by 2025. $1 trillion to $1.3 trillion per year in developing countries.
Internet of things	The use of sensors, actuators and data communication technology built into physical objects—from roadways to pacemakers—that enable those objects to be tracked, coordinated, or controlled across a data network or the internet.	$2.7 trillion to $6.2 trillion per year by 2025. $0.81 trillion to $1.86 trillion per year in developing countries.
Cloud	It brings computer architecture full circle, enabling network access to a shared pool of computer resources such as servers, storage, and applications that can be used as needed.	$1.7 trillion to $6.2 trillion per year by 2025. $1.19 trillion to $4.34 trillion per year in developing countries.
Advanced robotics	Advanced robotics have greater mobility dexterily, flexibility and adaptability, as well as the ability to learn from and interacting with humans, greatly expanding their range of potential applications.	$1.7 trillion to $4.5 trillion per year by 2025. $0.3 trillion to $0.9 trillion per year in developing countries.
Autonomous and near autonomous vehicles	An autonomous vehicle is one that can maneveur with reduced or no human intervention.	$0.2 trillion to $1.9 trillion per year by 2025. $0.04 trillion to $0.38 trillion per year in developing countries.
Next generation genomics	Next generation genomics can be described as the combination of next generation sequencing technologies, big data analytics, and technologies with the ability to modify organisms which include both recombinant techniques and DNA synthesis.	$0.7 trillion to $1.6 trillion per year by 2025. $0.14 trillion to $0.32 trillion per year in developing countries.
Energy storage	Energy storage systems convert electricity into a form that can be stored and converted back into electrical for later use, providing energy on demand.	$0.09 trillion to $0.63 trillion per year by 2025. $0.03 trillion to $0.25 trillion per year in developing countries.
3-D printing	3-D printing belongs to a class of techniques known as additive manufacturing. Additive processes build objects layer by layer rather than through molding or subtractive techniques.	$0.23 trillion to $0.55 trillion per year by 2025. $0.09 trillion to $0.2 trillion per year in developing countries.
Advanced materials	Any use or manipulation of materials with features at a scale of less than 100 nanometers (roughly molecular scale) can quality as nanotechnology.	$0.15 trillion to $0.50 trillion per year by 2025. $0.015 trillion to $0.05 trillion per year in developing countries.
Advanced oil and gas exploration and recovery	Unconventional oil and gas reserves are defined as reserves that cannot be extracted by conventional drilling methods.	$0.09 trillion to $0.46 trillion per year by 2025. $0.018 trillion to $0.092 trillion per year in developing countries.
Renewable energy	Renewable energy is energy that is derived from a source that is continuously replenished, such as the sun, a river, wind, or the thermal power of world oceans.	$0.16 trillion to $0.27 trillion per year by 2025. $0.12 trillion to $0.21 trillion per year in developing countries.

Source: UNIDO elaboration based on Manyika and others (2013).

> **Capability building, guided by economic policy, is decisive**

Capability building requires balance and must consider the combination of economic factors and incentives between firm and national capabilities that influence the development process.

Innovation and technology are not public goods and do not diffuse freely and instantaneously throughout the world. Although knowledge can be codified in blueprints or instructions, much of it is tacit and is embodied in people (Lundvall and Johnson 1994). So, technological knowledge is imperfectly imitated; it requires learning (Lee 2013b). And it is best developed in stages (Katz 1995; Kim 1980).

The technology gap theory is an overarching view of how economic development is fuelled by the international diffusion of technological knowledge, the development of capabilities by economic actors who adopt that knowledge and the institutions that facilitate that adoption. Applying it to technological capabilities, we look at the diffusion of knowledge and how it is linked to other economic factors, such as trade and participation in global value chains (GVCs), essentially since 1995, when GVCs begin acquiring significant importance.

The theory emphasizes structural change that accompanies the application of technological knowledge in developing economies. And it interprets innovation in a strict sense, referring to "hard" technological knowledge. Such technological innovation often is accompanied by organizational, marketing or even social innovation. But the emphasis here is on the technological part of innovation. Because of a lack of specific innovation indicators for the broad country sample and given the sectoral detail we are interested in, we also mainly look at the productivity side of technology and innovation.

The theory was pioneered by Abramovitz (1986), Fagerberg (1987) and Verspagen (1991). It focuses on technological knowledge, as generated by research and development (R&D) and other systematic investments by firms and public actors. The theory considers technological knowledge as the core engine of development because it increases productivity, creates new products and services and above all does that by creating economic externalities. Once created, that knowledge can be used widely, as by economic agents who did not originally develop the knowledge, if they have the capability to absorb the knowledge. Because wide use of productivity-enhancing technological knowledge is a key way for developing countries to climb the economic ladder, the lack of that knowledge in many countries explains why so little convergence has been seen in the international distribution of living standards (see Chapter 1).

The most famous examples of economies that have assimilated technology and moved out of a low level of development are the four Asian Tigers[4] (Box 2.2), while Japan also resorted to foreign technology to reconstruct. Their post-war economic history, well documented, shows how they gradually but deliberately assimilated foreign knowledge. Using that knowledge, they developed completely new manufacturing sectors and became competitive globally, as they transformed from largely agricultural societies into manufacturing and modern service economies. Many countries however, especially in Sub-Saharan Africa and Latin America, have been unable to follow that road, stuck at a low level of development.

The capabilities to assimilate international knowledge depend on earlier heavy investments in infrastructure, education, the political system, universities and other research institutes—"institutions" in a broad sense. Such investments are costly and require a high degree of state capacity. Economic strategy—particularly industrial and innovation policies—plays a crucial role in plugging the technology gaps. It determines whether countries can catch up with the global economic frontier (rare) or fall behind (which was the fate of the large majority of developing countries). Capability building, guided by economic policy, is also decisive. Countries that have taken that route to development have been characterized as "developmental states." But developing those capabilities, or becoming a developmental state, is very difficult when a country starts from a low base, which can easily become a low-development trap. Still, as the next section suggests, achieving that goal is not impossible.

> **Creating capabilities for catch-up requires substantial resources and a concerted effort**

Box 2.2
A tiger leaps from its lair: The Republic of Korea's structural transformation

The Republic of Korea became a high-income and highly technologized dynamic industrial economy within 50 years. Lacking natural resources, it focused on a technological growth paradigm. Robust science and technology capacity was enabled through the creation of a state-led research and educational capacity, as well as through corporate research and development efforts by the country's large conglomerates (OECD 2009).[5] Manufacturing was the focus of expansion, as were high-tech exports.

The country went through three transition phases: The 1960s and 1970s were characterized by technological learning, a well-educated workforce, selectively restricted foreign direct investment (FDI), imitation plus borrowing of foreign technology and technical agreements (Gupta and others 2013). The 1980s were marked by increasing technological requirements as industries became more complex and sophisticated (Chung 2011). The 1990s, the innovation phase, saw heavy R&D investments by the public and private sector, extensive government support for innovative small and medium enterprises (SMEs) and information and communications technology, and the launch of important technological institutions (Chung 2011).

Beyond boosting R&D spending, the government offered tax incentives for R&D (shifting from direct financial support) and R&D subsidies for SMEs in the innovation phase. Tax incentives were some of the most effective policy instruments, helping increase gross domestic expenditure on R&D as a share of GDP from 1.7 percent in 1991 to 4.1 percent in 2013 (Chung 2011). The R&D subsidies helped SMEs move ahead of large enterprises in value added for the first time: in the 2000s, SMEs accounted for 50.8 percent of value added and large enterprises for 49.2 percent (SMBA 2012).

Technological capabilities, exports and long-run economic growth

Structural transformation leading to sustained development take-off is very hard to achieve for developing countries because creating capabilities for catch-up requires substantial resources and a concerted effort by many economic and social actors. Countries that are already at high levels of development generally have high levels of capabilities, both because that is what enabled them to develop but also because being developed gives them the resources to develop and maintain capabilities. In other words, capabilities and development levels mutually influence each other (Fagerberg and Srholec 2008).

Capabilities are multifaceted, but measuring them often is inadequate because of the interlinkages among them, the multiple requirements at different development levels, and the intangible organizational strengths needed. Fagerberg and Srholec (2008) use principal components analysis on a large dataset to quantify capabilities, drawing on indicators such as trade, FDI, patents and scientific papers, educational attainment, the use of information and communications technology, corruption, banking and the regulatory environment.

Kaltenberg and Verspagen (2015) present a composite indicator for technological capabilities, much like the one by Fagerberg and Srholec, but updated with the most recent data (Figure 2.6).[6] They also present a similar (although not quite as tight) relationship for a single indicator related to technological capabilities: R&D spending in GDP, reproduced in Figure 2.7.[7] The latter measures investments in hard technology, reflecting the systematic discovery process that makes use of science and engineering methods. Given that narrow definition of capability, international comparisons are possible.

Figures 2.6 and 2.7 suggest that, because developing capabilities is hard when a country is at a low level of development, the technology gap that the country faces also represents a development trap. To break out of this trap, countries at low levels of development must develop capabilities that help them assimilate foreign technologies, and become competitive in international markets. Figure 2.8 illustrates this process. It makes use of recent literature on the capabilities and export performance of countries, suggesting that the specialization pattern of already developed countries holds information on what is necessary to escape the low-development trap. The products requiring

> **The technology gap that a country faces also represents a development trap**

Figure 2.6
Index of innovation system and institutional capabilities versus GDP per capita, 2013

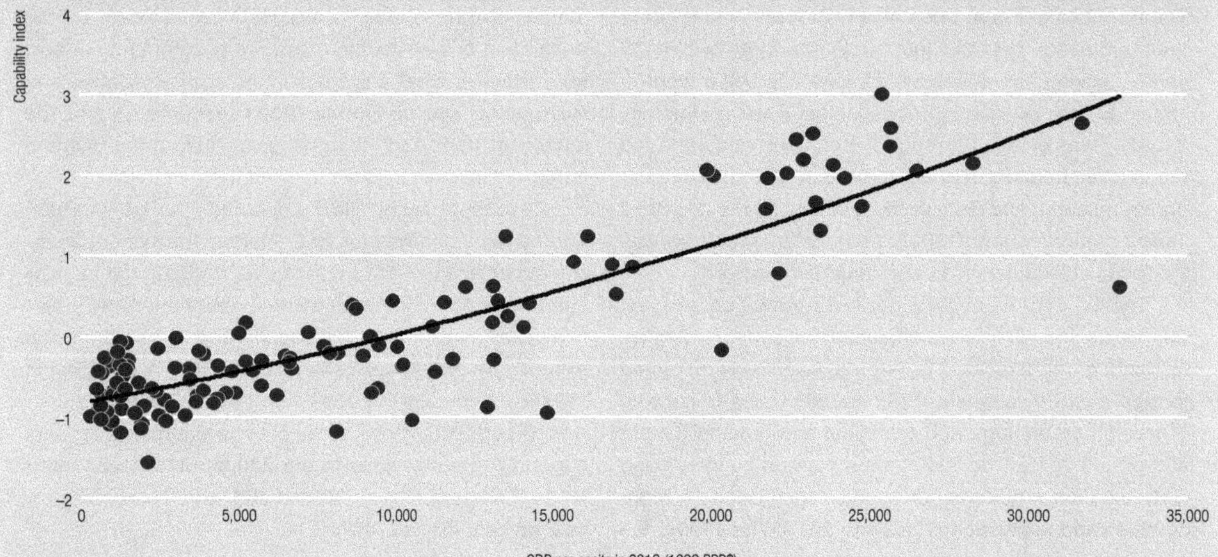

Note: GDP is gross domestic product; PPP is purchasing power parity.
Source: Kaltenberg and Verspagen (2015).

Figure 2.7
R&D spending (as share of GDP), 2011, and GDP per capita, 2013

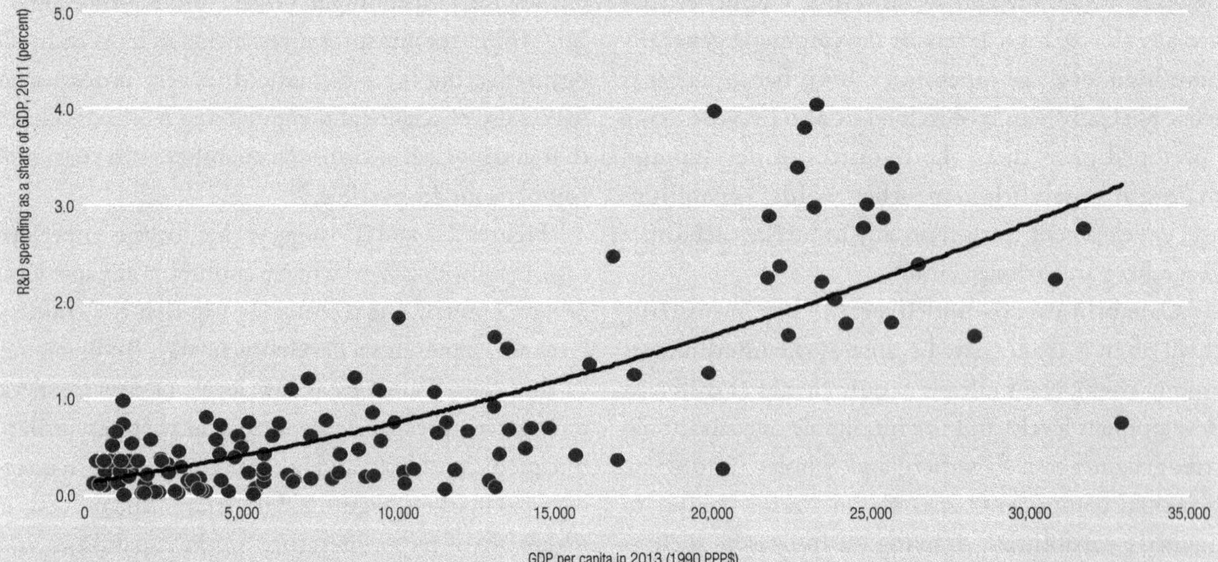

Note: GDP is gross domestic product; PPP is purchasing power parity. R&D is the total figure, including by business, by universities and by public laboratories.
Source: Kaltenberg and Verspagen (2015).

"When countries move closer to the technological frontier, they have to spend more resources on technology

complex production processes, and hence produced only in countries with high capabilities, are a reflection of "latent" capabilities, comprising technological, social and absorptive capabilities.

Figure 2.8 reproduces Figure 1.2 in Chapter 1 but adds the latent trade capabilities index constructed by Kaltenberg and Verspagen (2015),[8] which is displayed in the figure by the size of the dot representing each country. Dot size can clearly be seen to relate to the large variability in growth rates at low levels of development: up to around $10,000 on the horizontal axis, the countries with high capabilities have a much greater tendency to realize high growth rates (on the vertical axis) than countries with low capabilities.

Figure 2.9 depicts the latent trade capability indicator in relation to GDP per capita. The relation is displayed for 1998, so that the capability indicator can be related to post-1998 growth. The correlation is somewhat less tight than for R&D (see Figure 2.7) or the composite capabilities indicator (see Figure 2.6), but it is still high. The trade indicator is somewhat influenced by size, as for India and China, because trade capability is measured by diversification, and larger countries tend to be more diversified.

The relationships in the figure indicate the long-run relationship between technology and development predicted by the technology gap theory. In the long run, the only way to be highly developed (apart from resource riches connected to, for example, oil) is to become technology intensive (Fagerberg 1987). When countries move closer to the technological frontier, they have to spend more resources on technology, and their technological capabilities have to rise sharply to maintain their growth potential.

Whether countries move up in the development hierarchy seems much harder to explain than finding the long-run relationships. The basic premise of technology gap theory is that poor countries tend to have high potential for rapid growth, represented by the reservoir of global technological knowledge that is available for them to tap. The evidence suggests, however, that the tendency to realize that potential varies greatly within the group of poorer countries (see Figure 1.1). What explains that variance? Capabilities

Figure 2.8
GDP per capita and average rate of GDP growth conditioned on capabilities, 1998–2013

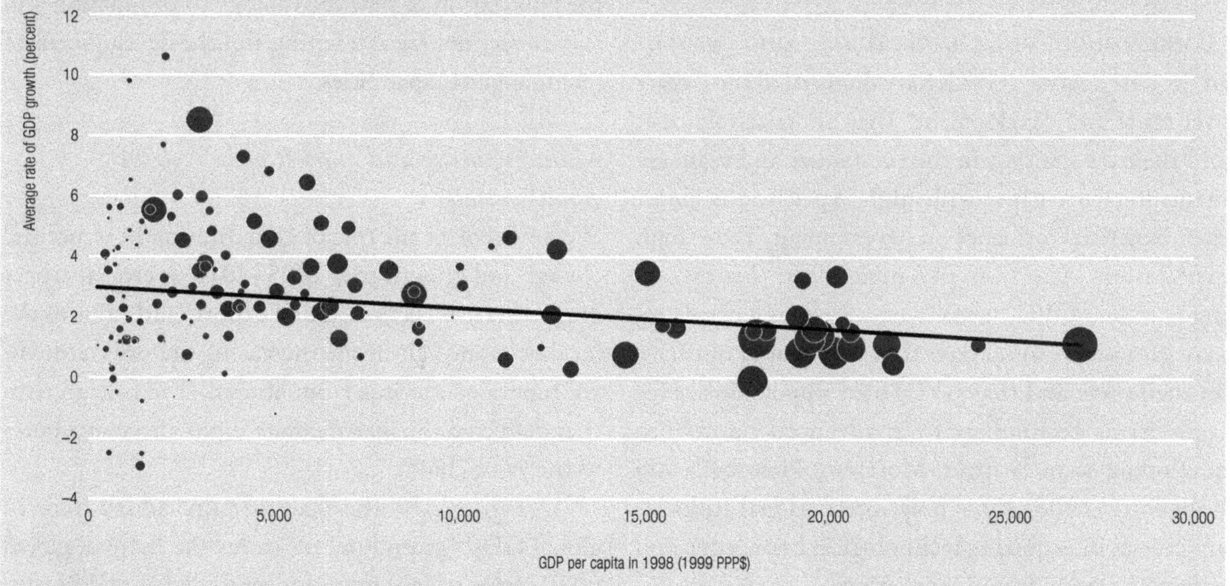

Note: GDP is gross domestic product. Based on a sample of 108 countries. Greater capabilities are marked by larger dots.
Source: Kaltenberg and Verspagen (2015).

"Not all exports add value to the local economy

Figure 2.9
Latent trade capability and GDP per capita, 1998

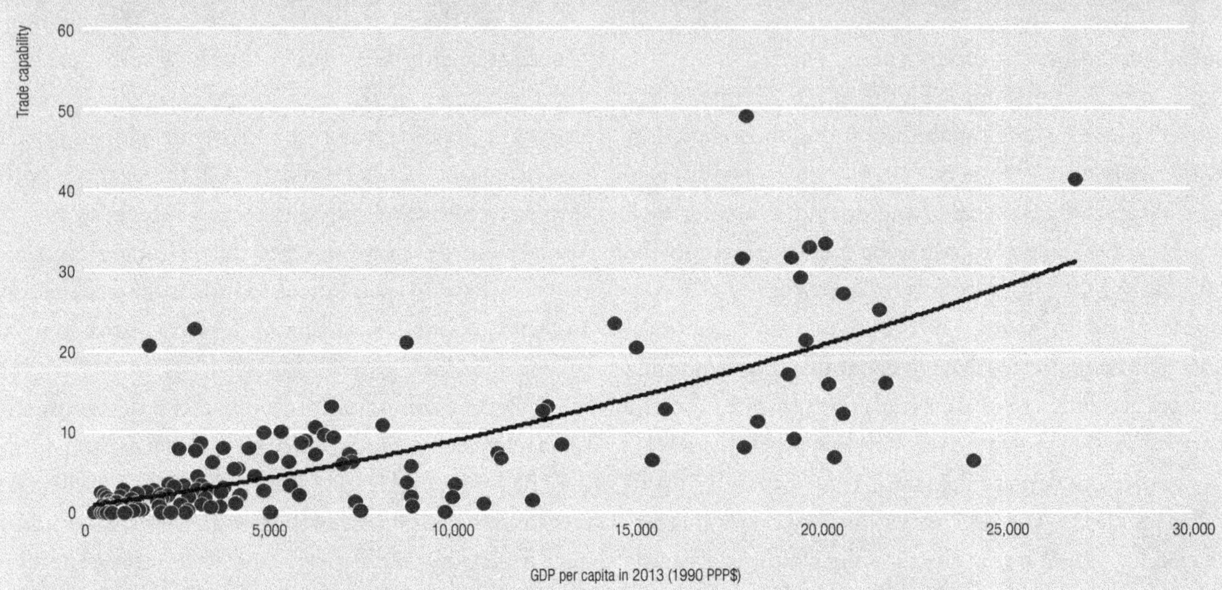

Note: GDP is gross domestic product.
Source: UNIDO estimate based on Gaulier and Zignago (2010).

to assimilate knowledge spillovers (see Figure 2.8 and Chapter 3).

Building technological capabilities in global trade and global value chains

Globalization, structural change and growth

In the 21st century, GVCs have dominated the debate over trade and development. They are relatively new, and their recognition in the economics and management literature came with high expectations about their beneficial influence on development. Those high expectations stem from two main ideas: that by taking part in a GVC, firms from developing countries may gain access to markets that they would not have on their own; and that GVCs offer opportunities for transferring technology from advanced-country to developing-country firms. Morrison, Pietrobelli and Rabellotti (2008) show how international linkages are crucial in acquiring technological knowledge and enhancing learning and innovation.

Essentially, GVCs are a global mode of production in which intermediate goods and services are traded in fragmented and internationally dispersed production processes. Goods are no longer produced in a single location; instead, their components are produced in various locations across the globe. Continuous division and specialization of labour internationally ease the application of new technology to production and lead to technological learning and the development of technological capabilities.

Manufacturing and market services drive globalization

We now look at the role of globalization in structural change and growth over 1995–2011. A crucial aspect of globalization is that not all exports add value to the local economy, given the increasing use of intermediate inputs or services from abroad. The foreign firm that delivered the intermediate input also contributes to the value chain.[9]

In Figure 2.10, the share of value added trade in global GDP (green line) measures the share of global GDP (value added) that services final demand (investment and consumption) in foreign countries.[10] That includes both direct trade in final goods (for example,

> **The largest part of the rise of globalization consists of derived demand, indicating the rise of global value chains**

Box 2.3

Bridging the technology gap depends on the type of value chain

De Marchi, Giuliani and Rabellotti (2015) conducted a study for this report, aimed at clarifying the role of GVCs in fostering innovation and learning by developing countries' firms that join GVCs. They collect evidence from case studies on topics such as the governance mode of the GVC, the degree to which GVCs produce innovations and the forms of learning inside and outside GVCs.

The authors discern three archetypal kinds of innovators in value chains. The one that puts GVCs most central in technology transfer (but present in only 9 cases of 50) is the group of *GVC-led* innovators, in which firms score high on innovation and use intensive learning channels within the GVC, such as face-to-face interaction between participants within the chain, training of the local workforce by the lead firm in the chain, direct knowledge transfer to local firms (usually confined to a narrow range of tasks), and pressure to adopt international quality standards. (An illustrative case of this group is the coffee GVC in Brazil, described by Cafaggi and others [2012]).

A second and slightly more common pattern (14 of 50 cases) is that of *independent* innovators. In this case, innovation is frequent and important, but learning takes place mostly in channels that are not directly related to the value chain. The important learning mechanisms in this category include within-firm and external learning mechanisms, including—but certainly not limited to—R&D. Other important learning mechanisms in this category are hiring skilled personnel and acquiring technology through licensing, joint ventures and mergers. Learning also takes place in local clusters, with close, short-distance interactions between firms. And learning is obtained from competitors (such as through reverse engineering) and from suppliers and buyers (from outside the value chain). The development of China's wind turbine industry exemplifies this group (Lema, Berger and Schmitz 2013).

The largest category consists of *weak* innovators (27 of 50 cases). In these value chains, the (non-lead) firms do not frequently innovate and they do not make extensive use of learning mechanisms. Among the numerous cases in this group are the clothing industry in Kenya and Madagascar (Kaplinsky and Wamae 2010). Local firms received some technical assistance from buyers, but it has not generated much product or process innovation.

That firms in developing countries, even when part of one or more GVCs, do not always use the GVCs as a privileged source of knowledge and technology implies that GVCs are not always crucial in closing technology gaps or spurring development. The prevalence of GVCs with weak innovators shows that the innovation road to development remains rocky and requires government policy to smooth it (see Chapters 3 and 6).

cars produced in Japan being delivered to Germany) and all indirect, value added streams associated with GVCs (for example, Chinese steel delivered to Japanese carmakers who sell cars on the domestic Japanese market). The other indicator—gross trade as a share of global GDP (red line)—looks at the gross value of all exports, including final goods and intermediates. Because not all trade is value added, this indicator is at a higher level than the value added trade indicator (Box 2.4).

In 1995, gross trade was about 20 percent of global GDP—in 2011, about 30 percent. Value added trade rose from about 15 percent to 20 percent (the trend includes two discontinuities: a marked increase in the rise after 2001, and the 2009 crisis). So, a substantial part of the global economy is "purely domestic"—domestic value added for domestic demand.

Which sectors were responsible for the rise in gross and value added trade since 2001?[11] Market services, closely followed by manufacturing, together account for about three-quarters. Almost half of the increase (48 percent) was attributable to final demand for manufacturing products (Table 2.1).

The discrepancy between the demand and production perspectives is a result of derived demand. Demand for manufacturing products leads to demand for services and for resources, and vice versa. In fact, the largest part of the rise of globalization consists of such derived demand, indeed indicating the rise of global value chains. Only 1.9 points (of the 4.8 point increase) occur in the same broad sector as final demand. The rest (2.9 points) is derived demand. The largest category in derived demand is market services production for manufacturing demand.

> **The rise of globalization since 2001 coincides with an uptake of the convergence of GDP per capita**

Figure 2.10
Gross and value added trade as a share of GDP, worldwide, 1995–2011

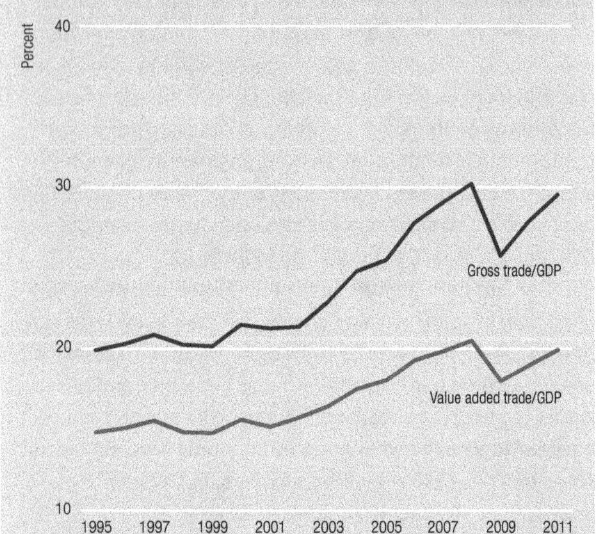

Note: GDP is gross domestic product.
Source: UNIDO elaboration based on Eora MRIO Database (Lenzen and others 2012; Lenzen and others 2013).

Manufacturing therefore plays a crucial role in increasing globalization. But market services also play a large role, both as parts of the services sector become more tradable and because of derived demand.

Globalization and convergence, 2001–2011

The rise of globalization since 2001 coincides with an uptake of the convergence of GDP per capita levels, although the causal relationship between the two is not clear. We use a broad geographical grouping of eight country blocks, so that the matrix that splits global GDP by source of demand and source of production (Table 2.2 below) can be presented with the same country groups (doing this for more groups, or even for countries, would make the matrix unmanageably large).

Figure 2.11 displays the growth rates of the shares of global GDP and GDP per capita for each country block, over 2001–2011. The analysis is in current dollars, so the GDP data are not comparable with the data that

Box 2.4
Macroeconomic accounting of production's fragmentation

Analysing the impact of GVCs using input–output tables can break down global GDP into the producing location and the demand location, in a matrix form. The diagram that follows uses a simplified example, with the global economy consisting of two countries, North and South.

In the diagram, if we look row-wise, we break down the GDP of a country by the demand location that it serves. In that way, GDP in North consists of value added produced for the local North market plus value added exported to South. A column breaks down total demand in a location by the source of production. The column for South sums to total final demand in South and breaks it down between value added imports from the North and domestically produced value added.

Taking the example of an iPod, the matrix works as follows. Suppose an iPod is assembled in South, using a license owned by North and using some parts (silicon chips) produced in North. The iPod is delivered to a final consumer in North. In this case, the sum of the demand column for North will be the value (selling price) of the iPod. Within this column, the cell for South will display the value added (wages and profits) of the firm in South that assembled the device. The cell for North will give the value added of all North firms involved in the knowledge embodied in the license and in the production of the silicon chip. The method takes account of all indirect linkages between the countries.

		Demand located in		
		North	South	
Production located in	North	Domestic value added in North	Value added exported from North to South	GDP of the North
	South	Value added exported from South to North	Domestic value added in South	GDP of the South
		Total demand of the North	Total demand of the South	

Source: UNIDO elaboration.

“Obtaining growth by being competitive in export markets has indirect effects on domestic growth

Table 2.1
The rise of globalization, by sector, 2001–2011

Traded value added increased by 4.8 percentage points over 2001–2011, and of this increase—	
1.9 percentage points are produced by market services sectors	2.2 percentage points are due to final demand in manufacturing sectors
1.8 percentage points are produced by manufacturing sectors	1.3 percentage points are due to final demand in other sectors
0.9 percentage points are produced by resources sectors	1.1 percentage points are due to final demand in market services sectors
0.2 percentage points are produced by other sectors	0.2 percentage points are due to final demand in resources sectors

Traded value added increased by 4.8 percentage points over 2001–2011, and of this increase—
1.9 percentage points are produced in the same broad group of sectors as where final demand occurs
0.7 percentage points are demand for market services derived from final demand in manufacturing sectors
0.5 percentage points are demand for resources derived from final demand in manufacturing sectors
0.5 percentage points are demand for manufacturing derived from final demand in "other" sectors
0.5 percentage points are demand for market services derived from final demand in "other" sectors
0.8 percentage points are other derived demand (smaller than 0.25 points per sector combination)

Source: UNIDO elaboration based on Eora MRIO Database (Lenzen and others 2012; Lenzen and others 2013).

Figure 2.11
Growth rates of share of global GDP, and GDP per capita growth, by country block and origin of demand, 2001–2011

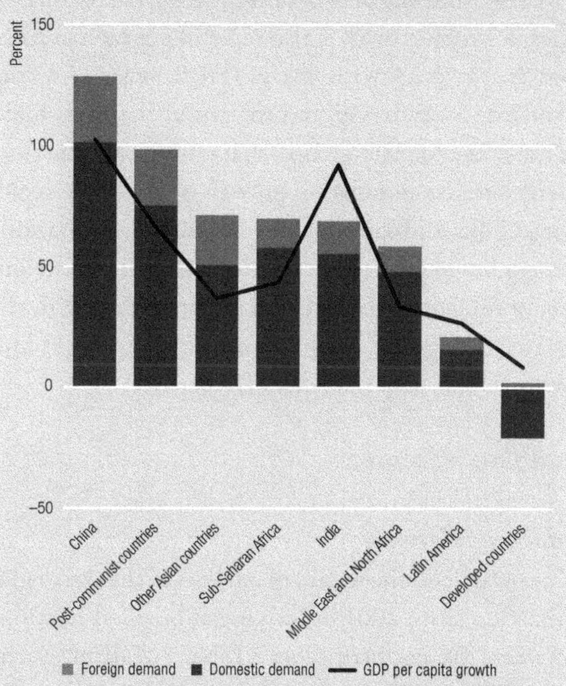

Note: GDP is gross domestic product. The eight country blocks are aggregations within the total sample of 189 countries. They are based on Annex A1, Table A1.2 with the following differences: "Developed countries" include also Greece, Portugal and the Republic of Korea. "Other Asian countries" are "East Asia", "South Asia" and "South-East Asia." "Post-communist countries" are "Former USSR" but also include Romania, Bulgaria, Slovakia, Czech Republic, Poland, Hungary, Slovenia, Bosnia and Herzigovina, Albania, Macedonia (Former Yugoslav Rep. of), Croatia and Serbia. "Latin America" is "Central America and Caribbean" and "South America." "Middle East and North Africa" includes Qatar, United Arab Emirates, Kuwait, Iran and Israel. The category "Other"includes Oceania and the following countries: Former USSR, Bhutan, Afghanistan, Maldives, Suriname, Cape Verde, Macao SAR (China), Montenegro, Sao Tome and Principe, Guyana, Cyprus, Brunei, the Democratic People's Republic of Korea, Djibouti, Belize, Andorra, Malta, Bermuda, Liechtenstein, Cayman Islands, Antigua, Luxembourg, Cuba, Barbados, British Virgin Islands, Seychelles, Iceland, Aruba, Bahamas, Greenland, Monaco, French Polynesia, Netherlands Antilles and San Marino. China and India, because of their size, are separate. This classification is also valid for Table 2.2 and Figures 2.12–2.18.
Source: UNIDO elaboration based on Eora MRIO Database (Lenzen and others 2012; Lenzen and others 2013).

are normally used for GDP per capita levels. Although we correct for global inflation by expressing all calculations in dollars and taking shares of global GDP, the results are not always indicative of the actual trends in real GDP per capita over the period, and therefore those growth rates are also included (the black line).

As expected, China is the fastest grower, with its share of global GDP rising from 4.2 percent in 2001 to 9.7 percent in 2011. The post-communist world's share almost doubles, which puts that bloc in second place. Those countries are indeed the ones catching up rapidly also in GDP per capita, and they have a high rate of structural change. The other country groups show slower convergence, and that is also the case for their shares of global GDP (see Figure 2.11). India is something of an exception, with GDP per capita growth that is relatively rapid but its share of global GDP increasing slower. Latin America is a slow grower. The developed bloc shrinks about 20 percent.

Final-demand or domestic demand?

The figure also shows the breakdown of the total growth rate into one part due to domestic demand and the other part due to foreign demand. The domestic part is always by far the largest. But the simple decomposition used here does not point out the role that foreign demand plays in catalysing domestic growth. This is connected to export-led growth: obtaining growth by being competitive in export markets has indirect effects on domestic growth, because of the spending that occurs out of income

> **The role of the 'foreign' part of growth should not be underestimated**

earned by workers or because of investment. So, the role of the "foreign" part of growth should not be underestimated.

China and the post-communist world are the two cases in which growth is also rapid in per capita terms, that is, catch-up with the global frontier of living standards is occurring, and structural change is high. In the other country groups, that is much less the case, partly because population growth is also fairly rapid. Both China and the post-communist bloc have sizable foreign-based growth. In relative terms, the contribution of foreign demand–based growth is largest in the Asia–Other group (29 percent of total growth) and lowest in Sub-Saharan Africa (17 percent).

Building relations

Bilateral relations?

We now extend the details of analysis of the rise of globalization since 2001 by looking at bilateral relations between the country blocs (Table 2.2 displays the changes over 2001–2011 in the basic matrix, providing the bilateral view).

The rise of globalization is indicated by the positive values shown in all but one cell. The exception is GDP produced in developed countries for foreign markets in the same group of countries, which declines by about half a point. The largest values in the table are, as expected, associated with China. GDP produced in developed countries for Chinese final demand is slightly more than GDP produced in China for demand in developed countries. The row for developed countries also attracts high values, indicating that those countries participate strongly in globalization. Post-communist countries also have high values in that area, and also within the block itself. The within-Latin America cell also contains high values.

On the production side (rows), the developed world and China take slightly more than half of the total increase of 4.8 points, with the developed world leading China by about 0.4 points. The post-communist world and other Asian countries rank next

Table 2.2
The rise of globalization, by country groups, demand and supply, 2001–2011 (percentage points)

		Demand									
		Developed countries	China	India	Other Asian countries	Post-communist countries	Latin America	Middle East and North Africa	Sub-Saharan Africa	Other	Total
Supply	Developed countries	-0.496	0.872	0.151	0.167	0.308	0.109	0.288	0.076	0.085	1.560
	China	0.785	0.000	0.038	0.081	0.062	0.061	0.078	0.023	0.017	1.146
	India	0.082	0.024	0.000	0.020	0.010	0.009	0.041	0.011	0.003	0.199
	Other Asian countries	0.178	0.120	0.021	0.044	0.015	0.013	0.027	0.007	0.004	0.429
	Post-communist countries	0.220	0.078	0.016	0.012	0.202	0.015	0.055	0.007	0.006	0.612
	Latin America	0.074	0.046	0.006	0.007	0.012	0.122	0.013	0.006	0.036	0.323
	Middle East and North Africa	0.140	0.057	0.025	0.019	0.029	0.014	0.058	0.010	0.004	0.356
	Sub-Saharan Africa	0.038	0.030	0.007	0.004	0.007	0.006	0.006	0.022	0.002	0.122
	Other	0.000	0.005	0.001	0.001	0.004	0.000	0.001	0.001	0.002	0.015
	Total	1.021	1.233	0.267	0.355	0.650	0.349	0.566	0.162	0.158	4.762

Note: As Figure 2.11. Numbers are percentage points. A value of "1.000" would indicate that the share of this cell in global gross domestic product increases by 1 percentage point. The sum of all cells in the matrix is 4.8 percent points, or the increase of the green line in Figure 2.10 over the period. The diagonal of the matrix refers to international flows of gross domestic product within the block—all (changes to) purely domestic flows have been kept out of the table.
Source: UNIDO elaboration based on Eora MRIO Database (Lenzen and others 2012; Lenzen and others 2013).

"Of the rise of globalization, 40 percent is produced in market services, 37 percent in manufacturing

in their row-totals. The contrast between India and China is large, with India participating much less in globalization. Sub-Saharan African countries, which have a low share of global GDP to begin with, take least advantage of globalizing production networks.

The demand side (columns) is somewhat different, but not very much. Developed countries and China take the bulk of the total. The developed world takes a larger share of the production total than the demand total; the reverse is true for China.

Manufacturing and market services or other sectors?

Figure 2.12 provides the overall sectoral breakdown of the rise of globalization. The numbers are the same as those in Table 2.1, but they are expressed as sectoral shares (by production). Of the rise of globalization, 40 percent is produced in market services, 37 percent in manufacturing; 19 percent in resources, and 4 percent in other sectors. Because a large proportion of globalization takes place in manufacturing, countries involved in international trade are probably undergoing radical changes in their use of equipment and in the production processes (Alcorta 1994). How does that distribution compare with the distributions at the country level, or even at the level of the (inner) cells of the matrix in Table 2.2? Those data are displayed in Figure 2.13, which gives totals per country group, for production of GDP (row totals).

Within those broad country groups, the distribution differs quite a lot from the global total. The developed world has a larger share of services, at the expense of resources, not manufacturing. As expected, China has a large share of manufacturing, but only 7 percentage points above the global average. China also has a larger share in resources and a correspondingly smaller share of market services. The post-communist world also has a large share of manufacturing, as large as China's. Other Asian countries have the same share of manufacturing as the global average; all other blocs (India, Latin America, Middle East and North Africa, and Sub-Saharan Africa) have a much lower share of manufacturing.

Figure 2.12
The rise of globalization, by producing sectors, 2001–2011

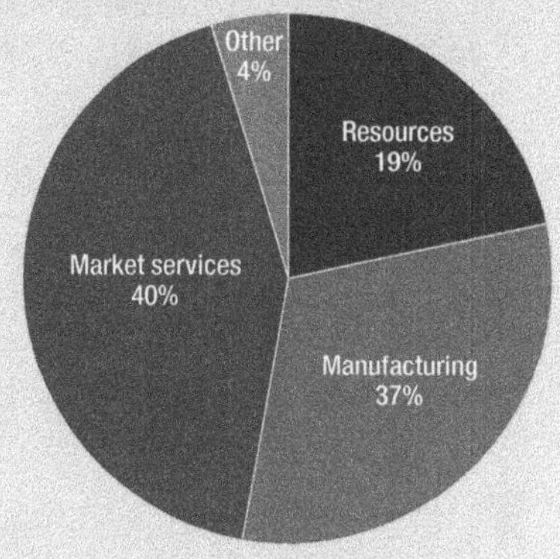

Note: Classification as indicated in the note to Figure 2.11.
Source: UNIDO elaboration based on Eora MRIO Database (Lenzen and others 2012; Lenzen and others 2013).

India has a comparatively large share of market services, but its resource share is also large. In Other Asian countries, besides manufacturing, the resources share is large. As expected, Middle East and North Africa has the highest resource share (mostly crude oil), but Latin America and Sub-Saharan Africa also rank high on resources. The latter two groups also have fairly large market services shares (though not much higher than the global average).

Relationships contributing most to the rise in globalization

We also look at a few of the bilateral relationships, focusing on the ones that are responsible for the larger part of the rise of globalization. The bilateral relationships (the off-diagonal cells in the matrix) are responsible for slightly more than the 4.8 percent points increase, as the developed-to-developed nations cell declines as a share of global GDP over 2001–2011. Among the bilateral relationships, those in which the developed world is involved are most important, in quantitative terms. Those economic relationships account for exactly 75 percent of the increase in globalization—that is, "South to South" relationships are only one-quarter of the increase of globalization.

> **"The post-communist world has a large share of manufacturing**

Figure 2.13

Growth of value added produced for foreign markets, by sector and country group, 2001–2011

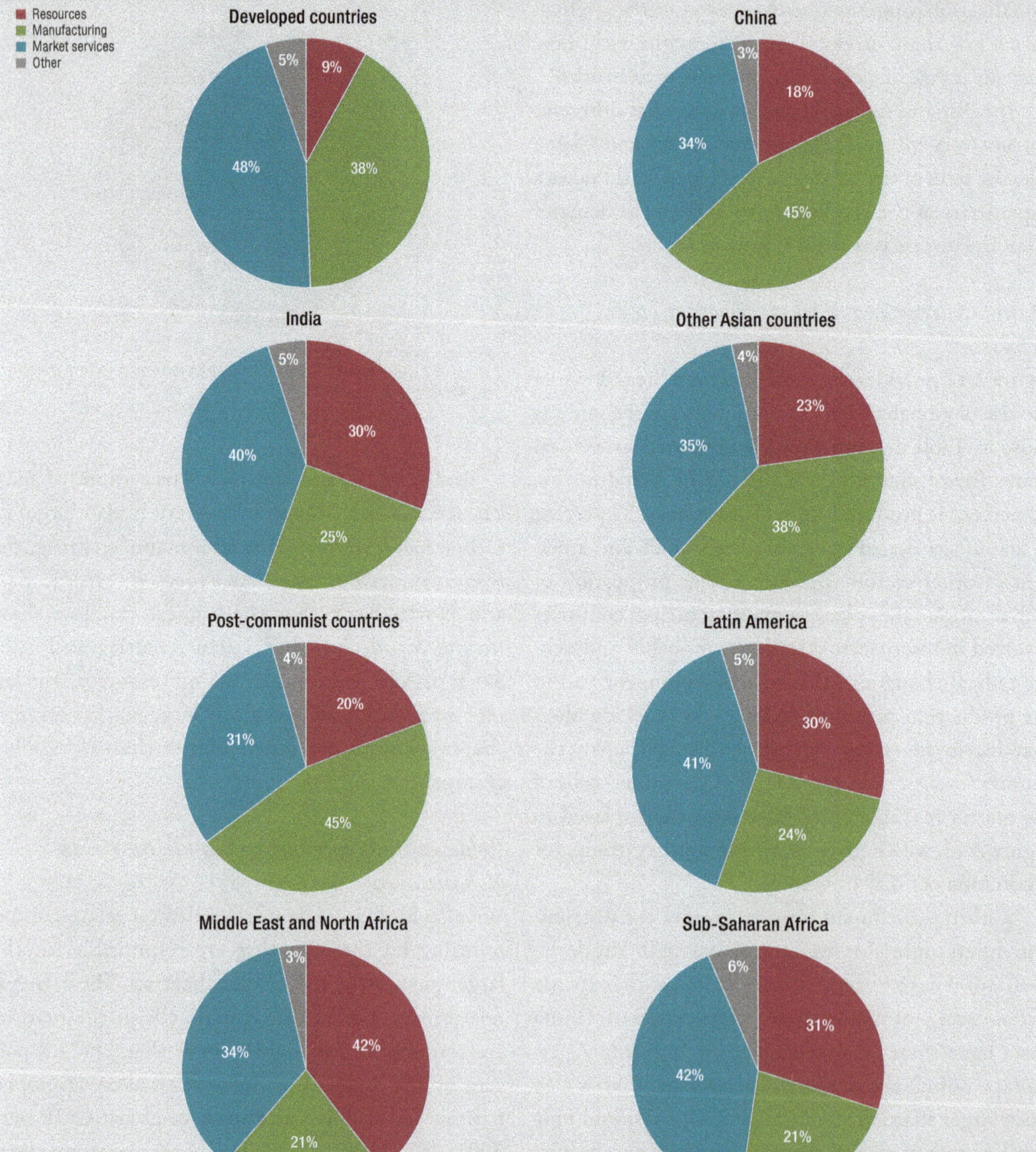

Note: Classification as indicated in the note to Figure 2.11.
Source: UNIDO elaboration based on Eora MRIO Database (Lenzen and others 2012; Lenzen and others 2013).

GDP produced by developing countries for demand in China consists of a larger share of manufacturing than it does in the other direction

The China–developed countries (and vice versa) relationship is responsible for slightly more than one-third of the increase in globalization (1.7 of 4.8 points), dominated by manufacturing, in both directions (Figure 2.14). In fact, GDP produced by developed countries for demand in China consists of a larger share of manufacturing than it does in the relationship in the other direction, though in both cases, manufacturing is the largest share. Resources are a larger part in China's production for developed-country demand than vice versa.

The same structure is found for the relationship between the developed countries and the post-communist world (Figure 2.15). This pair of country groups is responsible for slightly more than 10 percent of the total change. Again, manufacturing has the largest share in both directions, followed by market services, and the share of manufacturing in developed-country GDP is larger than in the other direction. The value added share of resources in demand from developed countries is much larger in post-communist countries than vice versa. These two cases indicate that, in net terms, the developed world mainly imports resources from the rest of the world, but in manufacturing, trade is far more balanced.

Figure 2.16 shows a further selection of other bilateral cells. The share of manufacturing in Sub-Saharan Africa is small, and that of resources is large, but that differs between Chinese and developed markets. In the Chinese case, the share of manufacturing is twice as large as in the case of developed markets, a difference that comes at the expense of resources.

India is generally a case of relatively low globalization, and in the case of Chinese demand, is heavily biased towards resources. The most manufacturing-intensive case is the fourth panel in Figure 2.16, GDP in post-communist countries for Chinese demand.

The results show that the increase of globalization over 2001–2011 has indeed been a process of structural change, with different parts of the world specializing in different broad sectors. Manufacturing and market services play the leading roles at a global level, but resources play a very important role in some parts of the world: resource-based growth is important to Sub-Saharan Africa, India and Middle East and North Africa. The China–developed nations economic relationship is very manufacturing-intensive in both directions.

Figure 2.14

Growth of value added produced for foreign markets, by sector, bilateral relationship—China and developed countries, 2001–2011

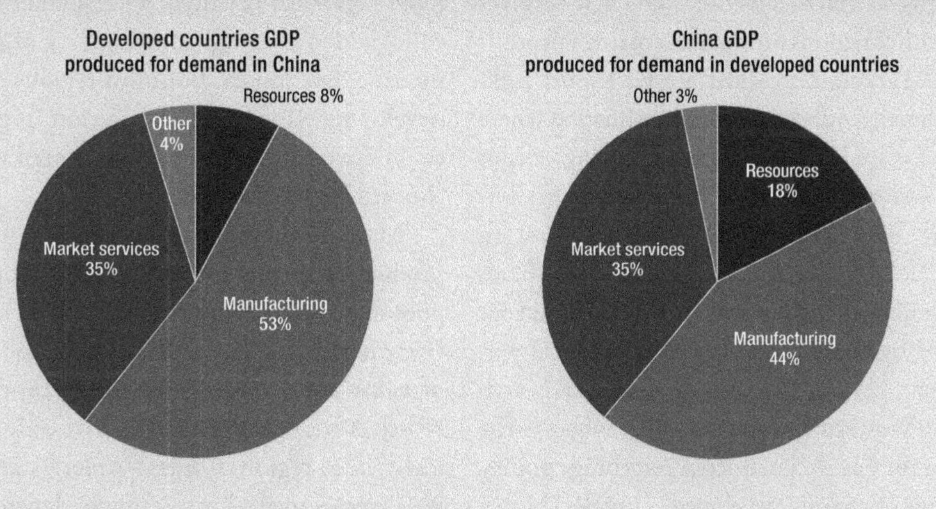

Note: GDP is gross domestic product. Classification as indicated in the note to Figure 2.11.
Source: UNIDO elaboration based on Eora MRIO Database (Lenzen and others 2012; Lenzen and others 2013).

> **"The developed world mainly imports resources from the rest of the world, but in manufacturing, trade is far more balanced**

Figure 2.15
Growth of value added produced for foreign markets, by sector, bilateral relationship—post-communist and developed countries, 2001–2011

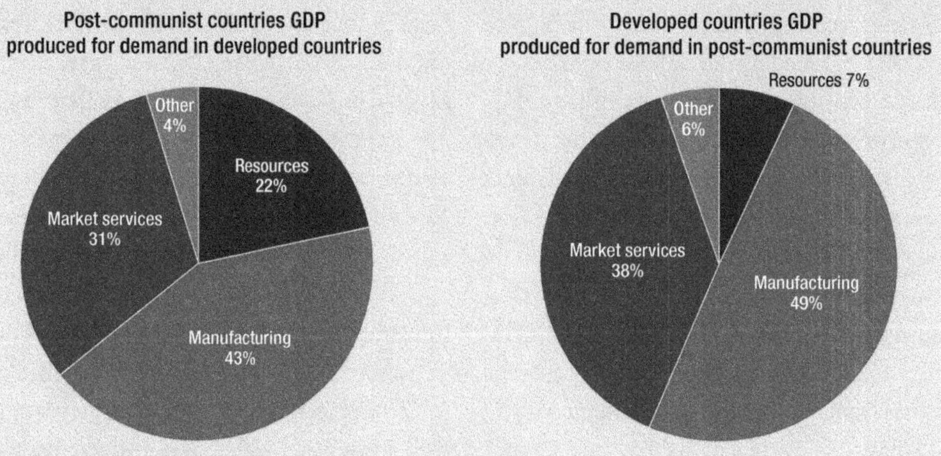

Note: GDP is gross domestic product. Classification as indicated in the note to Figure 2.11.
Source: UNIDO elaboration based on Eora MRIO Database (Lenzen and others 2012; Lenzen and others 2013).

The regional dimension

Regional production linkages and GVCs

In the process of globalization discussed in the previous section, the spread of production networks across countries has increased interdependence in manufacturing between regions. This section illustrates the evolution of regional production linkages by looking at how each region generates value added through its manufacturing activities and how much it receives inputs from other regions for manufacturing output.[12]

Figures 2.17 and 2.18 illustrate backward linkages in the global production system—showing where value added for each region's manufacturing output comes from—in 1990 and 2011. The bubble sizes show that the three regions of North America, Western Europe and East Asia account for the dominant shares of world manufacturing output: in 1990, 80 percent was produced by those regions, and in 2011, 73 percent. Over the period, the output shares of North America and Western Europe declined, whereas the output shares for East Asia gained 3 percentage points.

The regions of, mainly, developed countries (North America, Western Europe and East Asia) tend to have higher intra-regional value added contributions for their manufacturing output (as indicated in the higher numbers in their bubbles). Only about 10 percent of their total manufacturing value added in 1990 and 15 percent in 2011 came from the imports of intermediate inputs from other regions. In contrast, developing regions, especially Africa and South-East Asia, had higher shares of value added coming from other regions. Among developing regions, the manufacturing production of South America and South Asia are more regionally oriented, with smaller shares of their value added coming from imports of other regions' intermediate inputs. But most regions developed and developing—increased integration in global production networks over the period (as seen in decreases in the numbers in the bubbles).

Many developing regions have strong backward production linkages with Western Europe, which supplies intermediate inputs to various regions. Even relatively remote regions of Asia receive substantial shares of value added from Western Europe. In contrast, North America and East Asia have only one geographically close region, Central America and South-East Asia, respectively, both strongly dependent on value added inputs from North America and East Asia. Over the past 20 years, those two developing regions

“ The increase of globalization over 2001–2011 has been a process of structural change

Figure 2.16
Growth of valued added produced for foreign markets, by sector and selected bilateral relationships, 2001–2011

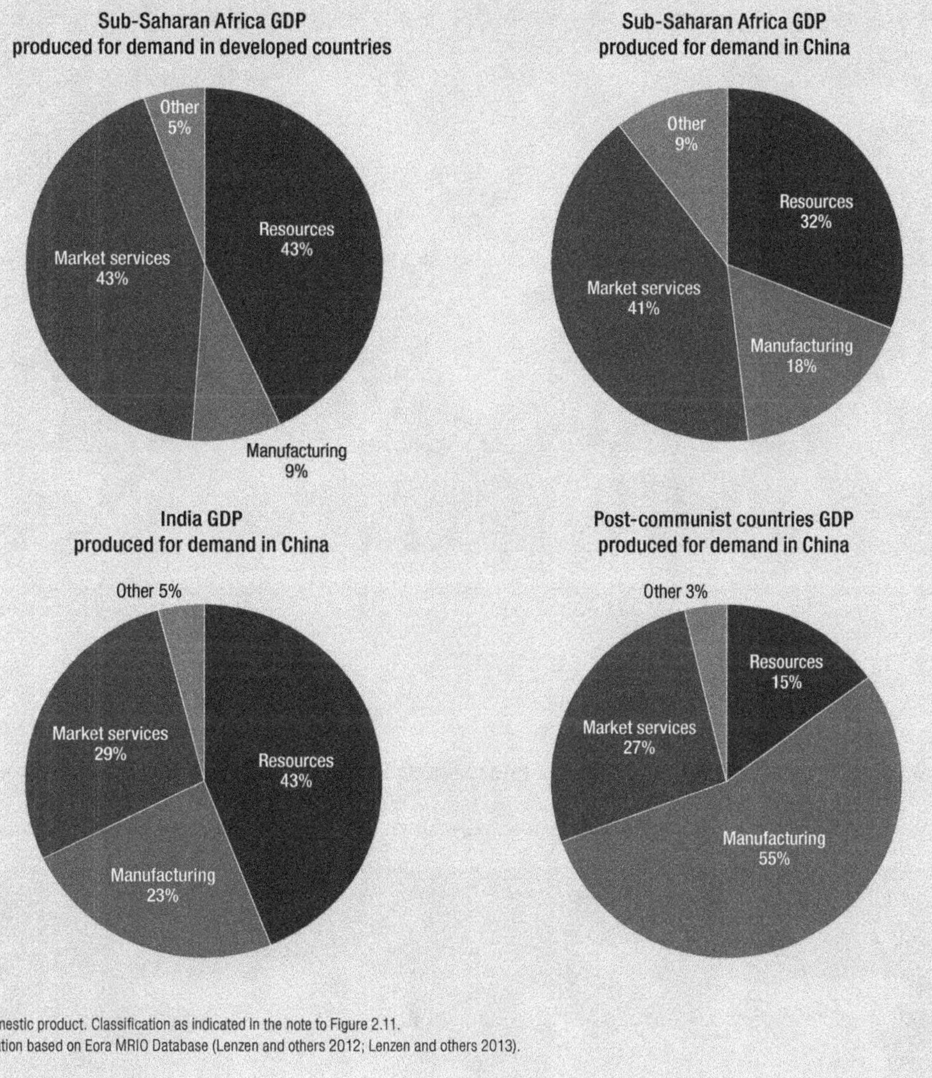

Note: GDP is gross domestic product. Classification as indicated in the note to Figure 2.11.
Source: UNIDO elaboration based on Eora MRIO Database (Lenzen and others 2012; Lenzen and others 2013).

have reduced their dependence on the major manufacturing hubs in the north and increased the share of value added generated within their own regions.

Finally, a comparison between Figures 2.17 and 2.18 and Table 2.3 indicates that globalization of manufacturing production and greater participation in global production networks do not automatically translate to technological development and to increased shares of a region's output in the world. To make global production linkages and GVCs work for sustained growth, countries need to make conscious efforts to generate the conditions conducive to technological learning. (Chapter 3 discusses the relevant policies in detail.)

The previous two figures showed how each region produced its total manufacturing output, which contains value added came from different regions. Figures 2.19 and 2.20 show regions' forward linkages. They map how each region generated its manufacturing value added. They show the region's value added by destination —whether value added is generated for manufacturing production within the region or for intermediate

> **“ Many developing regions have strong backward production linkages with Western Europe**

Figure 2.17
Backward linkages—where value added for each region's manufacturing output came from, 1990

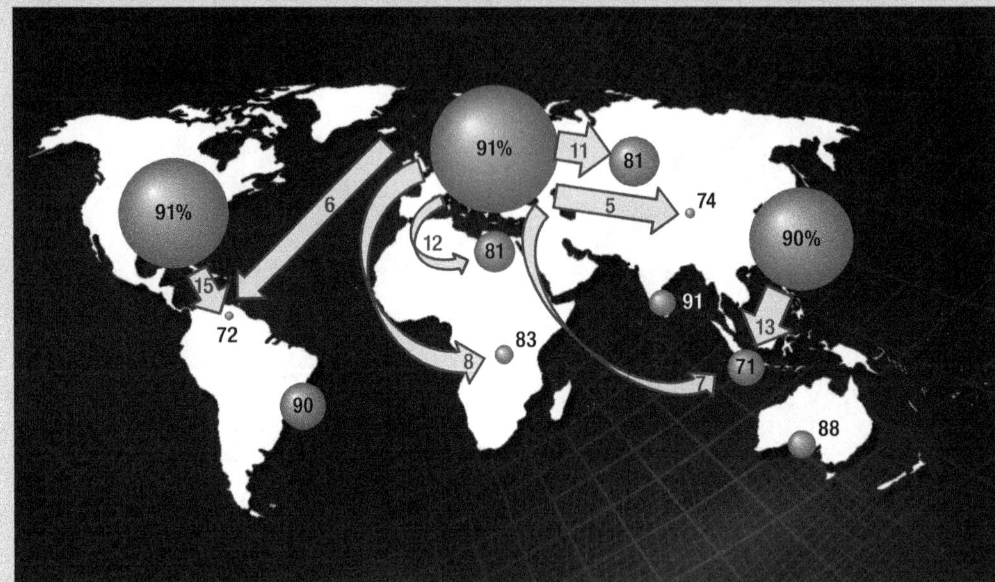

Note: Arrows show where value added came from for the production of manufacturing outputs at destination. The area covered by the bubbles represents the relative size of total value added (both generated within each region and from other regions) for a region's manufacturing output; the number in or next to a bubble (in black or white) indicates the share of manufacturing value added within the region for the production of final output, which can either be used within the region or exported to other regions. Arrows show the main sources of value-added contributions for manufacturing output in each region. Red numbers indicate the percentage of the value added that came from these transactions (that is, backward linkages with global value chains) in the region's total manufacturing value added. (Only transactions of 5 percent or more of the region's value added are shown.) Regional classification based on Annex A1, Table A1.2.
Source: UNIDO elaboration based on Eora MRIO Database (Lenzen and others 2012; Lenzen and others 2013).

Figure 2.18
Backward linkages—where value added for each region's manufacturing output came from, 2011

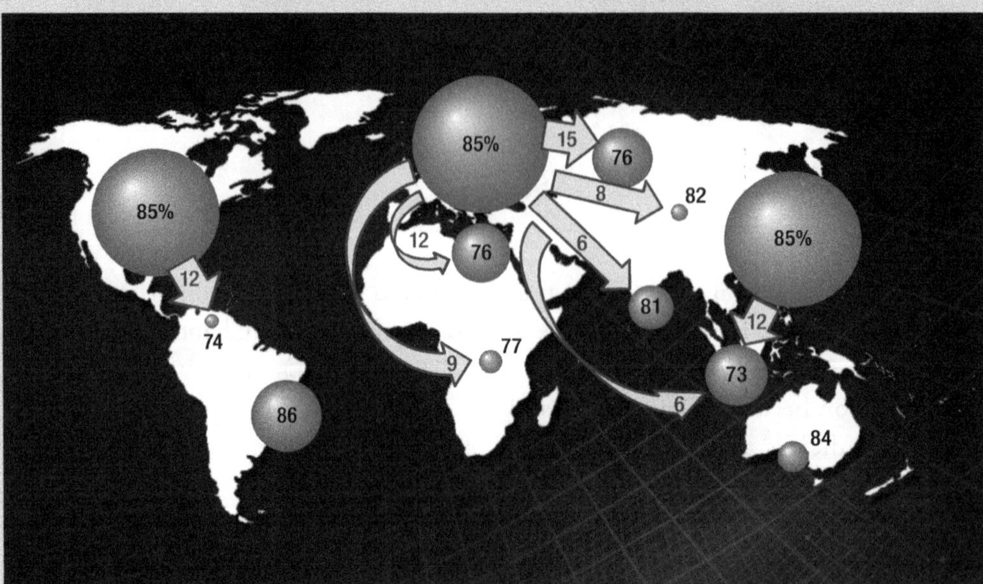

Note: Arrows show where value added came from for the production of manufacturing outputs at destination. See Figure 2.17.
Source: UNIDO elaboration based on Eora MRIO Database (Lenzen and others 2012; Lenzen and others 2013).

“Greater participation in global production networks does not automatically translate to increased shares of a region’s output in the world

Table 2.3
Manufacturing output shares in world total, 1990 and 2011, and gain or loss (percent)

	1990	2011	Gain or loss
South-East Asia	2.71	5.20	2.49
South America	4.59	6.90	2.32
North Africa and the Middle East	3.41	4.77	1.35
South Asia	1.39	2.52	1.13
Central Asia	0.19	0.37	0.18
Central America	0.13	0.24	0.10
South Africa	0.57	0.64	0.07
Sub-Saharan Africa	0.67	0.69	0.02
Eastern Europe	4.98	4.75	−0.22

Note: Regional classification based on Annex A1, Table A1.2.
Source: UNIDO elaboration based on Eora MIRO Database (Lenzen and others 2012; Lenzen and others 2013).

exports to other regions.[13] Compared with backward linkages, the ways that developing countries added value are more diversified. For example, in 2011, the significant shares of Sub-Saharan Africa’s manufacturing value added came from exports of intermediate goods to Western Europe, North America, North Africa and the Middle East, and East Asia. North Africa and the Middle East and Central Asia also had strong dependence on three regions for intermediate exports. Along with the diversification of the demand for their value added, they greatly increased the share of value added coming from the exports of intermediate goods relative to the value added generated for manufacturing output sold within the region. Sub-Saharan Africa and Oceania increased the share of value added coming from intermediate exports by 16 percentage points over the period. South Asia, Central Asia and Eastern Europe increased the share more than 10 percentage points.

Of the global trade in intermediate goods and the value added generated by such trade, the importance of East Asia has increased significantly over the past 20 years. In 2011, one quarter of global manufacturing value added was generated by that region. Its value added share was the third largest after North America in 1990, and it became the second largest after Western Europe in 2011 (as indicated by the

Figure 2.19
Forward linkages—how regions generated manufacturing value added, 1990

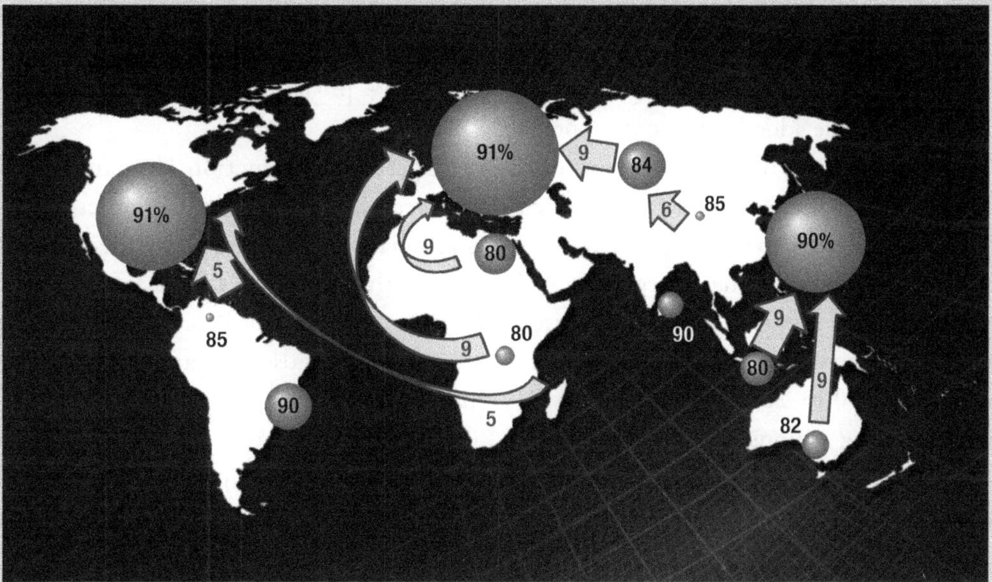

Note: Arrows show the value added generated from the export of intermediate goods by the region of origin. See Figure 2.17.
Source: UNIDO elaboration based on Eora MRIO Database (Lenzen and others 2012; Lenzen and others 2013).

> **Compared with backward linkages, the ways that developing countries added value are more diversified**

Figure 2.20
Forward linkages—how regions generated manufacturing value added, 2011

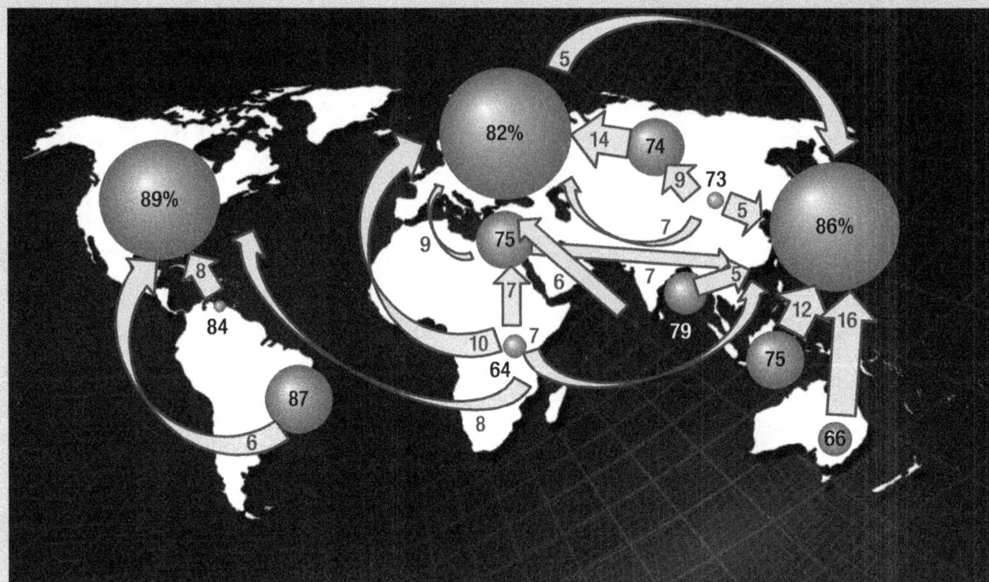

Note: Arrows show the value added generated from the export of intermediate goods by the region of origin. See Figure 2.17.
Source: UNIDO elaboration based on Eora MRIO Database (Lenzen and others 2012; Lenzen and others 2013).

bubble sizes). Further, seven regions came to have high dependence on East Asia for generating their value added through intermediate exports to that region.

The maps show that the world's manufacturing production increased its participation in global production networks and GVCs and integration into supply chains, led by North America, Western Europe and East Asia. Although Sub-Saharan Africa increased its share of value added coming from other regions in its manufacturing output (backward production linkages) and the share in its total value added derived from intermediate exports to other regions (forward production linkages), that integration did not see the region rapidly industrializing. It increased its global manufacturing value added share by a mere 0.13 percent from 1990 to 2011, one of the lowest increases in developing regions. The following section looks into Sub-Saharan Africa's participation in GVCs in greater detail.

The presence of Africa in GVCs

Sub-Saharan Africa generally has the lowest living standards on the planet, and the largest challenges of human development. Foster-McGregor, Kaulich and Stehrer (2015) confirm that in volume, Africa as a whole trades less than other regions—notably, in trade in intermediate goods. However, African exports are heavily involved in GVCs, particularly in upstream production. Downstream involvement in GVCs is low, although some individual countries have positioned themselves downstream in GVCs.[14]

Upstream production in manufactures usually involves high value added activities, but in Africa it is associated with low value added primary production, with few chances for learning and upgrading. Exports of primary products dominate the intermediate exports of most African countries, with negative implications for upgrading (Figure 2.21).

Primary sectors tend to make up 20–30 percent of upstream GVC participation (25 percent on average). Low- and high-tech manufacturing contribute on average 10 percent and 9.5 percent, respectively. That implies that the major contribution to upstream GVC participation comes from low-tech (28 percent) and high-tech (27.5 percent) services.

> **Export sophistication and product discovery are generally lower in Africa**

Figure 2.21
Upstream global value chain involvement, African countries, by sector, 2010

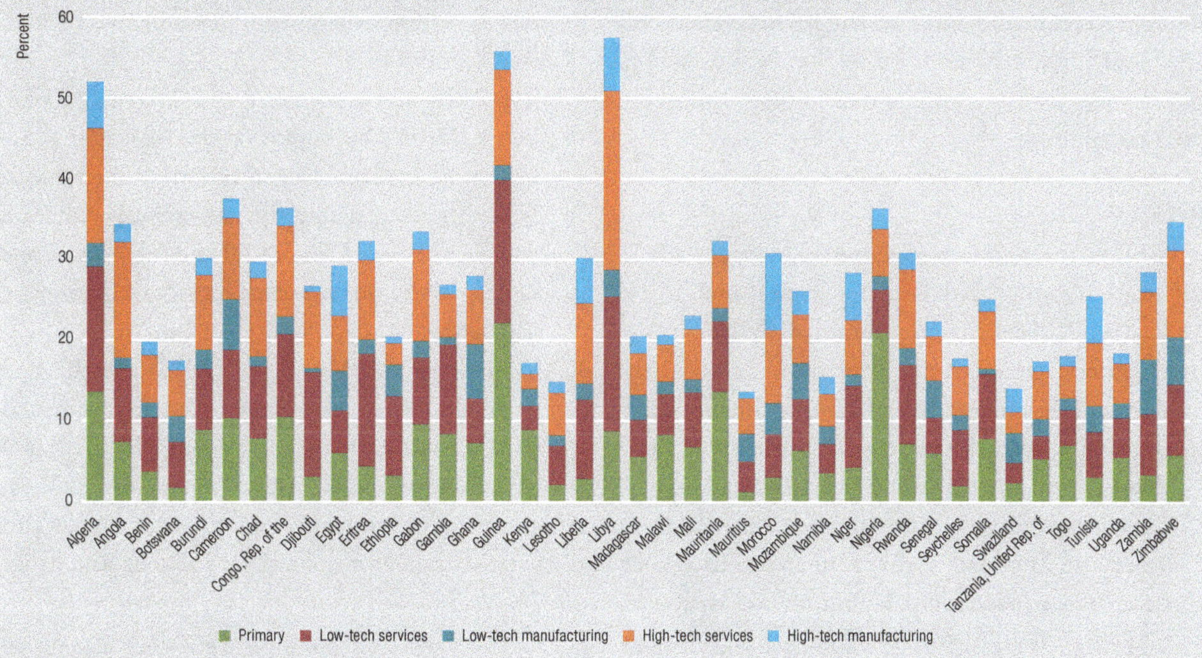

Note: For technology classification please see Foster-McGregor, Kaulich and Stehrer (2015).
Source: Foster-McGregor, Kaulich and Stehrer (2015).

Export sophistication and product discovery (intended to capture the extent of upgrading within GVCs) are generally lower in Africa than in other developing regions. Still, Africa has made some progress in upgrading, and African countries have exported their products at higher prices while maintaining market shares. Such economic upgrading was observed across a broad range of sectors, particularly in electrical and machinery, transport, and other manufacturing industries.

Although for Africa as a whole, GVC participation and upgrading are limited (particularly in upstream production), some countries have moved into downstream production, acting as assembly hubs for motor

Box 2.5
High-tech global value chain development in Morocco—the automotive industry

The Moroccan government has followed consistent sectoral strategies to promote new industries, such as the manufacture of automobiles, now a driver of growth and a crucial area of innovation (Mansour and Castel 2015). In the National Pact for Industrial Emergence 2009–2015, the automotive sector is seen as a "competitive advantage industry" (Elharouni, Benmoussa and Ouahman 2013). Morocco benefits greatly from this value chain, because of Morocco's geographical proximity to Europe, low labour costs, financial incentives, skilled labour, good infrastructure and political stability. Two special economic zones were created for the automotive industry.

By 2012, Morocco was producing automobiles and spare parts through the Société Marocaine de Constructions Automobiles, which has the capacity to manufacture 90,000 vehicles a year, the majority for export. And Renault's new €1 billion plant in Tangiers can assemble up to 400,000 vehicles a year, bringing 6,000 direct jobs and many more indirect jobs. Morocco assessed the potential of more than 600 automotive parts before selecting around 100 parts in which to compete through local production (Fine and others 2012).

The sector[15] employed 14,466 persons in 2011, almost double the 7,690 persons in 2009 (UNIDO 2015b). The export value of parts and accessories for motor vehicles and their engines rose from $27 million in 2000 to $151 million in 2012.[16]

vehicles (Box 2.5) and other manufacturing output. Those countries report relatively large exports of parts and components and some of the highest shares of exports of high-tech products in the region, raising their chances for knowledge and technology spillovers and for upgrading.

Notes

1. TFP is the amount of value added per "K&L unit," in which K stands for capital and L for labour. A K&L unit is obtained by weighting together capital and labour by the shares of their payments in value added. This is a well-established procedure, based on the economic theory of production, which we do not explain in detail here.
2. Social capability loosely includes current processes of knowledge diffusion, conditions of competence (education), labour market structure and migration, organization of firms (business environment), political stability, macroeconomic conditions affecting investment and effective demand, and financial institutions to mobilize capital (Abramovitz 1986; Fagerberg, Srholec and Verspagen 2010).
3. At the firm level: investment capabilities—the ability to plan, acquire technology, provide human resources and establish a new project; production capabilities—the skills to operate, maintain and improve process or product innovations; and linkage capabilities—the ability to coordinate inputs, information, services and institutions. At the national level: physical investment, human capital and technological effort (Lall 1992).
4. The Four Asian Tigers are Hong Kong SAR China, the Republic of Korea, Singapore and Taiwan Province of China.
5. The *chaebol* have moved from safe technology investments and incremental innovation towards cutting-edge science-based innovation, and are some of the country's main investors in R&D (Gupta and others 2013).
6. The capability index applied by Kaltenberg and Verspagen (2015) is based on an update of the data in Fagerberg and Srholec (2008). However, the index combines Fagerberg and Srholec's innovation system capability index and institutional quality index.
7. Kaltenberg and Verspagen (2015) source R&D data from UNESCO (United Nations Educational, Scientific and Cultural Organization). If the value for 2011 was not available, the nearest year with available data was used. Figure 2.7 and subsequent ones (unless otherwise indicated) include only countries with a population of more than 1 million.
8. Kaltenberg and Verspagen (2015) use the Base pour l'Analyse du Commerce International (BACI) dataset for trade (see CEPII 2010), and have close to 5,000 product classes and 229 geographical units in the underlying analysis. They use the indicator defined in Tachella and others (2012).
9. Classic cases in the literature are electronics products such as mobile phones or music players. Many are assembled in Asia, but much of their value added associated with their export is produced by firms that deliver parts or knowledge (in the form of licenses).
10. We use input–output data from the Eora MRIO database, 26-sector version (Lenzen and others 2012; Lenzen and others 2013). The database has data for 189 countries. We manipulated the basic input–output table to ensure consistency. Despite those manipulations, several countries show rather strange trends, but we aggregated those countries into larger groups so that the impact of those apparent anomalies is limited.
11. To keep the results presentable, we group the nine earlier sectors into four broad groups of sectors: resources (agriculture, mining and utilities); manufacturing; market services (trade, hotels and restaurants; transport and telecommunications; finance and business services); and other (also including re-exports, which is a separate sector in the underlying database).
12. The global regions used here are North America (which belongs mostly to the developed countries

group, as used previously), Western Europe (exclusively part of the developed group), Eastern Europe (mostly part of the post-communist world), Middle East and North Africa, Sub-Saharan Africa, Central Asia (part of the post-communist world), East Asia, South-East Asia, South Asia, Oceania, Central America and South America.

13. Again, the figures show only those intermediate exports that contributed more than 5 percent of a region's manufacturing value added.
14. The level of upstream production is measured as a share of a country's value added that enters as an intermediate input into the value added exported by all other countries in the country's total exports, whereas that of downstream production is defined as a share of foreign value added used in a country's own exports in the country's total exports.
15. ISIC Rev. 3, division 3410: motor vehicles + 3420 automobile bodies, trailers and semi-trailers + 3430 parts/accessories for automobiles.
16. ISIC Rev. 3, division 3430: manufacture of parts and accessories for motor vehicles and their engines.

Chapter 3

Sustaining economic growth

Ongoing changes in economic structure are important to sustain growth. At lower per capita GDPs, the diversification of economic structures reduces volatility and makes positive growth episodes last longer. As economies develop, services become much more important, but even at higher per capita incomes, manufacturing interacts with services to maintain growth's momentum, as outlined in Chapter 2.

What is clear in all this is that the structure of the economy matters for the ability to sustain growth. Technological change drives structural transformation by generating a continuous flow of new products and productive activities, as a result of which new sectors emerge and old ones shrink that have become technologically obsolete. Technological changes in production processes result in rapid productivity increases, which—in interaction with changes in demand—result in dramatic transformations in the structure of production. The productivity gains driven by technology reduce employment in some sectors and liberate resources that can be employed elsewhere in the economy. The classic example is agriculture, of which the employment share has been declining across the globe. Productivity gains—with product innovations—can also increase a sector's shares in employment and value added, when demand is increasing rapidly as in the case of modern information and communications technology (ICT). But in advanced economies, productivity gains in manufacturing have reduced the shares of manufacturing and increased the shares of services in value added.

Thus, technological change drives structural transformation, but the effects of technological change are mediated by various conditions including investment in human capital, the improved functioning of innovation systems, increased absorptive capacities, upgrading in domestic clusters and upgrading in the context of global value chains. Structural transformation enhances the ability of economies to sustain growth over longer periods of time. (Figure 3.1 provides a schematic conceptual framework of the relationships between the key concepts in this chapter.)

Specialization or diversification—don't put all your eggs in one basket

What all "structuralist" theories have in common is that the structure of the economy is important for economic growth because some sectors have more growth potential than others. So, when an economy increases the share of sectors with high growth potential, this enhances overall growth, and vice versa. This opens the search for structural characteristics that are growth enhancing or growth reducing. The first level of the analysis focuses on a general characteristic of economic structure—the degree of specialization or diversification.

Specialization and diversification as alternatives

Trade theories of comparative advantage argue that successful economic development is associated with specialization in a narrow range of activities, but some other theories posit that economic development involves a process of diversification of sectors, activities and exports (for a review see Kaulich 2012). Too much specialization leaves an economy vulnerable to shocks and changes in the terms of trade (Osakwe 2007). According to this argument, an economy needs a broad portfolio of activities.

In a seminal contribution, Imbs and Wacziarg (2003) take an intermediate position. They found a non-linear inverted U-shaped relationship between gross domestic product (GDP) per capita and sectoral diversification. For low-income countries, the relationship is positive between diversification and levels of GDP per capita. The underlying assumption is that developing countries that diversify the structure of their production or the structure of their export package will grow more rapidly because diversification makes them more resilient to external

> **"Low-income countries can overcome their economic marginalization by acquiring skills and knowledge to diversify their economic portfolio**

Figure 3.1
Conceptual framework: Technological change for sustained economic growth

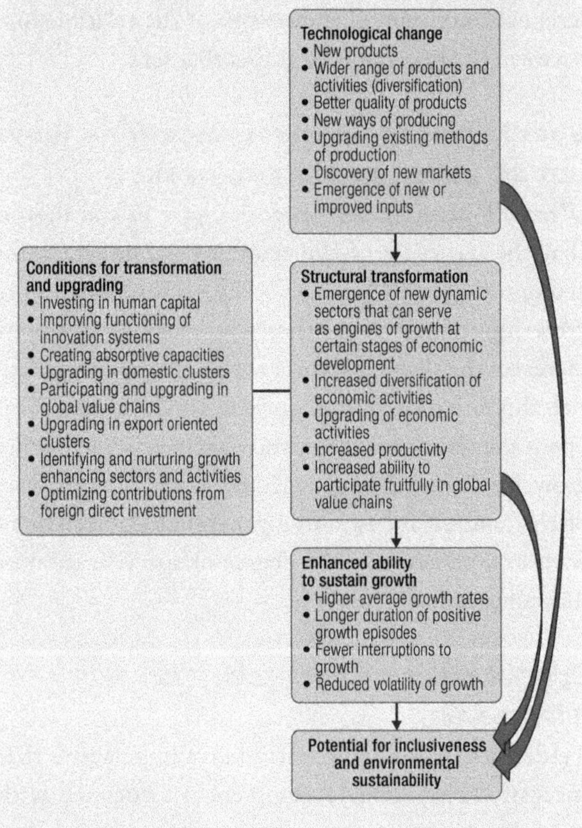

Source: UNIDO elaboration.

shocks. But beyond some threshold level of GDP per capita, the opposite relationship comes to dominate. As GDP per capita continues to increase, it is associated with increasing concentration and specialization, both in the total economy and within manufacturing. The turning point at which specialization sets in is quite high ($16,500 in 2000, purchasing power parity [PPP]).

From the perspective of low- and lower-middle-income economies, that observation implies that diversification from agriculture and diversification within manufacturing are associated with increases in GDP per capita. The implication for low-income countries in particular is that they can overcome their economic marginalization by acquiring skills and knowledge to diversify their economic portfolio rather than by focusing on "what they do best." High-income countries seem to benefit from specialization (Kaulich 2012).

Diversification also is linked to sophistication.[1] It is the ability to competitively produce a wider range of increasingly sophisticated goods that drives diversification. Low-income economies are typically specialized in a limited range of products. As their per capita incomes increase, they become more diversified, and the range of products broadens. At even higher incomes, specialization again comes to predominate.

Analysing export and sector structure data, UNIDO (2012) examines the inverted U-curve relationship and finds support for the notion that at lower levels of GDP per capita, a positive relationship exists between the degree of diversification and the level of per capita income. Regarding specialization at higher income levels, UNIDO (2012) has mixed results, although some interesting interactions between manufacturing and services emerge, which we will examine below.[2]

Specialization and sustaining growth, 1960–2010

Combining a new dataset on economic structure 1960–2010 with data on GDP per capita,[3] we analysed the relationships between structural diversification or specialization and different dimensions of economic growth: average growth, duration and volatility (Foster-McGregor, Kaba and Szirmai 2015). We find that diversification of the economic structure is related to the ability to sustain growth. Increased specialization of an economy has negative effects on the average rate of growth, the duration of positive growth episodes and the volatility of the growth path.

Based on a panel analysis with the growth rate in five-year periods as the dependent variable and the degree of specialization or concentration at the start of each five-year period as the independent variable, we find a consistent negative relationship between the degree of specialization (as measured by the Herfindahl index, the Gini index or the Theil index[4]) and the rate of growth. The more specialized an

“As an economy becomes less diversified (or more specialized), the rate of growth tends to slow

economy, the lower its average growth rate. The partial correlation scattergram between per capita growth and the (normalized) Theil index is reproduced in Figure 3.2.[5] This partial correlation shows the effect of the specified variable (in this case, concentration as captured by the Theil index) on per capita growth, holding constant other relevant determinants such as the initial level of GDP per capita relative to the United States, the ratio of exports to GDP, capital formation as percentage of GDP and population.

Another way to connect specialization and growth is to look at changes in specialization (rather than levels) and growth during the five-year periods. That relationship is also highly significant: as an economy becomes less diversified (or more specialized), the rate of growth tends to slow (Foster-McGregor, Kaba and Szirmai 2015).

The next step is to relate specialization to the length of uninterrupted growth episodes (duration).[6] The more specialized an economy is (irrespective of the indicator chosen), the shorter its growth episodes will tend to be (Figure 3.3).[7]

A very similar relationship is found between changes in specialization during a growth episode and the length of that episode. If there is an increase in specialization, the growth episode will tend to be shorter; if the economy diversifies, the growth episode will last longer (Figure 3.4).[8]

The degree of specialization as measured by the Theil or Gini indexes also has a significant effect on volatility in the five-year panel analysis. The more specialized or concentrated the economic structure at the start of a five-year period, the higher the volatility of growth in that period (as measured by its standard deviation). So, specialization affects not only the duration of growth episodes but also the volatility of the growth path.

Finally, we examine the chances that a growth episode will end. In any given year, the chances of a country ending a positive growth episode will be significantly higher in an economy with a more specialized structure of production. For instance, a 0.1 increase in the Theil index would result in 23 percent higher chance of a growth episode coming to an end.

Figure 3.2
Specialization and the rate of growth

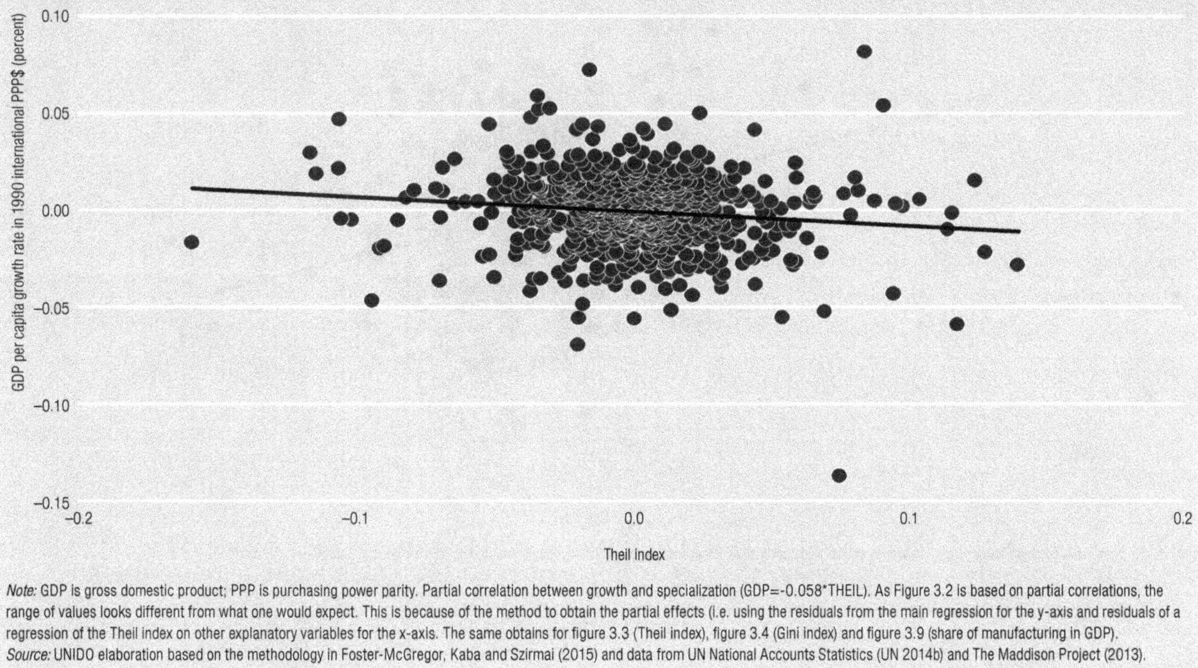

Note: GDP is gross domestic product; PPP is purchasing power parity. Partial correlation between growth and specialization (GDP=-0.058*THEIL). As Figure 3.2 is based on partial correlations, the range of values looks different from what one would expect. This is because of the method to obtain the partial effects (i.e. using the residuals from the main regression for the y-axis and residuals of a regression of the Theil index on other explanatory variables for the x-axis. The same obtains for figure 3.3 (Theil index), figure 3.4 (Gini index) and figure 3.9 (share of manufacturing in GDP).
Source: UNIDO elaboration based on the methodology in Foster-McGregor, Kaba and Szirmai (2015) and data from UN National Accounts Statistics (UN 2014b) and The Maddison Project (2013).

The more specialized an economy is, the shorter its growth episodes

Figure 3.3
Specialization and the duration of growth episodes

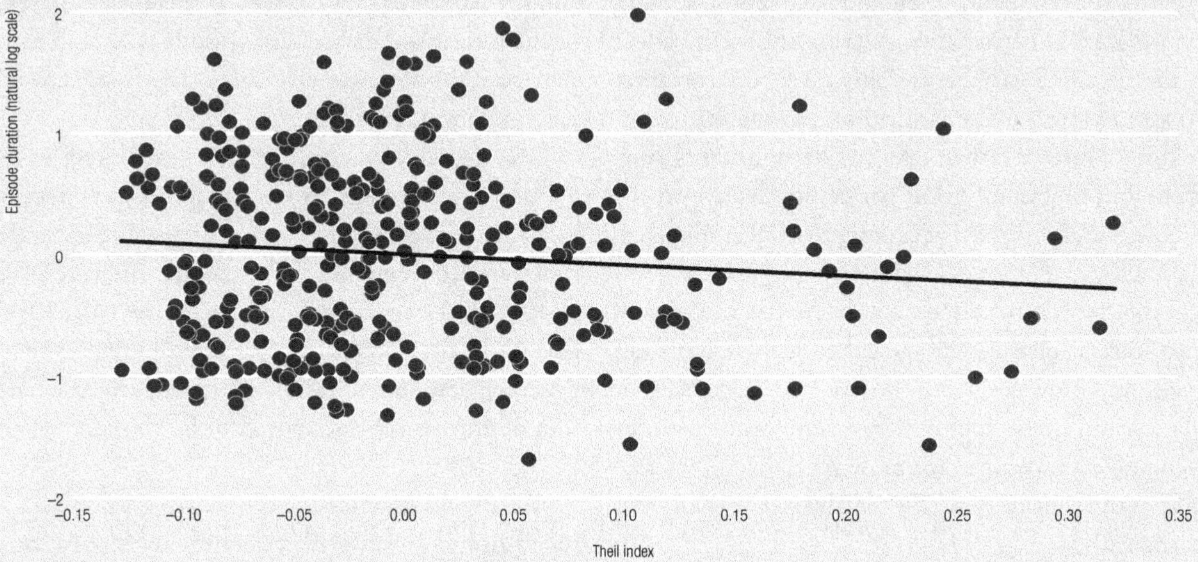

Note: Partial correlation between normalised Theil index and the natural logarithm of the duration of positive growth episodes.
Source: UNIDO elaboration based on the methodology in Foster-McGregor, Kaba and Szirmai (2015) and data from UN National Accounts Statistics (UN 2014b) and The Maddison Project (2013).

Figure 3.4
Change in specialization and the duration of growth episodes

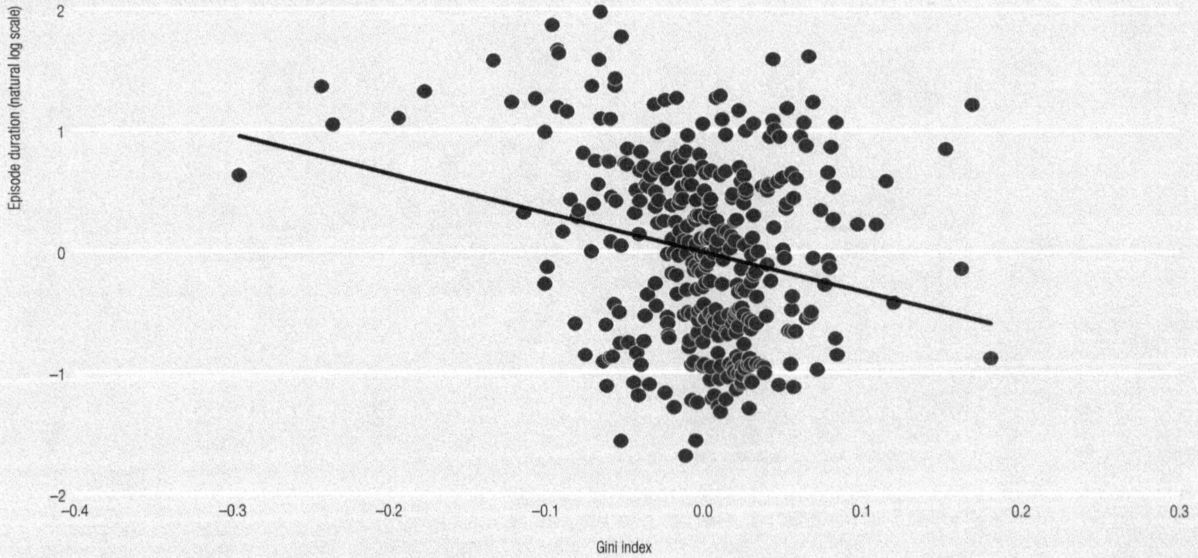

Note: Partial correlation between the annual change in specialization during an episode (measured by the normalised Gini Index) and the natural logarithm of its duration.
Source: UNIDO elaboration based on the methodology in Foster-McGregor, Kaba and Szirmai (2015) and data from UN National Accounts Statistics (UN 2014b) and The Maddison Project (2013).

> **“The share of manufacturing in value added and employment tends to increase when developing countries start growing**

Latent capabilities

As we saw in the previous chapter, a new index measures the latent capabilities of an economy—that is, the complexity and diversity of its export products. The more complex and diverse the export package, the stronger a country's latent capabilities (Kaltenberg and Verspagen 2015). Those capabilities had a significant influence on average GDP growth per capita in 108 countries from 1998 until 2013 (see Figure 2.8).

The low-income countries that grow rapidly and converge towards high-income countries are typically those with greater capabilities. Low-income countries with low capabilities tend to grow slowly and fall behind. Of course, high-income countries have the greatest capabilities, but they also tend to grow slowly because growth at the technological frontier is more difficult than growth at lower per capita incomes.

Manufacturing is still vital for sustaining growth

The second structuralist debate focuses on the role of sectors in economic development, whether agriculture versus industry, or industry versus services, or specific sectors that can act as engines of growth, such as ICT hardware, ICT software services, the automobile industry, capital goods sectors or high-tech sectors. The role of sectors can change over time, and different sectors can play key roles in different type of economies. Still, since 1950, an important hypothesis in developing economies has been that manufacturing is the key to development in low-income economies.

A substantial part of the literature—summarized in Lavopa (2015a), Szirmai (2012a), Szirmai and Verspagen (2015) and Tregenna (2015)—supports this engine-of-growth hypothesis.[9] But it is also contested. Several modern service sectors—such as ICT services, financial services, transport and logistics—are engines of growth in a manner similar to that of manufacturing in the past (Timmer and De Vries 2009). India is often mentioned as a case of service-led growth since the 1990s.

As discussed in Chapter 1, an important stylized fact of economic development is that the share of manufacturing in value added and employment tends to increase when developing countries start growing at low levels of per capita income. It peaks at intermediate per capita incomes and later declines as services become more important at high per capita incomes. An example of such an inverted U-curve is reproduced in Figure 3.5.

The interpretation of the relationship between GDP and manufacturing is that manufacturing plays a special role as an engine of growth and catch-up at lower levels of economic development, due to its special characteristics (Kaldor 1966, 1967; Szirmai 2012a):

- Productivity in manufacturing is higher than in other sectors, and productivity growth is faster. Structural changes involving a shift of resources to manufacturing thus provide static and dynamic productivity bonuses. The assumption is that the service sector provides less of a bonus.
- Manufacturing provides special opportunities for capital accumulation, spatial concentration, agglomeration economies and dynamic economies of scale.
- Manufacturing goods are internationally tradable, so the sector can profit from both domestic and global demand (Kaltenberg and Verspagen 2015).

Figure 3.5
Inverted U-curve: Manufacturing and GDP per capita, 2009

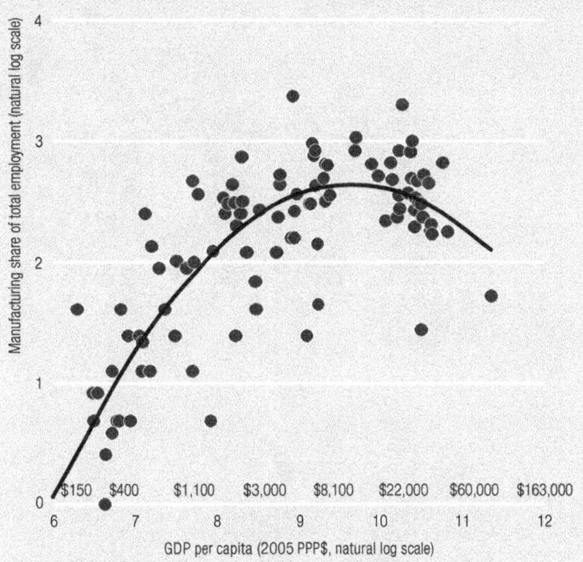

Note: GDP is gross domestic product. Based on a sample of 103 countries.
Source: Adapted from Tregenna (2015).

> **The relationship between the share of manufacturing and the rate of growth is positive and significant**

- Manufacturing has a special role as a driver of technological change, and it has more opportunities to profit from global technology and knowledge flows (Cornwall 1977; Kaldor 1966).
- Finally, its spillovers and linkages are stronger than those in other sectors (Lavopa and Szirmai 2012).[10]

The effect of manufacturing shares on growth, 1950–2005

In a detailed examination of the engine-of-growth hypothesis for a large sample of countries since 1960, Szirmai and Verspagen (2015) conclude that there is support for it, particularly in low-income developing countries that have invested heavily in education. But they also conclude that this route has become more difficult since the 1990s.

This research is based on a panel dataset with shares of manufacturing at current prices for four major sectors (agriculture, manufacturing, industry and services) in 89 countries from 1950 to 2005. The data relate the share of manufacturing at the start of each of 11 five-year periods (1950–1955 and onward) to the growth rate in that five-year period.

For the whole period 1950–2005, the relationship between the share of manufacturing and the rate of growth is positive and significant. But when the relationship is estimated for sub-periods, only 1970–1990 shows a significant relationship. Of special interest are two interactions: the share of manufacturing with GDP per capita as a share of the level of the United States, and the share of manufacturing with a measure of education.[11] The first relationship yields a significant negative effect. The interpretation is that industrialization matters for growth, especially at lower levels of per capita income. As countries become richer, industrialization becomes less important. Regarding the second relationship, manufacturing contributes

Figure 3.6
The marginal effect of manufacturing shares on growth, 1950-2005

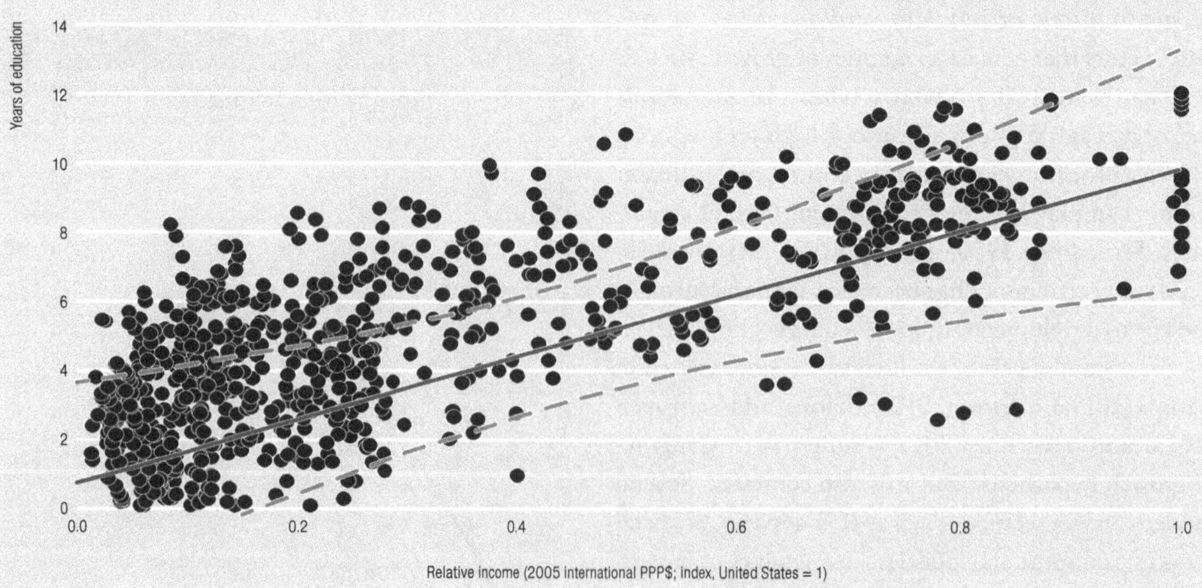

Note: The graph depicts the relation between relative income and average years of education for the population above 15 years of age. Relative income is defined as a percentage of United States income. The red line represents combinations of education and relative income at which there is no significant effect of manufacturing on growth. All points (country observations) above and below the red line represent combinations with positive and negative marginal effects, respectively. The farther from the line the points lie, the greater the chance of a significant effect. Beyond the dotted lines, which represent significance boundaries, there are significant effects of manufacturing on growth.
Source: Szirmai and Verspagen (2015).

> **"The second relationship, manufacturing contributes most to growth precisely in the low-income economies that have invested the most in education**

most to growth precisely in the low-income economies that have invested the most in education.

The effects of manufacturing on growth, together with its indirect effects through relative income and education, are depicted in Figure 3.6. Of particular interest are the dots above the top dotted line on the left-hand side. These represent countries with high educational investment and low levels of income per capita, where manufacturing plays a significant role in their catch-up. All the well-known examples of catch-up through industrialization—such as the Republic of Korea, Taiwan Province of China, Singapore and China—are in that area of the graph.

When this analysis is repeated for three subperiods—1950–1970, 1970–1990, and 1990–2005—the results are similar. But the red line shifts upward and becomes steeper in the latest period. To stay above the confidence interval, a country has to invest much more in education than it has previously to profit from the growth-enhancing effects of manufacturing. This route to growth has become more difficult (Figure 3.7).

Manufacturing and the ability to sustain growth, 1960–2010

We continue the analysis of the relationship between sectoral shares and dimensions of growth performance using the new dataset of sector shares 1960–2010 already discussed. This analysis builds on the results of Szirmai and Verspagen (2015), but for a larger sample of countries and a longer period. As before, the focus is on the share of manufacturing in GDP as the explanatory variable. But whereas the emphasis in the previous section was on average growth rates, the focus now is on the relationships between sectoral structure and the duration and volatility of growth.

Figure 3.7
The marginal effect of manufacturing on growth, selected periods, 1950-2005

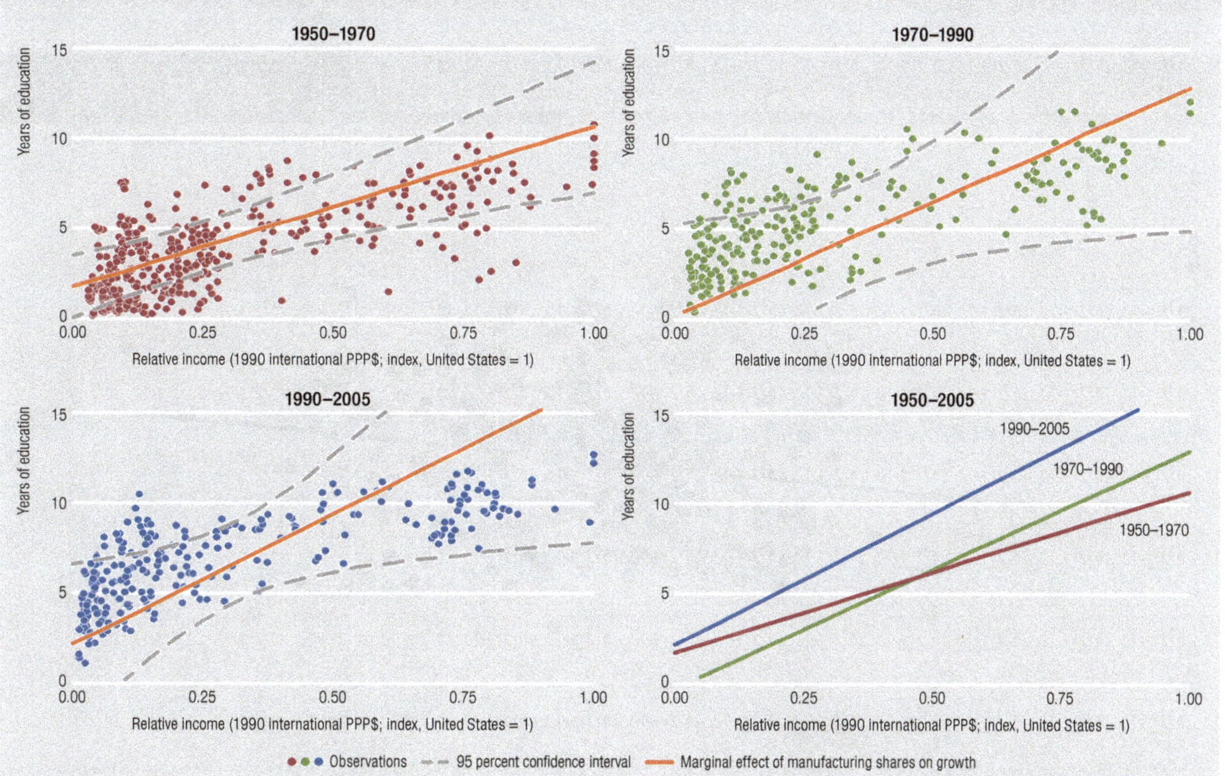

Note: See note for Figure 3.6.
Source: Szirmai and Verspagen (2015).

> **The larger the share of the manufacturing sector at the start of a growth episode, the longer growth continues**

One of the most striking and robust findings concerns the relationship between the share of manufacturing and the duration of growth episodes. The larger the share of the manufacturing sector at the start of a growth episode, the longer growth continues (Figure 3.8).[12]

The share of manufacturing within the modern sector defined in Annex A3 yields similar results, and they have significant positive effects on duration.

In line with the effects on duration, the chances of ending a growth episode are substantially reduced as the share of manufacturing in GDP at the start of the episode increases (Figure 3.9). Obviously, the longer an episode lasts, the greater the chances of it finally ending. But clearly the risk is much lower in every year in which the share of manufacturing at the start of the episode is higher.

One of the most interesting findings is that the effects of industrialization on the duration of growth episodes are more robust than its effects on average rates of growth.[13] Duration—the ability to sustain growth—really seems to matter.

Modern market activities and technological change as growth drivers

Critics of the manufacturing-as-engine-of-growth hypothesis argue that in the modern economy, some service sectors have dynamic growth-enhancing characteristics similar to those of manufacturing sectors in the past. Lavopa and Szirmai (2014) and Lavopa (2015a) attempt to broaden the discussion of drivers of growth by defining the modern sector to include other sectors with dynamic productive potential, such as financial services, business services, transport and logistics along with manufacturing (Annex A3).

The two studies show that sustained catch-up depends on the share of the modern sector in employment increasing at the same time as the productivity gaps between a country's modern subsectors and world best practice narrow as a result of technological upgrading. To capture that, consider an index of "structural modernization," defined as the relative labour productivity (as compared to the world

Figure 3.8
Initial shares of manufacturing in GDP and duration of growth episodes

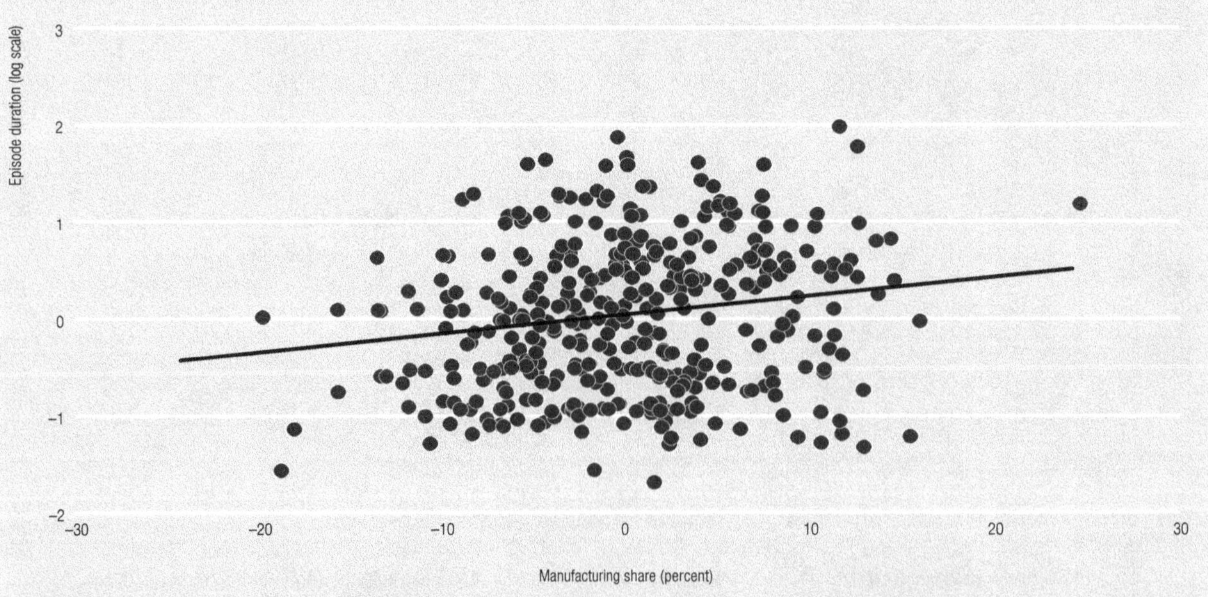

Note: GDP is gross domestic product. Partial correlation between the share of manufacturing at the start of a growth episode and the natural logarithm of the duration of the episode, controlling for population size, export-to-GDP ratio and GDP per capita as a percentage of the level in the United States in the initial year. GDP values are measured in 1990 international PPP$.
Source: UNIDO elaboration based on the methodology in Foster-McGregor, Kaba and Szirmai (2015) and data from UN National Accounts Statistics (UN 2014b) and The Maddison Project (2013).

Different countries can follow different trajectories in their process of modernization

Figure 3.9
The effect of the size of manufacturing on the hazard rate for ending a positive growth episode

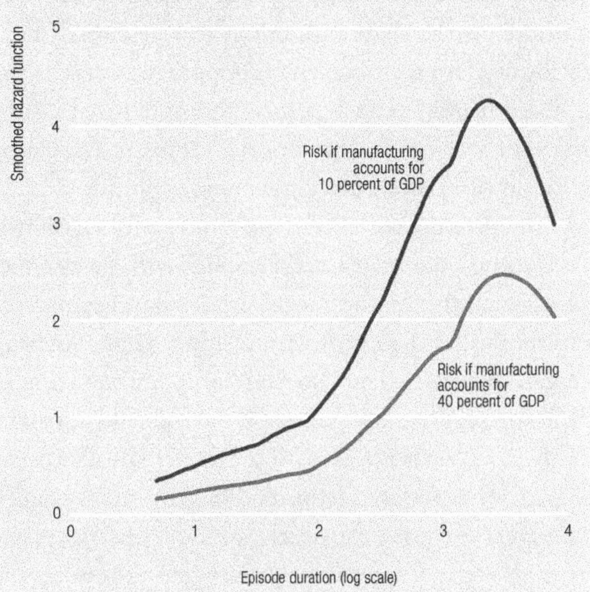

Note: The horizontal axis marks the log of time since the start of a growth episode. The vertical axis marks the hazard rate (the chance of an episode ending in a given year). The lower line represents the risk of growth coming to an end if manufacturing accounts for 10 percent of GDP at the start of an episode. The top line represents the risk in the hypothetical situation of a manufacturing share of 40 percent of GDP. Regression based on Cox Proportional Hazards Model.
Source: Foster-McGregor, Kaba and Szirmai (2015).

Figure 3.10
Structural modernization landscape

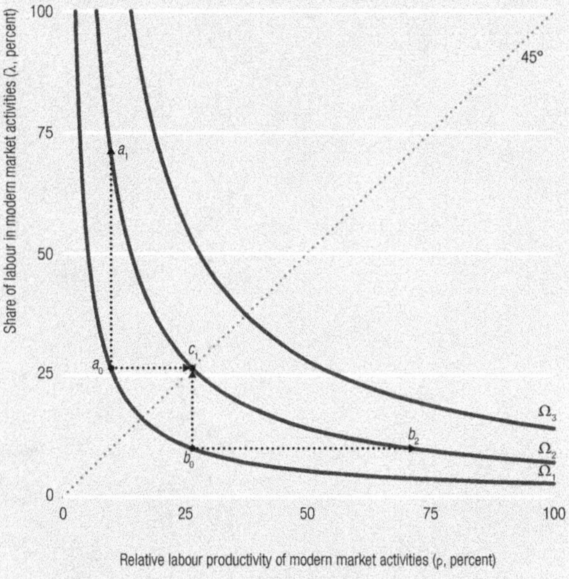

Note: The vertical axis represents the percentage of the labour force in the modern sector (λ); the horizontal axis represents labour productivity relative to world best practice (ρ). The three omega lines represent different combinations of the two components which result in a given level of structural modernization (ρ * λ).
Source: Lavopa and Szirmai (2014).

frontier) of the modern sector, weighted by the share of the modern sector in the total labour force.

Figure 3.10 depicts the structural modernization landscape. The same degree of structural modernization can be achieved with a large modern sector with lower relative productivity, or a smaller modern sector with higher relative productivity. A sector that usually has high absolute and relative productivity, but a very small share of employment, is mining. Success in economic development would entail a movement towards the upper-right corner of the landscape, where the modern sector expands and the technological gap with the world frontier shrinks. The figure also indicates that different countries can follow different trajectories in their process of modernization. Thus the movement from a_0 to c_1 involves reducing the technology gap, while the movement from a_0 to a_1 involves increasing size of the modern sector, while having no change in the productivity gap.

The structural modernization landscape can be used to chart the trajectory of a developing economy over time, as for the Republic of Korea in Figure 3.11. Korea is one of the very few that moved from low-income (I) to high-income status (IV) between 1960 and 1985. Its trajectory combines an expansion of modern employment and increases in relative productivity. Between 1985 and 2005, the modern sector ceased to expand as a share of labour force, but relative productivity was still improving. An alternative trajectory can be defined for economies that remain trapped at low-income levels, failing both to increase their modern sector and to bridge technology gaps.

Figure 3.5 provided a stylized depiction of the inverted U-shaped relationship between the share of manufacturing and GDP per capita. A similar figure can be derived for the modern sector (Figure 3.12). But there are differences. At low levels of income, the relationship between the modern sector and GDP per capita is much steeper, pointing to a strong connection

> **As income levels increase, the growth-enhancing effect of structural transformation erodes**

Figure 3.11
Modernization trajectory of the Republic of Korea, 1960–2009

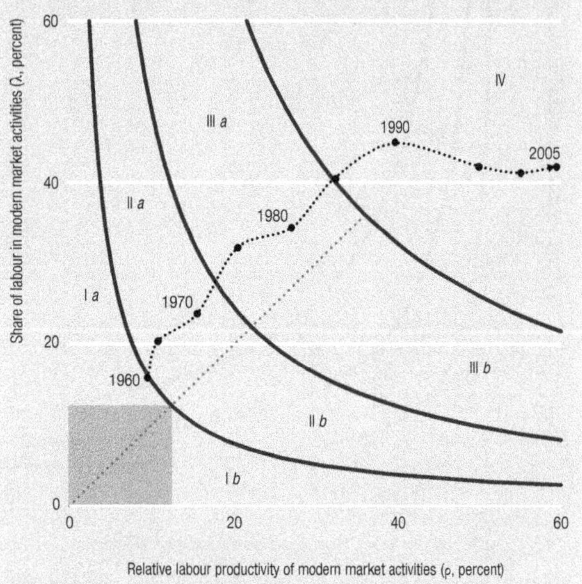

Note: Five year averages. The dotted isoquants refer to the levels of income that distinguish low, lower middle, upper middle and high income economies according to the World Bank classification (For more details see Annex A1, Table A1.3). These income boundaries are associated with the lowest levels of structural modernization found within each of the income categories.
Source: Lavopa and Szirmai (2014).

between structural modernization and per capita income. Next, the peak value is reached at higher levels of per capita income. The most important difference is that no decline occurs after the peak value is reached. The share of the modern sector as a whole remains stable. But in the modern sector, in advanced economies, the share of industry declines, while the share of modern services increases.

Lavopa (2015a) analyses the contributions of the two components of the modernization index—relative productivity and the share of the modern sector in employment—to growth. The relationships are negative, indicating that the contribution of structural change and technological upgrading is powerful at low levels of income and becomes weaker as countries start approaching the global frontier and start being similar in economic structure to the most advanced economies.

One of the most important and unexpected findings has to do with the share of manufacturing employment within the modern sector. When that term is entered in the growth regressions, it always yields positive and highly significant results. A similar exercise for the share of modern service employment in the modern sector gives the opposite results. Its significant negative effect on growth implies that a large share of services in the modern sector is associated with slower growth, even in mature economies.

The interpretation of those findings is as follows. As income levels increase, the growth-enhancing effect of structural transformation erodes because mature economies tend to grow more slowly than catch-up economies. But when the share of manufacturing in modern sector employment is introduced into the analysis, the countries with a larger manufacturing share in the modern sector tend to grow more rapidly. At higher incomes, the deceleration of growth typical of high-income economies is reduced. Those results are consistent with the virtuous interactions between modern manufacturing and dynamic market services.

The value-added share of exports in GDP is one of the most important determinants of economic growth (Lavopa 2015a). But the composition of exports matters hugely. Manufactured exports are far more important for growth than any other type of exports with a consistent positive effect of manufacturing exports.

Technological opportunities within manufacturing and beyond

Some service sectors play the same dynamic role as manufacturing has done in the past, but they do so when interacting with sophisticated manufacturing activities (Tregenna 2015; UNIDO 2013a). At more advanced levels of development strong linkages exist between manufacturing and service sectors, comparable to the relationship between agriculture and industrialization at early stages of development (Szirmai and Verspagen 2015). Service activities depend heavily on manufactured inputs. Manufacturing is also an important source of demand for modern intermediate service inputs such as financial services, transport and logistics, and business services, as illustrated in the global input–output analysis for this report (Park

“ Manufacturing is still the most important recipient of technological progress

Figure 3.12
Share of the modern sector and industry in GDP, by per capita income, 1950–2009

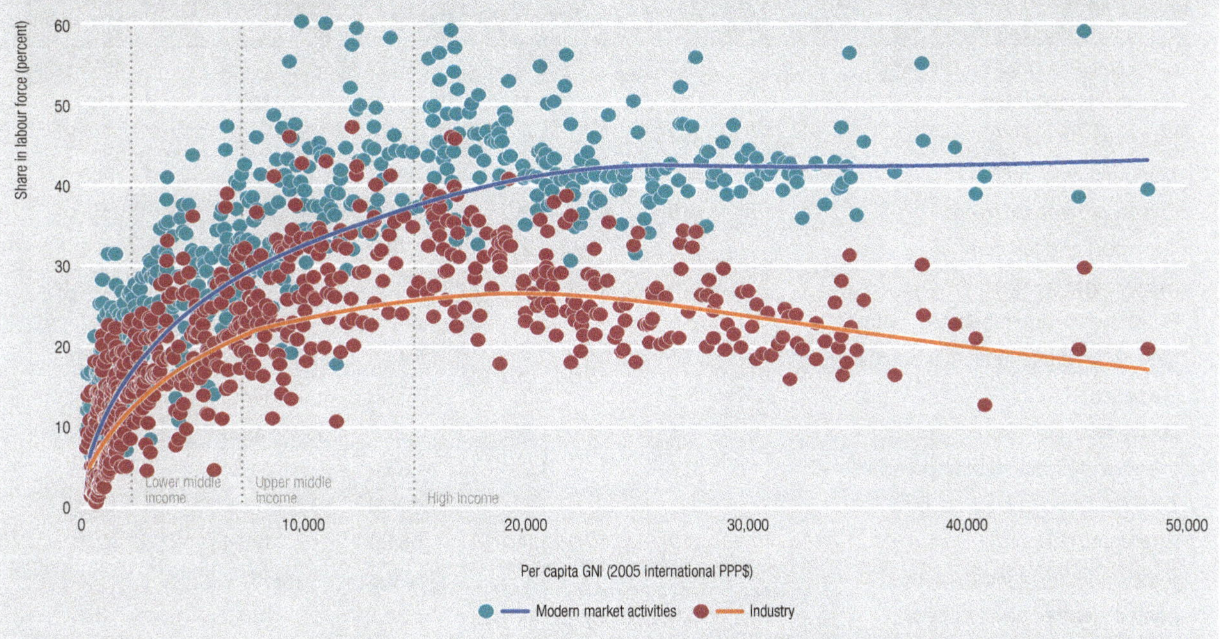

Note: GNI is gross national income. Five year averages for 97 countries. For income classifications please see Lavopa and Szirmai (2014).
Source: Lavopa and Szirmai (2014).

1989; Park and Chan 1989). The sectors use each other's inputs, but services depend more on manufacturing than manufacturing does on services. In addition, the emergence of modern service activities depends on the structure of manufacturing. Knowledge-intensive manufacturing sectors—such as office and computing machinery, electrical apparatus or industrial chemicals—are the main users of producer services (Guerrieri and Meliciani 2005).

Some sectors matter more than others

Table 3.1 provides summary statistics of the rates of growth of total factor productivity (TFP) over 1995–2007 for a sample of 37 countries.[14] The sectors are ranked by their median growth rate of TFP. The manufacturing sector as a whole has relatively high TFP growth rates. The top five sectors in the table are manufacturing and four of its subsectors. The highest-ranking service sector is transport and telecommunications, at rank 6. Of the service sectors, trade, restaurants and hotels is the only other sector that has median TFP growth higher than the median rate for the aggregate economy of the 37 countries.

Manufacturing is still the most important recipient of technological progress in those economies. Where catching up has been the norm in the 21st century—primarily in China and the post-communist world—manufacturing is crucial. The economic relations between those parts of the world and the developed part of the world are very manufacturing intensive in both directions, meaning that the most recent globalization trend is not a process in which manufacturing is just relocated from the developed world to other parts of the world—but one in which both sides of the trade relation are becoming more specialized in specific areas of manufacturing. The parts of the world where resources or market services take a leading role in globalization tend to also be the parts where growth is slower, and catching up rarer—primarily India, Sub-Saharan Africa and Latin America. They are globalizing in a very resource production–intensive way, but they are not catching up with the developed world in any strong and systematic fashion.

> **Differences in technological opportunities are fairly large**

3

SUSTAINING ECONOMIC GROWTH

Table 3.1
Distributions of growth in total factor productivity within sectors, 1995–2007 (percent)

Sector	Average	Median	Kurtosis	Skewness	Standard deviation
Electrical and optical equipment	15.3	9.5	600.1	240.9	22.6
Transport equipment	7.1	4.4	235.7	122.1	11.5
Machinery, n.e.c.	6.2	4.0	28.1	42.0	6.5
Manufacturing, total	4.7	3.4	–25.4	62.0	4.6
Other non-metallic mineral	4.6	3.2	117.4	113.6	5.1
Transport and telecom	4.0	3.2	79.9	97.0	5.4
Rubber and plastics	7.8	3.1	715.2	277.2	15.5
Basic metals and fabricated metal	5.0	2.7	669.9	229.8	9.2
Chemicals and chemical products	3.1	2.5	84.0	43.2	4.6
Trade etc.	2.9	2.2	502.8	144.8	3.5
Textiles and textile products	2.5	2.0	1479.3	324.9	3.9
Wood and products of wood and cork	2.7	1.9	63.9	17.8	7.1
Manufacturing, n.e.c.; recycling	3.7	1.9	694.0	79.0	14.2
Aggregate economy	3.0	1.9	564.9	188.4	2.7
Paper, printing and publishing	1.9	1.7	1,320.8	–211.2	7.2
Leather, leather and footwear	0.8	1.4	1,504.7	–326.5	3.7
Agriculture	4.0	1.3	2,144.6	442.9	10.1
Food, beverages and tobacco	1.3	1.0	182.0	73.7	3.7
Finance and bus services	2.2	0.9	145.5	82.7	4.8
Utilities	1.0	0.4	309.3	65.7	6.2
Public and other services	0.8	0.1	1,361.2	321.6	3.2
Mining	1.2	–0.2	1,325.0	144.5	12.2
Construction	–0.2	–0.4	305.4	8.5	4.1
Coke, refined petroleum and nuclear fuel	3.0	–0.7	1,264.2	278.6	23.1

Note: n.e.c. is not elsewhere classified. Based on a sample of 37 countries including all but three EU countries as well as the United States, Japan, the Russian Federation, the Republic of Korea and large developing countries such as Brazil, India, China, Turkey and Mexico.
Source: Adapted from Kaltenberg and Verspagen (2015) based on World Input-Output Database (Timmer and others 2015).

Ranking the sectors by the average growth rates, instead of the median rates, reveals much the same picture, but some differences emerge. Agriculture has a high average TFP growth rate but a much lower median. The same holds for mining and refined oil and, to some extent, for business services and finance.

In summary, differences in technological opportunities are fairly large. Manufacturing provides strong opportunities, as do some market services. The TFP growth rates of the service sectors of transport and telecommunications and trade are more similar to those of manufacturing than of traditional services.

Sectoral distribution of R&D and foreign direct investment

Table 3.2 provides an overview of the sectoral allocation of research and development (R&D) expenditures in high-income economies (Jacob and Sasso 2015) and provides strong support for the view that manufacturing is the most knowledge-intensive sector of the economy. On average, 60.3 percent of all R&D takes place in manufacturing, with services in second place at 35 percent. On R&D intensity (R&D as a share of value added), the sectoral differences are even more pronounced: R&D intensity in manufacturing is close to 6 percent on average but only around 0.5 percent in other sectors.

> **Manufacturing is the most knowledge-intensive sector of the economy**

Table 3.2
R&D expenditures, by income group and sector, 2011

	Agriculture			Manufacturing			Mining, construction and utilities			Services		
	Billion PPP$	Shares of total R&D expenditure (percent)	Sectoral R&D intensity (percent)	Billion PPP$	Shares of total R&D expenditure (percent)	Sectoral R&D intensity (percent)	Billion PPP$	Shares of total R&D expenditure (percent)	Sectoral R&D intensity (percent)	Billion PPP$	Shares of total R&D expenditure (percent)	Sectoral R&D intensity (percent)
Low- and middle-income countries												
China	0.48	0.26	0.04	162.47	86.56	3.78	12.69	6.76	0.64	12.05	6.42	0.21
Poland	0.01	0.77	0.06	0.95	49.15	0.70	0.05	2.76	0.05	0.91	47.32	0.19
Turkey	0.01	0.27	0.01	2.59	53.34	1.23	0.07	1.36	0.06	2.19	45.04	0.30
High-income countries												
Australia	0.13	1.04	0.56	2.98	24.57	4.29	3.51	28.94	1.85	5.51	45.46	0.86
Austria	0.00	0.03	0.04	4.34	63.69	7.02	0.09	1.29	0.27	2.38	34.98	1.04
Belgium	0.03	0.46	1.09	4.21	62.93	7.13	0.14	2.14	0.40	2.30	34.48	0.75
Canada*	0.11	0.83	0.55	6.03	46.62	4.41	1.05	8.11	0.48	5.75	44.44	0.63
Czech Republic	0.01	0.33	0.13	1.46	56.23	2.19	0.04	1.50	0.11	1.09	41.94	0.66
Denmark	0.01	0.14	0.21	2.48	51.94	9.34	0.05	0.99	0.21	2.24	46.92	1.43
Finland	0.01	0.10	0.11	4.27	76.83	12.06	0.12	2.17	0.64	1.16	20.90	0.90
France	0.18	0.52	0.44	17.00	49.75	6.82	0.81	2.37	0.44	16.18	47.36	0.94
Germany	0.16	0.25	0.65	55.77	85.62	7.93	0.35	0.53	0.15	8.86	13.60	0.42
Italy	0.00	0.03	0.01	10.36	73.60	3.43	0.15	1.07	0.09	3.56	25.30	0.25
Japan	0.03	0.02	0.05	100.35	87.87	12.35	1.56	1.37	0.47	12.26	10.74	0.39
Korea, Rep. of	0.04	0.09	0.12	39.11	87.54	8.81	1.57	3.51	1.58	3.96	8.85	0.47
Norway	0.08	3.14	1.98	0.98	37.07	4.66	0.35	13.08	0.37	1.23	46.70	0.78
Portugal	0.00	0.24	0.09	0.70	35.48	2.16	0.05	2.42	0.21	1.21	61.86	0.64
Slovenia	0.00	0.08	0.07	0.76	72.08	7.07	0.02	1.69	0.35	0.27	26.15	0.81
Spain	0.14	1.32	0.40	5.85	55.70	3.10	0.68	6.44	0.42	3.84	36.54	0.38
Sweden	0.02	0.25	0.38	6.59	71.86	9.91	0.05	0.51	0.13	2.51	27.38	0.98
United Kingdom	0.02	0.07	0.13	9.18	36.90	4.39	0.35	1.40	0.16	15.33	61.63	0.95
United States	–	–	–	201.36	–	10.56	3.79	–	0.30	88.95	–	0.73

Note: PPP is purchasing power parity. (1) Billion PPP$; (2) Sectoral shares of total R&D expenditure (percent); and (3) sectoral R&D intensity (percent). * Values are from 2010. The sectoral R&D intensity is computed dividing the sectoral R&D expenditure (in current local currency) by the sectoral value added (in current local currency), i.e. R&D intensity$_{ij}$ = R_{ij}/Y_{ij} where R_{ij} is the R&D expenditure of sector *i* in country *j*.
Source: Jacob and Sasso (2015).

For the sectoral distribution of foreign direct investment (FDI), a similar picture emerges (Jacob and Sasso 2015). Between 2003 and 2011, manufacturing attracted 42 percent of total greenfield investment, followed by 29 percent in mining, construction and utilities combined. Services accounted for 28.6 percent. However, in least developed countries and other developing economies, mining attracts the bulk of FDI. Within manufacturing, 47 percent of all greenfield investment projects were in electronics, electrical equipment, machinery, and motor vehicles (Table 3.3).

Together, those figures provide strong support for the proposition that manufacturing is indeed the locus of technological change in the wider economy and that sectoral differences matter.

High-tech sectors generally drive growth, but not in low-income countries[15]

Do technology-intensive sectors make a greater contribution to growth than other manufacturing sectors? Structuralism identifies manufacturing as having a special role as an engine of economic growth

> **High-tech value added growth has a significant positive effect on the rate of growth**

Table 3.3
Inward greenfield foreign direct investment projects in manufacturing sectors, by group of economies, 2003–2011 (percent)

	Emerging industrial countries	Industrialized countries	Least developed countries	Other developing countries	Total
Electronics, electrical equipment, machinery, motor vehicles	48.9	47.0	28.1	38.0	47.0
Chemicals, rubber, plastics, fuel and minerals	26.8	26.5	28.7	29.1	26.8
Metals	9.2	8.2	12.9	10.4	8.8
Food, beverages, tobacco	6.9	6.6	21.3	13.7	7.4
Furniture, repair and installment, other	3.9	5.6	2.4	3.1	4.7
Textiles	2.2	2.9	4.9	3.7	2.7
Paper, wood, printing	2.3	3.1	1.7	2.1	2.7
Total	100.0	100.0	100.0	100.0	100.0

Source: Jacob and Sasso (2015).

and social development—not only through a direct effect but also through greater dynamic economies of scale and positive externalities across the whole economy (Kaldor 1966, 1967). The underlying idea is that GDP growth is correlated to the manufacturing growth rate. Using a panel dataset on GDP and manufacturing value added in 146 countries between 1970 and 2011, that idea is analysed for manufacturing as a whole and for technology-intensive sectors in particular (Lennon, Cantore and Clara 2015).

The most important finding is that specialization in high-tech manufacturing sectors provides an important growth premium. When growth of high-tech value added is included in models that explain the growth of aggregate value added, the high-tech value-added growth has a significant positive effect on the rate of growth. In contrast, the coefficients for low- and medium-tech activities are not significant. Holding the rate of growth of manufacturing constant, countries specializing in high-tech sectors tend to grow more rapidly than those specializing in low-tech sectors. Those results appear even more robust when one focuses specifically on middle- and low-income countries and on the most recent period, 1991–2011.

But for low-income countries, this is a cause for concern, given their modest share of high-tech activities. In the past two decades, high-tech sectors accounted for almost one-third of total manufacturing value added in high-income countries, one-fifth in middle-income economies but only one-tenth in low-income economies. There is also a structural shift towards more high-tech activities in high-income economies (an increase of 4.4 percentage points) and middle-income economies (an increase of 3.2 percentage points), but no change in low-income economies.

Short-cycle technologies—or long

As seen, some economic sectors play a special role in economic development and are better suited to sustaining economic growth, but so far we have not explored the technological reasons for that. Technology-based specialization is necessary to sustain economic growth at middle incomes and to avoid falling into the middle-income trap (Lee 2013a). Short-cycle technologies lead to the highest degree of intra-national knowledge diffusion and creation and the advantages of established producers are the least.

Long-cycle technologies use old knowledge, but in short-cycle technologies, knowledge becomes obsolete faster and therefore changes faster, and so latecomer emerging economies will find it easier to enter a sector with short-cycle technology because they will not need

> **Involvement in short-cycle technologies leads to indigenous knowledge creation**

to master the old knowledge, increasing their chances of catch-up. The frequent emergence of new knowledge that characterizes short-cycle technologies also provides higher growth prospects and lower barriers to entry. And involvement in short-cycle technologies also leads to indigenous knowledge creation because it implies less reliance on the old technologies dominated by industrially advanced countries.

What sectors have this type of technologies? Detailed patent-data analysis suggests that ICT sectors such as semiconductors, TV/displays, information storage and telecommunications tend to have extremely short-cycle technologies, while textiles, wearing apparel and food and beverages generally have longer-cycle technologies (Lee 2013c).

Sectors with short-cycle technologies are exactly those where the most successful catch-up countries of Asia (the Republic of Korea and Taiwan Province of China) have based their recent specialization. The Republic of Korea began with labour-intensive (long-cycle technology) industries such as apparel and shoes in the 1960s, then moved to shorter- (medium-) cycle technology sectors such as consumer electronics and automobile assembly in the 1970s and 1980s. In the 1990s, it moved to shorter-cycle technology sectors such as telecommunications equipment, memory chips and digital TV. This kind of technology-driven structural change allowed the Republic of Korea to sustain growth over a very long time, sweeping it past the middle-income trap.

In contrast, less successful middle-income countries (such as Argentina and Brazil) have failed to move into shorter-cycle technologies, instead adopting a direct replication strategy focused on technologies with longer cycles such as wells, pumps, hydraulic engineering, food and beverages, and metal working. This is put forward as a seminal reason why these countries did not maintain their growth rates and fell into the trap that the Republic of Korea has avoided (Lee 2013a).

This analysis is consistent with the patterns of structural change in Chapter 1. Low-tech sectors (with long-cycle technologies) grow fast at initial stages of development but tend to have growth decline as countries get richer. Only high-tech manufacturing industries manage to sustain high growth rates, even at upper middle and high incomes.

The technological characteristics of different sectors thus seem to lie at the core of the relationship between structure and capacity to sustain growth, so policies to facilitate this transition (by building up the capabilities to enter short-cycle technology sectors) are important (Chapter 6).

Creating the conditions for technology to sustain economic growth

Globalization presents a moving target of competition. Standing still condemns an economy to falling standards of living. Second, upgrading and the policies required to deliver it are contextual. They also vary over time. So while the Republic of Korea and Japan could upgrade while keeping foreign firms at arm's length, in the present international order FDI and multinational enterprises (MNEs) are key to accessing global technology and cannot be circumvented (Szirmai 2015). Third, the acquisition of technological capabilities in many cases is closely related to the capacity of producers to insert themselves productively into global value chains (GVCs). Fourth, given the structure of GVCs, success in technological upgrading is determined both by shifts in positioning within sectors and by diversification into new sectors.

As has been argued extensively in this chapter, developing the right kind of growth-enhancing sectors is a key element of structural transformation. At low levels of economic development, this involves building up productive capacity in industry, often of a labour-intensive kind, but also in labour-intensive service sectors with the potential to evolve. At higher levels of per capita income, successful development involves technological upgrading within sectors shifting or even leapfrogging towards technologically more dynamic activities (Lee 2015). Identifying the appropriate sectors for given countries at given stages of development is an important challenge for firms and for industrial policy.

> **Technology-specific skills are typically needed to absorb new technologies**

Technological capabilities for sustained growth

Successful development is not only a matter of reallocating resources (mainly labour) towards *good* sectors; it is also—and quite fundamentally—tied to the capacity of countries to reduce their gap with the technological frontier in these sectors. This is the main idea behind Lavopa and Szirmai (2014) modernization index, introduced earlier in this chapter. The Republic of Korea (Figure 3.11) clearly illustrated a successful path for the modern sector to expand and reduce the technological gap with the world frontier. A crucial question, then, is how to improve or upgrade the national level of absorptive and technological capabilities to reduce the technological gap.

Technological capabilities are mainly related to the education of the population and the specific allocation of human capital and other resources to undertake R&D. The relative importance of each of these elements depends on a country's level of development. At early stages of development, technological gaps create the potential for accelerated structural transformation through access to global technological knowledge, but the extent to which such transformation will be realized depends heavily on the absorptive capacities of countries, sectors and firms (Lall 2000, 2002). Among the most important determinants of absorptive capacity are sustained investments in human capital. Strong basic and secondary education and specialized human capital are fundamental to absorb new technologies: basic education and new skills are needed to use new technologies, whereas a more educated population tends to adopt new technologies faster.

But basic literacy is not enough. Certain technology-specific skills are typically needed to absorb new technologies. In some cases, these skills can be provided by an improved basic education curriculum. In other cases, these skills have to be provided through specialized training at vocational centres. At middle ranges of development, the creation of new indigenous knowledge also becomes very important, in addition to the continued absorption of global technological knowledge. A strong tertiary education system in science and engineering and larger formal R&D play a key role at this stage (Dahlman 2010). In fact, the transition towards more technology-intensive manufacturing and service activities depends on a "hi-tech infrastructure," which includes universities and polytechnics capable of generating skilled technicians, engineers and scientists (Narula 2004) (Box 3.1).

Ultimately, technological capabilities are embedded in domestic firms. So, the specific conditions to achieve technological upgrading are also closely tied to the various channels for firms to acquire technological knowledge to upgrade their capabilities. Such channels include informal learning, learning from FDI

Box 3.1
Human capital and the aerospace industry in Querétaro, Mexico

The aerospace industry in Querétaro has grown rapidly. Bombardier—one of the leading companies in the sector, based in Canada—arrived in 2006. The French group Safran and Spanish airframe manufacturer Aernnova established operations in 2007. By 2012, more than 30 foreign firms were operating in the state. Mexico's aerospace exports reached $4.5 billion by 2011, up from $1.3 billion in 2004 (Gereffi 2015).

Growth was supported by the state government, partly through the creation of the National Aeronautics University of Querétaro (UNAQ) in 2007. Various technical programmes emerging from public–private initiatives were offered to introduce the country's first aerospace engineering programme. By 2009, state investments in UNAQ had totaled $21 million. Many teachers working in regional aerospace firms were hired to teach at the university and offer training of teaching staff in both Canada and Spain. By 2012, UNAQ counted 488 technical and professional students. UNAQ is further credited with adding to an already strong engineering training base by contributing to human capital development (Gereffi 2015). Forty-one percent of undergraduate degrees were earned in engineering, while 65 percent of all master's degree programmes offered in the country were in engineering-related fields (Casalet and others 2011). In 2007 an aircraft maintenance programme was established in Querétaro by the National Mexican Technical Training Institute, which graduates 90 technicians annually (Gereffi 2015).

> **Developing domestic technological capabilities stands as one of the most important elements to sustain growth**

partners, licensing, strategic alliances and co-development (Lee 2013b). They vary with a country's development level. At early stages, technological knowledge is mainly embodied in imported machinery, and the main channel for capacity building relates to learning by doing. In an intermediate stage, domestic firms recognize the need for more systemic learning and technological development and tend to resort to technological licensing, or seek for knowledge transfers from FDI partners. This tends to be complemented with increasing in-house R&D capacity. At a later stage, once the channels of licensing and learning from foreign partners have reached their limit, domestic firms rely on public-private R&D consortia, existing literature, overseas R&D outposts, co-development contracts with foreign R&D firms and international merges and acquisitions.

Learning and technological absorption may take place in firms, but the success or failure of individual firms occurs within a "system" (Lall and Narula 2004). So, the scope for countries to upgrade their technological capabilities also depends on the functioning of the national systems of innovation.[16] From this perspective, learning and innovation involve complex interactions between firms and their environment, which includes not only the firms' network of customers and suppliers but also the institutional and organizational framework, the technological infrastructure and knowledge-creating and -diffusing institutions. As innovation systems improve, countries are better able to tap into international sources of knowledge and technology, while the domestic flows of knowledge improve. In an effective innovation system, knowledge does not remain limited to a few modern firms—it circulates rapidly from firm to firm and actor to actor.

Technological upgrading needs broad dissemination of knowledge throughout the whole economy. Such dissemination requires strong public policies for diffusing the use of new technologies and an institutional infrastructure that includes, among other things, technical information services, extension services, productivity organizations, metrology standards, quality control institutions and industrial clusters. Upgrading technological capabilities also requires an appropriate technological commercialization infrastructure that can put into practice the new knowledge created—as in government research labs and universities. This infrastructure includes adequate intellectual property rights protection systems, technology-transfer offices at universities and research institutes, science and industrial parks, business incubators, early stage technology finance and venture capital (Dahlman 2010).

It follows that developing domestic technological capabilities stands as one of the most important elements to sustain growth. Take China. In the early 1980s, it had low levels of secondary and tertiary education, but massive investments in basic, secondary and tertiary education modified this picture radically. By 2007 it had more students at the tertiary level than the United States, more than 40 percent of them in engineering and sciences.[17] This had tremendous importance for acquiring and using new knowledge. In parallel, huge investments increased the number of scientists and engineers carrying out R&D and the expenditure on R&D as percentage of GDP. This led China to be by 2007 the third largest investor in R&D at the world level, devoting 1.42 percent of its GDP to R&D.

Major efforts were also devoted to the creation of an appropriate technological commercialization infrastructure, developing an impressive number of science parks and business incubators, and implementing an aggressive strategy to promote spin-off activities in universities. All this has been complemented by important efforts to achieve broad diffusion of new technological knowledge, including specific programmes such as the Spark programme for rural innovation[18] and the Torch programme for high-tech innovation (Dahlmann 2010).[19]

China has also extensively exploited international knowledge resources in an "open national system of innovation" (Fu 2015). It combined fostering indigenous capabilities and opening to external knowledge sources. In doing so, China has used unconventional channels that are not often used in developing countries, such as outward direct investment, international

> **Dynamic clusters can provide a stepping stone in the growth trajectory of developing economies**

innovative collaboration and attracting highly skilled migrants (Fu 2015). Furthermore, specific policies were implemented to strengthen the linkages between foreign and local firms to make effective technology transfer possible. Over a certain period, China required joint ventures as a condition for FDI, negotiated local content requirements in certain industries and imposed training requirements on MNEs, as for Motorola (Fu, Pietrobelli and Soete 2011).

These elements cannot be considered in a vacuum, looking exclusively at domestic conditions. In the general context of increasing fragmentation of production, technological upgrading is very much related to upgrading both in clusters and in GVCs.

Technological upgrading in industrial clusters

An important condition for achieving technological upgrading relates to the functioning of industrial clusters.[20] As mentioned above, industrial clusters foster broad dissemination of technological knowledge throughout the economy and bring important benefits to the domestic economy. These benefits can be broadly defined along three dimensions (Ketels and Memedovic 2008). First, clusters enable higher productivity. Firms can operate with a higher efficiency, drawing on more specialized assets and suppliers with shorter reaction times than when working alone. Second, firms and research institutions can build connections to better learn and innovate, as tacit information and knowledge are best developed and exchanged locally. Knowledge spillovers and the close interaction with customers, other firms, venture capitalists and knowledge-intensive service providers create more new ideas and provide intense pressure to innovate, while the cluster environment lowers the cost of experimenting. Third, business formation tends to be higher in clusters. Start-ups are more reliant on external suppliers and partners—all to be found in a cluster. Clusters can spread the cost of failure, as entrepreneurs can fall back on local job opportunities in the many other firms in the same field.

In developing countries, industrial clusters bring two additional important effects. Most (but not all) clusters in low and middle-income markets use labour-intensive and small-scale technologies and meet the needs of domestic and low income consumers. Thus, they not only represent an important source of economic growth, but they also provide important opportunities for social inclusion. But most of these economies suffer from a "missing middle" between small and informal enterprises and the large-scale formal sector. Dynamic clusters can thus provide a stepping stone in the growth trajectory of these economies. But if this dynamism is merely a form of extensive growth—the replication of what exists—it is unlikely to deliver sustained and sustainable growth. Hence the capacity to upgrade within industrial clusters also represents a key condition for sustaining economic growth.

To upgrade within clusters requires overcoming challenges in final markets, process technology, organizational technology, and the interfirm division of labour, which includes positioning in the value chain (Box 3.2).

Extending final markets. Meeting new demands from consumers and overcoming the offerings of competitors are often the prime drivers for cluster upgrading, whether the extended market is at home or abroad. Strengthening user–producer interactions and extending markets are routes to upgrading.

Upgrading processes. Most clusters—survivalist and dynamic—are small and use basic technologies, often second hand. Their small final markets do not allow the purchase of large and scale-intensive technologies, but it also may be a reflection of the high acquisition costs of more sophisticated equipment. The upgrading challenges in these clusters, particularly in the informal sector, are complex. In some cases, the solution to process upgrading lies in buying new equipment, or improving equipment. A further solution might be to search for new sources of capital goods, which represent an improvement in what they use, but may be of lower quality than equipment bought from more established capital goods suppliers. But the prospects may be good for South-South technology transfers.

> **"Once enterprises begin to participate in global value chains, they also need to upgrade functionally**

Box 3.2
Upgrading in Kenya's furniture industry

A good example of these upgrading challenges can be observed in Kenya's furniture industry (Attah-Ankomah 2014). In recent years, many of the clustered furniture manufacturers have switched from using industrial-country machinery to much cheaper equipment from China, which is generally less robust and produces to a lower quality. The production process using these machines is more labour-intensive and generally has lower unit costs.

Kenyan machinery manufacturers—also operating in the informal sector—have responded and either produce their own cheaper versions of furniture-manufacturing equipment, with even lower-quality output, or help to "blend" Chinese and industrial-country machines, using motors from high-income countries. The overall impact is an improvement in productivity and in product quality and at least one of the furniture clusters is venturing into much higher-quality final markets using a mix of Chinese and blended equipment.

Similar benefits from using Chinese and Indian equipment are seen in agricultural mechanization in Tanzania (Ageyi-Holmes 2014) and in Uganda's apparel sector (Botchie 2014).

Source: Kaplinsky (2015a).

Organizational upgrading. Many informal sector clusters especially offer wide scope for upgrading, and with lower barriers to entry with disembodied organizational technologies than from the introduction or adaptation of embodied technologies. This may involve workflow, quality procedures, storage of materials, maintenance of machines and business strategy. These "soft" elements of process technology are often reduced to writing business plans and securing finance from governments and non-governmental organizations. But they meet only a restricted part of the organizational technology upgrading agenda. In each of the East African clusters using Chinese and Indian equipment (Box 3.2), there is no evidence of structured attempts to facilitate cluster upgrading by addressing skill development, machine maintenance and workflow. Each of these arenas was the sole responsibility of the individual entrepreneurs, and there has been very little change in these clusters.

Interfirm division of labour and functional upgrading. One of the major drivers of productivity growth is specialization within firms and the division of labour between firms. This is often a natural outcome of cluster dynamics, as Schmitz documented in the Sinos Valley footwear industrial district in Brazil (Schmitz 1995). An increase in the inter-firm division of labour poses multiple upgrading challenges for clusters. It reflects a drive towards the specialization of components manufacture and their disassociation from assembly. But increasingly it also involves the development of specialized business services providers, for example, in the extension of standards in value chains and in the provision of support for finance and marketing.

Once enterprises begin to participate in GVCs, they also need to upgrade functionally. That is, an upgrading strategy may involve the capacity to change position in the chain, perhaps moving from low-skilled assembly to more skill-intensive component manufacturing, or beginning to design, brand and market products independently. The drive towards functional upgrading may only have broader economic benefits if the cluster as a whole changes its position in the value chain. If individual firms merely swap their position in the chain, they may gain or lose as separate economic agents, but there may be little cluster upgrading in the chain as a whole.

Technological upgrading in GVCs

Entering GVCs. In the "governance" (coordination) of GVCs, multinationals play a key role.[21] Thus the rise of GVCs goes hand in hand with a substantial increase in global FDI flows, from an average of $65 billion (1995 dollars) a year in 1970–1974 to $1,42 billion in 2011–2014 (UNCTAD 2015). The share of FDI going to developing countries increased, from 21 percent in 1970–1974 to 45 percent in 2009–2011 (Szirmai 2015). So, upgrading in GVCs is closely tied to FDI.

From developing countries' perspectives, GVCs offer new opportunities for industrialization and technological change. Rather than having to build

> **For firms newly incorporated into a global value chain or new entrants into a sector, the strategy is one of 'thinning in'**

capabilities over the complete range of industrial activities, countries can enter slices of GVCs (Baldwin 2011; Szirmai, Naudé and Alcorta 2013). But if their activities remain limited to thin slices, they may become too specialized, with little impact on growth.

Many countries have deliberately followed policies to enter GVCs by establishing special economic zones (SEZs) with special facilities and incentives to attract foreign investment. Among the best known are China's (Box 3.3).

For the firm receiving outsourced activities (usually in a developing economy), this may involve two contrasting strategies. For firms newly incorporated into the chain or new entrants into the sector, the strategy is one of "thinning in." That is, they enter the chain by contributing a low proportion of the value added embodied in the final product. Examples are the firms newly established to assemble apparel on a cut-make-and-trim basis. The other strategy is for supplier firms that have long operated in the sector and for whom GVC entry involves a "thinning out" of activities, cutting the range of activities that they have historically undertaken. Keeping the apparel sector as an example, this would represent a firm that gives up its own design and brand names to assemble apparel for an outsourcing lead buyer.

Mechanisms affecting opportunities for upgrading. In some value chains, the lead firm limits opportunities for upgrading in others. So a key objective in GVC upgrading is for firms to enter chains that provide such scope. Different markets have different requirements and vary in their scope for entry-profit margins. Environmental and health standards in advanced country markets provide serious barriers to entry, but also provide challenges (and incentives) for quality improvement and technological upgrading. Increasing concentration of buyers and final retailers reduce the bargaining power of entrants and the conditions for upgrading. But the more deeply embedded foreign firms are in the local economy, the more they can help upgrade their local suppliers.

Vertically specialized and additive value chains. From an analytical perspective, it helps to distinguish two types of GVCs: vertically specialized value chains and additive value chains (Kaplinsky 2015a). Vertical chains stem from the fracturing of chains as firms specialize more in their core competences and outsource non-core activities. This fragments production into a plethora of subprocesses that can be undertaken in parallel and lend themselves to global

Box 3.3
Special economic zones in China

China's industrial clusters are distinctive for their history, size, external orientation and government role. While China has a long history of cluster development, the recent dynamism of their clusters has been driven by government policy (Enright, Scott and Chang 2005). Building on earlier experiences, a series of SEZs was established after 1978, providing tax and other incentives and designed to promote exports through inward FDI and (increasingly) joint ventures between Chinese and foreign-owned firms.[22]

The first 5 experimental SEZs were established between 1980 and 1984, 14 were created in 1984 and the number has since expanded. Support for clustering in general, and SEZs in particular, has not been limited to the central government. Provincial government was also active, as were city and township governments. China's township and village enterprises were the backbone of its industrial development until the end of the 20th century.

Initially these clusters concentrated on labour-intensive sectors, and although of diminishing relative importance in the economy, they continue to make a major contribution to output, exports and employment. Most of these industrial clusters were located in formally constituted SEZs. In total, these SEZs—labour intensive and high-tech—were estimated to account for 22 percent of China's GDP and 60 percent of exports—and to have created 30 million jobs by 2007. More recently, the SEZs have concentrated on high-tech sectors and between 1995 and 2010 these high-tech clusters accounted for half of the value of high-tech industrial output and one third of China's high-tech exports (Zeng 2014).

Source: Kaplinsky 2015a.

> **For supplier firms that have long operated in a sector, global value chain entry involves a 'thinning out' of activities**

dispersion. A well-known example is the Apple iPhone. Additive value chains involve a process of sequentially adding value to each stage of the chain, as in processing natural resources. An example of a sequential value chain is the cocoa value chain (Kaplinsky and Morris 2014). Other examples are in garment manufacturing.

In both types of value chain, upgrading involves countries or firms capturing a larger slice of the total value added embodied in the final product and specializing in increasingly technologically sophisticated activities that create opportunities for learning and technological spillovers (Box 3.4). The policy prescriptions, however, may vary between these two models.

Vertical value chains. Given that much of the outsourcing of non-core competencies in vertical GVCs was offshored primarily to developing economies, the policies that address the promotion of vertically specialized GVCs, particularly in the short run, relate primarily to trade. The objective is to reduce impediments to trade, such as quotas and tariffs on imports, introduce incentives to promote exports and remove "at the border" bureaucracy and obstacles. Complementing trade should be the smooth functioning of trade infrastructure. In economies where entry into export markets is provided by foreign lead firms, the vertical GVC policy agenda also usually targets SEZs.

Beyond trade are follow-on efforts to deepen presence in the rent-rich links in the chain. Primarily executed at firm level, they often follow an upgrading trajectory reflecting process, product, functional and chain upgrading (Humphrey and Schmitz 2000). This trajectory is widely evidenced in the Asian economies that have successfully pursued a vertically specialized GVC path. Simply by assembling the lead firm's designs, a local firm can enter the GVC (for example, the iPhone 4 in China). In this early phase, the firm's technology upgrading trajectory indicates process improvement. The firm's priority at the outset may focus on improving its efficiency, but as its capabilities progress, it may shift from assembly to manufacturing and begin to incorporate a higher share of self-produced or locally sourced components (Kaplinsky 2015a).[23]

As the firm's capabilities develop, its activities may turn to product upgrading and design of its own products. The firm may eventually even get involved in functional upgrading or build its own brand presence,

Box 3.4

SME growth in the Indian auto components industry: Hybrid filtration systems

Following a period of market liberalization after the early 1990s, the auto components sector in India expanded rapidly with employment increasing from 162,000 to 462,000 between 1998 and 2009 and output increasing by over 700 percent to $21 billion, contributing 2.1 percent of GDP (Borgave and Chaudhari 2010).

While a significant share of R&D is performed abroad and imported by MNEs, successful technological innovations among SMEs played a significant role in boosting sales and employment in the sector (Krishnaswamy, Mathirajan and Subrahmanya 2014). The SMEs mainly support larger auto manufacturers as subcontractors and also cater to the low-value aftermarket (India Brand Equity Foundation).

A close customer relationship was crucial in planning and implementing new product designs in filtration-system suppliers. Linkages between large enterprises and the SMEs were strengthened by specific design requirements, expert assistance, drawings and product samples, leading to a more targeted innovation process and higher market acceptability. Substituting aluminum for steel made the product lighter, optimized the filtration ratio and allowed filters to be more reusable. Product innovations promoted employment growth through increased production while some of the larger SMEs also developed small research units with up to four employees given experimental freedom.

Technological innovation in SMEs is a dynamic process in which the first priority is to meet industry standards, then to establish credibility through gradual product and process innovations (according to the customers' needs) and finally to explore new markets by furthering technological capabilities. With SMEs accounting for 45 percent of India's manufacturing output and providing a good source of income to lower-skilled workers, their growth is vital for sustained and inclusive development (Goyal 2013).

> **A firm may move to another global value chain once competitors catch up and begin to master chain capabilities**

3

SUSTAINING ECONOMIC GROWTH

either in its own right (as Samsung did) or by purchasing the brand name of an already well-established firm (for example, Lenovo acquiring IBM computers and marketing the Thinkpad). The firm may move to another chain once competitors catch up and begin to master chain capabilities. Intensifying knowledge and a growing share of disembodied activities are underlying characteristics of this trajectory (Kaplinsky 2015a).

Additive value chains. In additive GVCs the strategic focus for developing economies is on "thickening" by building linkages to capture a larger slice of value added in the sector. This is predominantly the case in the resource sector, although as observed above, it also applies to some manufacturing sectors, such as apparel (Box 3.5).

Traditionally the dominant perspective on resource extraction was seen as an enclave with few linkages to the domestic economy. There has though been a reappraisal on these linkages, first because linkages seem more common than previously recognized, and second because there is more potential for increasing and deepening linkages than previously thought.

Linkages from the resource sector have played an important role in the industrial development of many now industrialized economies (Wright and Czelusta 2004). This includes Canada and the United States in the 19th century, Norway and the United Kingdom in the 20th century and now the United States in the fracking era of the early 21st century. Recent studies on production linkages in Africa's resource sector find considerable "below the radar" linkage development, including some in rather surprising circumstances (Kaplinsky and Morris 2014; Morris, Kaplinsky and Kaplan 2012).

The obvious and expected linkage development was in South Africa's mining equipment and services (Kaplan 2012). It has a positive balance of trade and is prominent in global patenting. Less expected was the Nigerian oil and gas sector, where high-level knowledge-intensive backward linkages in services include those provided by local firms employing engineering and ICT graduates (Adewuyi and Oyejide 2012). Ghana is emerging as a West African mining services hub, replicating the role long played by South Africa in the continent's resource sector (Bloch and

Box 3.5

Nicaragua in the apparel manufacturing global value chain

Nicaragua's apparel exports nearly doubled from $716 million in 2005 to $1.4 billion in 2011 (Bair and Gereffi 2014). Nicaragua mainly participates in the low-value "cut-make-trim" stage of the value chain. Leveraging the country's competitive wage advantage, the industry employed more than 51,300 people in 2010 (ILO and IFC 2010; Portocarrero Lacayo 2010). The majority of apparel exports from Nicaragua (89 percent) were being sold to the United States by 2009; Nicaragua had been gaining market share in some segments in the United States since 2004 (for example, with woven trousers and cotton shirts) due to its preferential trade status within the Dominican Republic–Central American Free Trade Agreement (Bair and Gereffi 2014). Most Nicaraguan apparel manufacturers primarily produce denim jeans and twill pants and T-shirts.

The majority of apparel manufacturers are foreign-owned firms, with very few locally owned companies. Among the foreign firms, most are owned by companies from the Republic of Korea and the United States, but some are owned by companies from El Salvador, Honduras, Mexico and Taiwan Province of China. As a large share of these firms belong to global or regional networks (especially in Central America), they can offer full-package services to clients by directing the interactions of multiple country operations. Knit-based firms sell to Walmart, Target, and Ralph Lauren. Woven apparel firms, on the other hand, usually have a regional focus—their operations in neighboring countries including Honduras, Guatemala and Mexico. The main buyers are, among others, Kohl's, Levi Strauss and Cintas (Gereffi 2015).

Between 2005 and 2010, the volume of Nicaragua's apparel exports grew by 8.6 percent, but despite this increase the country has had limited success in moving up the apparel value chain and competes mainly on low-cost apparel assembly. It remains vulnerable because its apparel exports depend on United States trade policy, specifically, the Tariff Preference Level exception that allowed it to import textiles from East Asia.

> **The advantages of foreign direct investment are far from automatic and depend on mediating characteristics of the domestic economy**

Owusu 2012). Botswana is developing its diamond trading, polishing and cutting industry (Mbayi 2011).

Over time, there has been a market driven process of linkage development, beginning with the "easy hits" of "low hanging fruit" and then slowing as the more difficult stages are confronted. This market process can either be speeded up or deepened into the core-competence rent-rich territory of the lead resource firm, or slowed by the policy environment.

Africa's resource sector shows the emergence of linkage development not just backward but also forward and horizontal linkages driven by GVC operations (Kaplinsky and Morris 2014; Morris, Kaplinsky and Kaplan 2012). But there are specific policies and strategic interventions adopted towards the linkage sector—sometimes in partnership with international agencies or lead firms—that can promote a greater depth of value added.[24]

Optimizing the contributions of FDI

Since the 1970s trade policy has shifted from an inward to an outward looking export-oriented stance. Strategic insertion into international trade and into GVCs is a key ingredient in structural transformation (Westphal 2002). In this process of breaking into export markets, FDI can provide access to technology, to brand names, to global markets and has the potential to provide spillovers to the domestic economy.

Possible mechanisms for FDI to contribute to technological upgrading and international knowledge flows include:

- Relationships between foreign affiliates of MNEs and suppliers and clients in host countries (these are vertical knowledge flows, usually more transfers than spillovers).
- Knowledge spilling over to local competitors through imitation or reverse engineering (horizontal spillovers).
- Knowledge spilling over through the mobility of (skilled) workers.
- Knowledge spilling over when workers leave to start new firms.
- MNEs increasing competition in the sector they operate in.

However, the advantages of FDI are far from automatic and depend on mediating characteristics of the domestic economy (Jacob and Sasso 2015). Horizontal spillovers within sectors are limited. Sometimes, the effects on domestic firms can be negative, when more productive foreign firms out-compete the local firms. MNEs can generate vertical knowledge flows through backward linkages. In general, the positive effects of FDI and the presence of foreign MNEs depend to a large extent on mediating factors, which have to do with the absorptive capacities of domestic firms, sectors and the national economy. They are also affected by policies for foreign investment. In absence of sufficient absorptive capacities and appropriate policies, FDI and participation in GVCs can even have deleterious effects on domestic firms.

Important moderating factors include the absorptive capacities of domestic firms, sufficient investment in human capital and R&D, the size of technology gaps between foreign and domestic firms, protection of intellectual property rights in middle-income countries, the degree of involvement of foreign firms with local enterprises and the development of relationships with subcontracting domestic firms. The goal of efforts to optimize the contributions of FDI should be to strengthen absorptive capacities.

Notes

1. See Kaltenberg and Verspagen (2015).
2. UNIDO (2012) warns that the evidence is not conclusive. In particular, the findings are still driven by large time-invariant differences in degrees of specialization in cross-country datasets.
3. The principal source of data for Chapter 3 is a new dataset with value added at current prices broken down by nine sectors for 108 countries covering the period 1960-2010 (see details in Foster-McGregor, Kaba and Szirmai (2015)). This dataset derives from UN National Accounts Statistics (UN 2014b), Groningen Growth and Development Centre 10-Sector Database

(Timmer, de Vries and de Vries 2014) and World Input-Output Database (Timmer and others 2015). Data on per capita economic growth derive from The Maddison Project (2013).

4. These are the three most commonly used statistics for measuring the degree of concentration of the distribution of a given variable.
5. As Figure 3.2 is based on partial correlations, the range of values looks different to what one would expect. This is because of the method to obtain the partial effects (using the residuals from the main regression for the y-axis and residuals of a regression of the Theil index on other explanatory variables for the x-axis. The same obtains for Figure 3.3 (Theil index), Figure 3.4 (Gini index) and Figure 3.8 (share of manufacturing in GDP).
6. Positive episodes are defined as follows. For each country, a year is considered to be part of a positive growth episode, if its GDP per capita is higher than that of the previous year for two successive years (see also chapter 1).
7. See endnote 6.
8. See endnote 6.
9. The seminal contribution of Nicholas Kaldor on the causes of the slow rate of growth in the United Kingdom (Kaldor 1966) is typically pointed out as initiating this tradition.
10. The argument that manufacturing is key for technological change for the whole economy is perhaps the most important argument in favour of industrial policies.
11. Average years of schooling for population older than 15 years of age.
12. See endnote 6.
13. In fact, within growth episodes, the rate of growth during the episode is even negatively correlated with the share of manufacturing at the beginning, although as countries become richer, the relationship turns positive.
14. The sample includes all EU countries except for three small ones, but it also includes the United States, Japan, Russia, the Republic of Korea and large developing countries, such as Brazil, India, China, Turkey and Mexico.
15. The results are from Kaltenberg and Verspagen (2015).
16. Following Freeman (1987) a national innovation system can be defined as the network of institutions in the public and private sectors whose activities and interactions initiate, import, modify and diffuse new technologies.
17. Of course this is related to population size. Tertiary enrolment in China in 2010 was still far below levels of the advanced economies.
18. The Spark program started in 1986 and has the objective of promoting rural economic development through the application of science and technology. It supports a large number of projects that use rural resources, need small amounts of capital, provide early benefits and use appropriate technologies. See OECD (2015) for further details.
19. The Torch program started in 1988 and has the objective of promoting high-tech industry. It supports a large number of projects that use advanced technologies to produce high-tech products and that explore management systems and operation mechanisms suitable for hi-tech industrial development. See MOFCOM (2012) for further details.
20. This section draws on Kaplinsky (2015a).
21. This section draws on Kaplinsky (2015a).
22. SEZs and clusters are not necessarily identical. A firm can enter a SEZ for infrastructure and tax incentives without collaborating in a cluster. In practice, Chinese SEZs did result in the emergence of clusters.
23. For example, between 2011 and 2013, the number of firms selling iPhone batteries in China doubled from eight to 16, and local firms began producing inputs that used to be imported, such as acoustic components (Mishkin 2013). The value added of the iPhone 5 production in China is—albeit undocumented—almost certainly much higher than the $6.50 incorporated in the early versions of the iPhone 4 (Kaplinsky 2015a).

24. These include local content policies (Nigeria and Angola), building infrastructure specifically to meet the needs of the resource sector (Botswana diamond sector), marketing institutions to support domestic processing (cocoa sector in Ghana), export taxes to force local value addition (leather in Ethiopia), building industrial zones to facilitate linkages between lead firms and local manufacturers (copper suppliers in Zambia and leather tanning and footwear manufacturers in Ethiopia), restructuring government agencies with the support of the EU to upgrade local certification to meet EurepGap standards (Nile perch fish in Uganda) and introducing human resource programmes to meet the specific needs of the resource sector (Angola and Nigeria).

Chapter 4

Promoting social inclusiveness

Technology enables structural transformation and economic modernization for improving living standards. But it can also trigger disruptive forces that might concentrate the benefits of economic growth in only some parts of society. What, then, are the conditions to induce technology-driven structural change that promotes social inclusiveness?

During structural transformation, society becomes more technologically complex and economically productive. That improves incomes, wealth and subjective well-being. And the ensuing demographic shifts, facilitated by rising incomes and the uptake of modern technologies, improve outcomes in health, education and urbanization.

Manufacturing is fundamental to this process. It provides productive employment in the early stages and is a catalyst for further technological innovation. Over time a country's manufacturing typically evolves from being labour intensive to being more capital and technology intensive, creating demand for more-skilled labour. A better skilled workforce in turn provides incentives for technological innovation, which can enable a virtuous circle of education, innovation and productivity growth.

But not everyone can take advantage of the opportunities that arise. So, socially inclusive industrial development demands specific domestic capabilities and technologies better suited to match these conditions. Only then can it distribute more evenly the fruits of economic growth.

Inclusiveness and industrialization

General overview

Social inclusiveness is typically equated with people's "full participation in all aspects of life" and contrasted with the concept of "social exclusion" that refers to "the conditions (barriers and processes) that impede inclusiveness" (UNDESA 2009, p. 12). Socially inclusive industrialization is consistent with people having equal opportunities to share in the growth from industrialization and ultimately contributing to a socially inclusive society. Such a society transcends "differences of race, gender, class, generation, and geography, and ensures inclusiveness, equality of opportunity as well as capability of all members of the society to determine an agreed set of social institutions that govern social interaction" (UNDESA 2008).

Social inclusiveness is thus multidimensional (de Haan 2015). No simple single measure can capture whether a society is inclusive or not. At a minimum, social inclusiveness requires growth that reduces poverty and inequality and creates jobs for the poorest and most vulnerable groups in society. In assessing whether industrialization has been socially inclusive, one would be concerned not only with the inclusiveness of manufacturing industries (see the inclusive and sustainable development [ISID] index in Chapter 1) but also with changes in levels, averages and distributions of broader measures of well-being. These changes are captured in poverty rates (such as the percentage of the population living in households with incomes below the poverty line); income distribution measures such as the Gini index and the share of wages in total (or sectoral) gross domestic product (GDP); and how employment is distributed across various skill, age and gender categories.

An empirical examination of the links between technology-driven structural transformation and social inclusiveness thus needs to explore how new technologies contribute to the broad dimensions of inclusion. As in previous chapters, technological innovation is considered central in the structural transformation of societies, which can do much to improve social inclusiveness.

The World Summit for Social Development in 1995 (The Copenhagen Summit) recognized that "new information technologies and new approaches to access to and use of technologies by people living in poverty can help in fulfilling social development goals." And in recent years the so-called "Arab Spring"

> **“Socially inclusive industrial development demands specific domestic capabilities and technologies**

highlighted how social media can increase participation in governance.

New technologies can improve the living conditions of the poor—their health, consumption and access to information as well as the type and quality of their employment. But the benefits of technology might not be equally or fairly distributed. As UNESCO (2014) pointed out, there is a “digital divide” between and within countries. Technological innovations may also result in patterns of growth and structural change that increase income inequality if the appropriate technologies are not implemented.

The channels for technological change to increase social inclusiveness through the transformation of economic structures can be broadly divided into two major areas: employment creation and income distribution (Figure 4.1). On the first, will new technologies create or destroy jobs? On the second, will they improve or impair the distribution of income within society?

In industrially advanced high-income countries, technological change is typically related to the generation of new technologies—and in developing countries, to the absorption of technologies from abroad. What these developments mean for social inclusiveness depends on the specific conditions that characterize a country and its technological trajectory.

Underlying conditions are the characteristics of the country, its factor and skill endowments, its absorptive capabilities, the type and direction of its technological change and the international context. Some technologies are better suited to fully utilize the factor and skill endowments of the country, thus creating new jobs without impairing the income distribution. But if the conditions of the country do not properly match the requirements of the technology implemented, the final outcomes can be negative.

By the same token, the net effect of a particular innovation on the creation of jobs depends on the type of innovation. Broadly, product innovations have a positive effect on creating new jobs while process innovations have a negative impact (UNIDO 2013a). Again, the specific conditions of the country—market structure, investment behavior and degree of substitution between factors, and so on—determine the effectiveness of compensation mechanisms that can alleviate the negative impact of labour-saving process innovations.

In addition, new technologies (and the direction of structural change) can introduce important trade-offs between multiple objectives. The technologies that promote social inclusiveness might be at the expense of environmental deterioration. Or the technologies that improve environmental sustainability might hurt job creation and income distribution.

We now discuss in more detail the elements in Figure 4.1, focusing on differences between advanced and developing countries.

Structural transformation and social inclusiveness

If the right capabilities are in place, technology-driven structural change can trigger the expansion of the modern formal industrial sector and industry-related services, thus absorbing labour from the pool of underemployed workers in agriculture or informal services.

Figure 4.1
Conceptual framework: Technological change for inclusive structural transformation

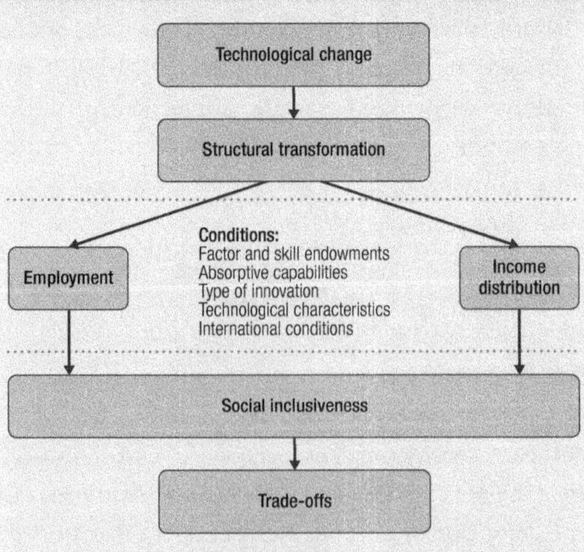

Source: UNIDO elaboration.

> **Manufacturing plays a key role, catalyzing the generation and diffusion of new technologies**

Manufacturing plays a key role, catalyzing the generation and diffusion of new technologies. Moreover, backward and forward linkages and spillover effects from manufacturing promote further regional and country development. Feedback loops can also be expected to accumulate human capital and improve institutions. The generation of direct and indirect jobs in manufacturing and manufacturing-related sectors leads to the inclusion of more people in the growth process. It also increases average productivity, wages and family income, thus reducing poverty (Box 4.1).

This process can temporarily lead to rising income inequality (Kuznets 1955; Kaldor 1957). Kuznets posited an inverse U-shaped relationship between inequality and GDP per capita (the "Kuznets curve"). In his view, income inequality would be low at early stages of development, when most people are employed in the traditional (agricultural) sector, where productivity and wages are equally low for everybody. However, as workers move to the manufacturing sector (inter-sectoral) inequality will rise, at least initially. Over time productivity and wages will also rise in agriculture, so that there will be an eventual decline in inequality. Inequality will also decline because the new "modern" sectors that come into being will generate a demand for the labour that has been made redundant by new technologies. In such a case, technological innovation will ultimately lead to job displacement instead of job replacement, over the long run. A typical example is how the invention of the internal combustion engine caused substantial job losses in the horse-drawn carriage industry but eventually resulted in substantial new employment in the automobile industry. So, technological innovation not only has static effects in the

Box 4.1

Woman empowerment through improved shea butter processing in Ghana

The production of shea butter is an important income-earning activity mainly for women in the rural areas of northern Ghana (FAO 2006b). More than 600,000 women in northern Ghana depend on shea butter production, and for many it is their key or only source of independent income and thus essential for their empowerment. Improved shea butter processing technologies can not only increase women's incomes and improve their health, but also save time and resources (UNDP 2015). Shea butter is produced from shea nuts and mainly used for cooking and cosmetic purposes. Ghana is one of the leading exporters in the region, with strongly increasing exports from 41,200 metric tons of shea butter in 2010 to 96,200 metric tons in 2012 (Ghana Ministry of Food and Agriculture 2012).

The traditional process of shea butter production requires large quantities of fuel wood, water and labour. Improved processing methods can be semi-mechanized with a nut crusher, an improved roaster, a kneader or a hydraulic screw press or mechanical bridge press. Semi-mechanized technologies can lead to an improvement of shea butter extraction from 20 to 35–40 percent, a water use of 1.9 head loads (instead of 2.5 head loads) and a processing time of 22 minutes (instead of 8 hours) for the extraction of 25 kilograms of shea kernel (Jibreel and others, 2013). The mechanical bridge press, another improved technology used in Ghana, reduces time from 9.5 to 3.25 hours, water usage from 90 litres to 1.7 litres and does not need any fuel wood for the production of 25 kilograms shea kernels, in comparison to the traditional processing method (Swetman and Hammond 1997).

Ghana's Technology Consultancy Centre has also developed improved technologies in collaboration with community-based women's groups in northern Ghana with similar advantages as the semi-mechanized technology (FAO 2006b). A study on the adoption of improved shea butter processing technologies on women's livelihood and women microenterprise growth in the northern region in Ghana has shown that fully mechanized technology adopters have improved their incomes, savings, employment, investment and credit levels as a result of improved technology over those who stayed with the traditional method. The adopters of improved technology registered a maximum income of GH¢ 160 before adoption and GH¢ 360 after adoption of the technology.

The mean income rose from GH¢ 23 to GH¢ 145 after adoption. Also quality improved significantly (Mohammed, Boateng and Al-Hassan 2013). Furthermore, the improved technologies significantly benefit the environment, due to less water consumption and less or no fuel consumption thereby preventing degradation of forest and water resources. The traditional method often caused dehydration and respiratory challenges due to indoor smoke pollution and exposure to fire, also reduced or even prevented the improved technologies (Jibreel and others 2013).

> **Product innovation creates jobs through the emergence of new markets**

once-off reallocation of labour—it also has dynamic effects facilitating the growth of productivity and output in modern urban-based industries (Frey and Osborne 2013).

The expansion of the modern formal sector gives the government a bigger tax base. And the greater revenues in the public sector can enable the government to improve economic, administrative and political institutions and widen social protection measures. It also induces the higher labour market participation of women. With better earning opportunities, parents will want their children to receive more education. Furthermore, the growing modern sector may induce fertility declines, allowing a shift of resources towards better education of children and enhancing human capital formation and labour productivity. A growing modern sector is thus a major determinant of fertility and the demographic transition (Galor 2012; Gries and Grundmann 2014).

So, from this perspective, even if new technologies have a negative impact on income distribution and employment creation, it is typically temporary. Persistent or rising inequalities would ultimately reflect institutional and policy failures that perpetuate technological gaps between sectors, and between regions and countries[1] or that fail to provide adequate social buffers in times of rapid change (Naudé, Santos-Paulino and McGillivray 2009).

Innovation and employment

On the relationship between technological innovations and employment, consider product and process innovations.[2] Product innovation creates jobs through the emergence of new markets, while process innovation typically creates a labour-saving effect, mainly related to introducing new machines that allow the same output to be produced with fewer workers (UNIDO 2013a).

In labour-saving process innovation, several economic forces can compensate for the reduction in employment (UNIDO 2013a):

- *New machines.* The same process innovations that displace workers in the product industries where the new machines are introduced can create new jobs in the capital goods industries where the new machines are produced.
- *Lower prices.* Although innovations involve the displacement of workers, these innovations reduce the unit costs of production, and in a competitive market this lowers prices, which stimulate new demand for products and so additional production and employment.
- *New investments.* The reduction in costs—due to technological progress—and the consequent fall in prices may allow innovative entrepreneurs to accumulate extra profits. If these profits are invested well, they will create new output and new jobs.
- *Lower wages.* Where there is demand for labour, the direct effect of job-destructive technologies may be compensated within the labour market. Assuming free competition and full substitutability between labour and capital, technological unemployment implies a decrease in wages, and this should induce a reverse shift back to more labour-intensive technologies.
- *Higher incomes.* Trade unions may redistribute part of the innovation rents back to the workforce, so a portion of the cost savings due to innovation can be translated into higher wage income and thus higher consumption. This increase in demand increases employment, which may compensate for the initial job losses.

None of these mechanisms is automatic, and to work they require some underlying conditions to be in place. Price and income mechanisms can counterbalance the direct job destruction caused by process innovation, but their effectiveness depends on such parameters as the degree of competition, the price and income elasticity of demand and the way business expectations are shaped (UNIDO 2013a; Vivarelli 2013, 2015).

Innovation and income distribution

Besides the labour-saving effects, technological change can also exhibit a skill-bias. Innovations tend to be skill-biased, replacing tasks traditionally carried out

> **"Entering manufacturing segments seems to deliver the most beneficial results**

by unskilled workers, with new jobs demanding qualified workers (Vivarelli 2013). So, if there are too few qualified workers, the adoption and diffusion of the technology might be checked and the pursuit of full employment constrained. Moreover, the benefits from adopting of the new technology might be restricted to a fraction of society, making incomes less equal.

Now consider the fragmentation of production. New technologies, particularly information technologies, can change the structure of the firm and the extent to which different types of workers share the dividends of productivity gains. Specifically, new technologies facilitate a separation of tasks at the firm level. Production processes have become networks of manufacturing nodes that can be located inside or outside the firm—and increasingly abroad. When these nodes are organized according to comparative advantages, labour-intensive processes are typically allocated to labour-abundant countries. Processes that are labour intensive and require low or medium skills will be produced in developing countries with low wages, reducing the prices for these processes. So, the globalization of tasks according to comparative advantage might contribute to growing wage inequality and job market polarization in high-income countries (Alderson and Nielsen 2002; Krugman 2008).

The story is similar for international trade, because it links not only goods markets (or markets for tasks) but also labour markets at the global level. Standard (neoclassical) trade theory predicts that the greater trade openness that characterizes globalization would, other things being equal, lead to rising income inequality in skill-abundant countries (high-income economies) and to narrowing income equality in skill-scarce countries (developing countries). Under free trade, countries specialize in production and trade of their relatively more abundant factors and that raises the relative demand for that factor. Reductions in trade barriers facilitate the "unbundling" of production processes across countries, causing the "offshoring" of low-skilled jobs from advanced economies (Baldwin and Venables 2013; Blinder 2009). When the jobs of low-skilled workers in advanced countries are offshored to low-skilled workers in developing countries, wages of low-skilled workers in advanced economies can decline, increasing income inequality in these economies. Many authors have argued that globalization has been a cause of "deindustrialization" in advanced economies (Wood 1998).[3]

In developing countries, the fragmentation of production is typically considered more positively—as an opportunity to access international markets and develop a modern (exporting) sector. In fact, global value chains (GVCs) have been particularly important in low- and middle-income countries, where the separation of unskilled and semi-skilled components across the world has led to massive employment creation, notably in China but also in Central America (Chataway, Hanlin and Kaplinsky 2014). Since much of this relocated labour has involved women who previously had little access to paid employment, this process also has important gender-distribution implications.

The final outcome relies heavily on the way developing countries integrate into GVCs. Entering manufacturing segments seems to deliver the most beneficial results. At early stages of development, countries are dominated by an agricultural or semi-agricultural structure and so face two important problems. On the demand side, there is insufficient income and productivity for a large share of the population to purchase goods produced in the modern industrial sector. On the supply side, even if there is a market for some goods, countries do not have the technology, product quality or organizational ability to offer a competitive product to domestic customers. Thus a simultaneous lack of markets and technological capabilities becomes a prohibitive barrier for developing a domestic modern sector.

Joining GVCs can solve both problems. First, world markets can provide almost unlimited demand for products conditional on the ability of the country to adjust to the necessary technological standards, qualities and requirements. Second, the ability to adjust to these requirements can improve domestic

> **The benefits of joining global value chains depend on the 'governance' of the process**

technological capabilities. It follows that linking to international value networks can drive modernization in developing economies (Gereffi, Humphrey and Kaplinsky 2001; Makki and Somwaru 2004).

The benefits of joining GVCs for developing countries also depend on the "governance" of the process. GVCs are not functioning in a fully competitive market environment. Pure rents are detectable in all kind of arrangements within these networks (Davies and Vadlamannati 2013). The exploitation of these rents is an important objective shaping the business model of GVCs, and thus the distribution of gains from trade and foreign direct investment. For instance, if a certain vintage of machines is depreciated in high-income economies and a newer vintage machine can substitute for the old vintage at lower cost, this machine may still be profitable in developing economies' lower wage conditions. Transferring technology to a developing country might therefore present a profitable investment at lower wages. In developing countries, market wages can even be lower than that required for making such investments profitable. So, this wage gap generates an extraordinary rent, and the inclusiveness of globalization depends on the distribution of these rents.

Through trade and labour policies, governments influence both technology transfers and rent distributions. For example, China's trade policy was aimed at encouraging international investors to transfer as much know-how as possible to local agents. Multinational enterprises (MNEs) were obliged to let local firms participate in the technology and rents. Governments can also influence technology transfers and rent distributions by enforcing rules on working conditions, such as minimum labour standards.

For these trade and labour-market policies, a trade-off can emerge between restricting international investors and obtaining know-how from them. So, competition among countries to attract foreign direct investment–led technological industrialization can lead to a "race to the bottom" on taxes, labour safety regulations or environmental rules, which could ultimately worsen inequalities and reduce social inclusiveness.

General trends in social inclusiveness

Indicators of inclusiveness

In this section we provide empirical evidence on general trends in terms of social inclusiveness that can be observed in the last few decades, relating them to our conceptual framework.

Lavopa (2015b) focuses on four indicators: the Non-Poor Ratio (NPR), defined as one minus the poverty headcount ratio;[4] the Human Development Index (HDI); the Equity Index (EI), defined as one minus the Gini index;[5] and the Inclusive Industrialization Development (IID), defined as the inequality-adjusted wage-share in manufacturing industries. Figure 4.2 presents the trends in these indicators between 1980 and 2014 for different developing regions.

It is clear that the indicators vary widely among developing regions. On poverty and human development, Eastern Europe and Latin America perform far better than Asia and Africa. On distribution, Latin America ranks as the worst both on the overall economy (that is, EI) and manufacturing (IID). Eastern Europe still ranks (just) as the most equal region.

Most indicators and regions show positive trends, albeit with variations. As expected, Asia shows the best performance on poverty and human development, with an impressive increase of the NPR and the HDI, especially after 1995. Its outcome on income distribution (the EI and the IID) is not so positive. Africa also shows solid gains in poverty, human development and overall income distribution, though the IID has dropped sharply. Latin America shows good achievements on poverty reduction and improved income distribution, especially from 2000. Its HDI has increased steadily.

A positive relationship between structural transformation and social inclusiveness

From Lavopa (2015b) it can be seen that the relationship between structural change (broadly defined as the expansion of manufacturing in total employment) and social inclusiveness is positive on the basic correlations for the social inclusiveness indicators (Figure 4.3).

Industrialization is associated with lower levels of poverty, better income distribution, and better Human Development Index rankings

Figure 4.2
Main trends in social inclusiveness indicators, by developing region, 1980–2014

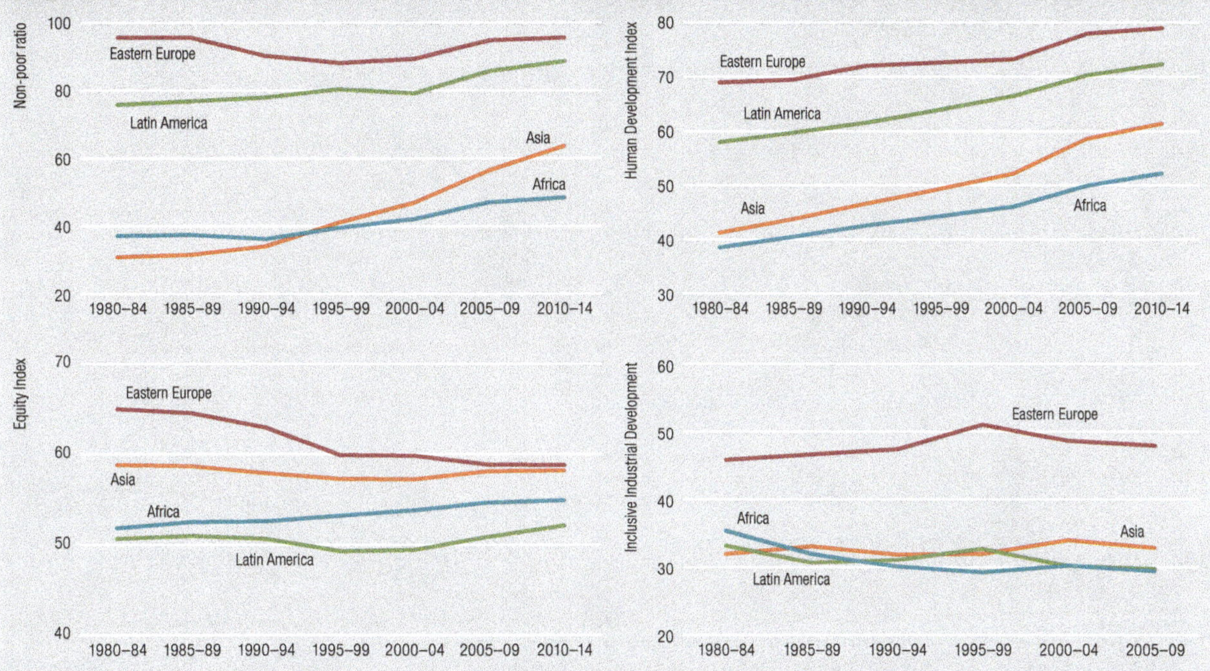

Note: Regional values are calculated as unweighted averages over countries with available data for the whole period. Developing countries are countries that in 1990 were not high income according to the World Bank's definition (see Annex A1, Table A1.2). Within these countries are four groups based on location: Africa (including the Middle East), Asia (excluding the former Union of Soviet Socialist Republics [USSR] and the Middle East), Eastern Europe (including the former USSR) and Latin America. Five-year averages are used to maximize the number of observations and minimize the potential effects of extreme years.
Source: Lavopa (2015b).

In all cases except EI this relationship seems to be decreasing with the share of manufacturing, but only in the case of HDI does it reach a turning point within the relevant ranges of manufacturing share. Industrialization is thus associated with lower levels of poverty, better income distribution, and better HDI rankings.

These basic correlations provide some preliminary evidence on the positive role of manufacturing in driving social inclusiveness. They might, however, also be indicative of other factors. One would be income: rich countries tend to have larger shares of manufacturing than very poor countries (though these are similar for middle-income economies), and their social inclusiveness indicators are, at the same time, much better than those in poor countries.

A more solid analysis that controls for the effects of other variables comes from a panel-data econometric exercise. Using this technique, Lavopa (2015b) analyses the main determinants of each of these social inclusiveness indicators in which one of the explanatory variables is the share of manufacturing in total employment (MEMPSH; Table 4.1). Other explanatory variables are per capita income (y), degree of openness (OPEN), investment share in GDP (INV), average years of schooling (EDU), population share older than 65 years (OLD), natural resources endowment (NNRR) and climate zones (CLIMATE).[6]

Each column presents the regression results for the corresponding inclusiveness index. In all cases the share of manufacturing in total employment has a positive and significant impact, even after controlling for other variables. It is the only explanatory variable that has a positive and significant coefficient across all models. As these regressions already control for income, there is something "special" about

> **Technological change is expected to have a positive impact on social inclusiveness in the long run**

Figure 4.3
Inclusiveness indices by share of manufacturing in total employment, 1970–2010

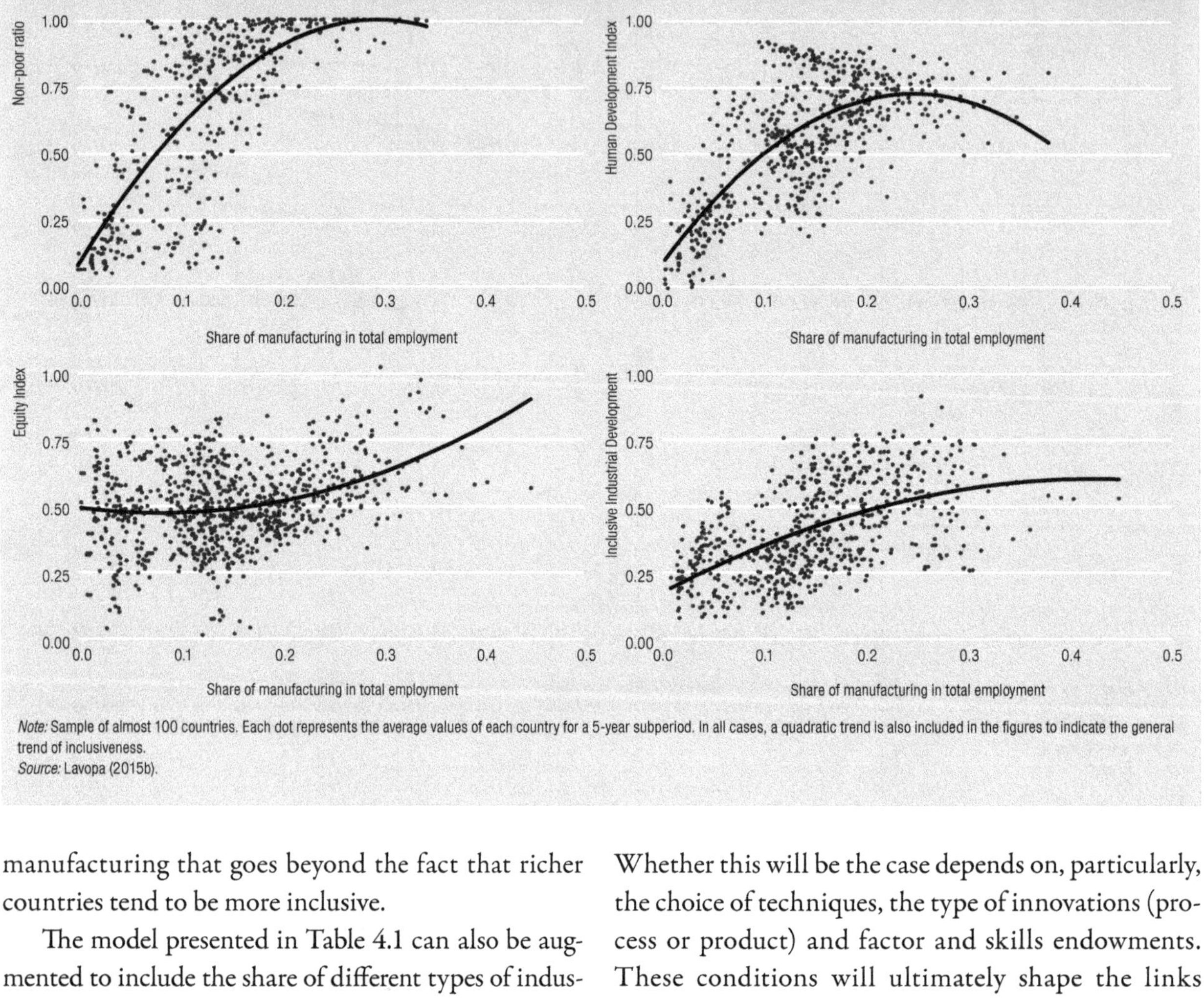

Note: Sample of almost 100 countries. Each dot represents the average values of each country for a 5-year subperiod. In all cases, a quadratic trend is also included in the figures to indicate the general trend of inclusiveness.
Source: Lavopa (2015b).

manufacturing that goes beyond the fact that richer countries tend to be more inclusive.

The model presented in Table 4.1 can also be augmented to include the share of different types of industries in total manufacturing employment in order to explore the role of industries with different technological characteristics in driving social inclusiveness. When an additional explanatory variable capturing the share of low-, middle- and high-tech industries in total manufacturing employment is included, the results[7] suggest that the larger the share of high-tech industries, the greater the positive impact of industrialization on social inclusiveness, in terms of income equity (EI) and broader social development (HDI). But for poverty (NPR) and industrial inclusiveness (IID) the share of high-tech industries does not show a significant impact.

As long as technological change leads to an expansion of manufacturing, it is expected to have a positive impact on social inclusiveness in the long run. Whether this will be the case depends on, particularly, the choice of techniques, the type of innovations (process or product) and factor and skills endowments. These conditions will ultimately shape the links between technological change, employment creation and income distribution.

Employment creation

Empirical studies of the links between employment and innovation have focused largely on higher income economies and tend to show that product innovations are usually associated with employment growth, while the effect of process innovations is often negative (e.g. UNIDO 2013a).

The analysis of this relationship is not, however, free of difficulties. It needs to take into account the different forms of technical change, their direct effect on labour, the various compensation mechanisms and the possible hindrances to these mechanisms

Research and development spending has a positive employment effect

Table 4.1
Role of industrialization in different inclusiveness indices, Hausman-Taylor estimates, 1970–2010

	Non-poor ratio	Human Development Index	Equality Index	Inclusive Industrial Development
Constant	−1.269*	0.000	−0.009	0.506
CLIMATE	0.035	0.029	−0.017	0.023
NNRR	0.049	0.069**	−0.005	0.039
Ln(y)	0.544***		0.125***	0.126
Ln(y)2	−0.028***		−0.008***	−0.008
Ln(OPEN)	0.029***	0.000	−0.009**	0.027***
Ln(INV)	0.040**	0.014***	−0.002	−0.015
Ln(EDU)	0.063		0.045***	−0.037
Ln(OLD)	0.195***	0.078***	−0.029*	0.087**
Ln(MEMPsh)	0.038**	0.024***	0.014*	0.028*
Europe (adv.)	0.000	−0.028	0.013	−0.054
Asia (adv.)	0.000	0.007	0.016	−0.140
Africa (dev.)	0.051	−0.216**	−0.050	−0.234**
Latin America (dev.)	0.126*	−0.106	−0.106*	−0.291***
Asia (dev.)	−0.109	−0.196**	−0.022	−0.277***
Europe (dev.)	0.000	−0.102	0.011	−0.164*
Oceania (dev.)				
d77	0.000	0.000	0.000	0.015
d82	0.003	0.968***	0.008	0.014
d87	0.000	0.989***	0.005	−0.009
d92	−0.028*	1.010***	−0.005	−0.014
d97	−0.048**	1.026***	−0.018	−0.018
d02	−0.062**	1.042***	−0.017	−0.022
d07	−0.041	1.077***	−0.010	−0.031
d12	−0.031	1.093***	−0.008	0.000
Rho	0.799	0.938	0.738	0.640
Obs.	434	577	798	644
Countries	69	89	97	95

Note: * $p < 0.05$, ** $p < 0.01$, *** $p < 0.001$. Calculations are five-year averages. Income and regional classification based on Annex A1, Table A1.2.
Source: Lavopa (2015b).

(Vivarelli 2013). Very few studies have tried to test the validity of the compensation mechanisms, and the evidence is inconclusive. Using data from the United States, Sinclair (1981) found strong evidence supporting the mechanism via decreasing wages but not that via decreasing prices. Vivarelli (1995), using data for Italy and the United States, found that the more effective compensation mechanism was lower prices. Most of the empirical evidence is thus based on sectoral or microeconomic studies and cannot fully disentangle the impacts of all the compensating mechanisms.[8]

Still, these studies have shown that research and development (R&D) spending has a positive employment effect, principally in high-tech sectors.[9] Net employment usually rises with innovation, and government support for product development is likely to lead to employment generation.

Work on a large database created from the World Bank Enterprise Surveys of 2002–2006 and covering more than 26,000 manufacturing establishments in 71 countries confirms the employment–innovation link (Dutz and others 2011). Firms that introduced

> **Innovators were found to employ more unskilled and female workers than non-innovators**

a product or process innovation (judged by their own responses in the survey) had employment growth of 2.9 or 2.1 percentage points, respectively, above that of non-innovating firms (controlling for all other factors). Process innovation had a positive employment effect overall, but not for the largest enterprises (more than 200 workers). Product innovation had a positive employment effect that was larger than that for process innovation and was significant for all size categories. Innovators were also found to employ more unskilled and female workers than non-innovators.

But the link between process innovation and unskilled employment growth was weaker than with product innovation, and there is some evidence that non-process innovation may have had a stronger link to employment of unskilled labour. Female employment in developing countries was strongly linked to innovation, while across the whole sample of countries it appeared to be linked to innovation only in new and medium-sized enterprises.

Income distribution

Technological direction.[10] The skill-biased technological change, including capital deepening and organizational changes, is a key driving force behind the increase in the relative demand for more skilled and more educated workers.[11] This is often found to be in line with direct evidence of capital-skill and technology-skill complementarity and with increases in the relative demand for skill within industries and within plants (Katz and Autor 1999).

The initial literature typically distinguished two broad categories (low and high skills), but more recent studies introduced the concept of job polarization,[12] looking at three skill levels—low, middle and high—and the distinction between routine and non-routine jobs (Autor, Levy and Murnane 2003). Non-routine jobs are associated with higher skills than routine jobs. When technological change is biased towards replacing labour in routine tasks, the demand for middle-skill segments will decrease relative to high- and low-skill segments. This, in turn, will polarize the labour market both within industries and between industries (Goos, Manning and Salomons 2014). Figure 4.4 illustrates some empirical data for the United States over the last three decades.

In the 1980s job growth was positively related to the skill level (see the green line). Middle-skill jobs (second third of the distribution) and high- skill jobs (upper third of the distribution) expanded broadly. Inclusiveness through education was made possible for a broad spectrum of the population. But this trend changed dramatically in the 1990s: job growth continued for the top three skill deciles and growth was stronger, the higher the skill level (the red line). Also, the bottom decile expanded its job share, but less than the top three deciles. In contrast and most important, the skill deciles in between saw a substantial hollowing out—middle-skill jobs started to disappear.

These changes became even more pronounced in the 2000s (the blue line), when the job share grew only in the bottom three skill deciles. Middle-skill jobs continued losing share and even the share of high-skill jobs stagnated. A similar polarization is seen in Europe (Figure 4.5).

Figure 4.4
Job growth by skill percentile (ranked by occupational mean wage), selected periods, United States, 1979–2007

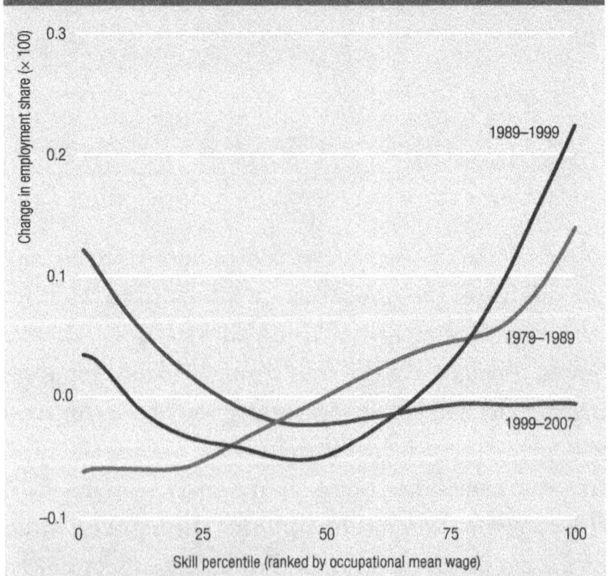

Note: The skill percentile is proxied by the occupational mean wage.
Source: Acemoglu and Autor (2011).

Skill-biased technological change seems to widen the income differential

Figure 4.5
Pattern of job growth, by labour income groups, selected European countries, 1993–2010

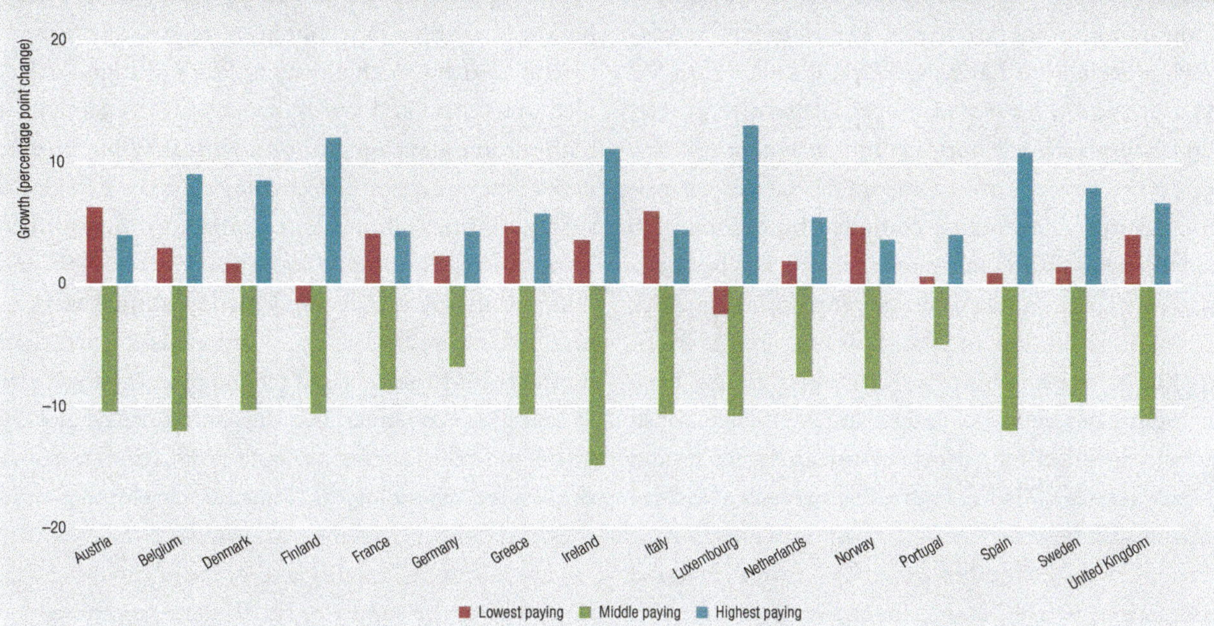

Note: Lowest-/middle-/highest-paying occupations are grouped according to the mean European occupational wage. See groupings in Goos, Manning and Salomons (2014).
Source: Gries and others (2015).

Skill-biased technological change seems to widen the income differential, but the evidence is stronger for the United States than for developing countries. The strong growth in the relative demand for high-skilled workers, in combination with fluctuations in the growth of relative skill supply, can explain major aspects in the evolution of educational wage differentials in the United States, such as the rise in the college premium from 1980 to 2005 (Goldin and Katz 2009). Changes in the wage structure very much reflect changes in the job structure (Figure 4.6).

Figure 4.6
Change in income for various skill groups, United States, 1964–2008

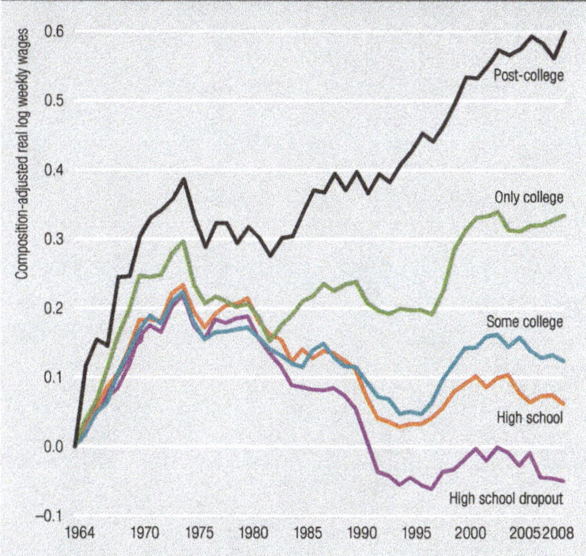

Source: Acemoglu and Autor (2011).

Until the early 1970s all skill levels in the United States benefited from growth. Since then, only college and upper education have further increased income growth rates. High school dropouts show an absolute decline (relative to the beginning of the sample period) from the beginning of the 1990s. The increasing competition in a shrinking market with job up- and downgrading may thus be driving wage movements at both ends of the distribution.

> **"Extending the quality and scope of higher education to raise the supply of skills is the recommended policy response**

For developing countries, the evidence base is less solid. In many non-OECD (Organisation for Economic Co-operation and Development) countries, especially in Latin America, it seems that the skill premium, measured by private returns to tertiary education, has been declining in recent years. A recent cross-country comparable dataset on private returns to schooling compiled by Montenegro and Patrinos (2014) indicates that this has been the case in a large sample of developing economies. But there are exceptions, and the skills premium as measured by private returns to tertiary education has been increasing in some developing countries such as South Africa while declining in others such as Argentina and Brazil (Figure 4.7). The figure also suggests a positive relationship between the skills premium and income inequality (the notable exception is the Russian Federation, where income inequality has increased significantly while the private returns to tertiary education dropped substantially).

Los, Timmer and de Vries (2014), however, use the World Input-Output Database[13] to quantify the contribution of changes in technology, trade and consumption to changes in employment levels of high-skilled, medium-skilled and low-skilled labour in a sample of high-income and emerging countries. In doing so, they derive a new measure of technological change in global supply chains, and find that demand for high-skilled labour increased in most countries between 1995 and 2008, and that the demand for low-skilled workers declined, except for India and Indonesia. Hence, their results point to skill-biased technological change, also in emerging economies. It is thus hard to tell if globalization (and the fragmentation of production) transmit skill-biased technological change to developing countries, given the mixed empirical evidence.

Extending the quality and scope of higher education to raise the supply of skills is the recommended policy response to inequality caused by skill-biased technological change. Many have argued that the

Figure 4.7
Rates of return on tertiary education and income inequality in selected emerging economies, 1992–2012

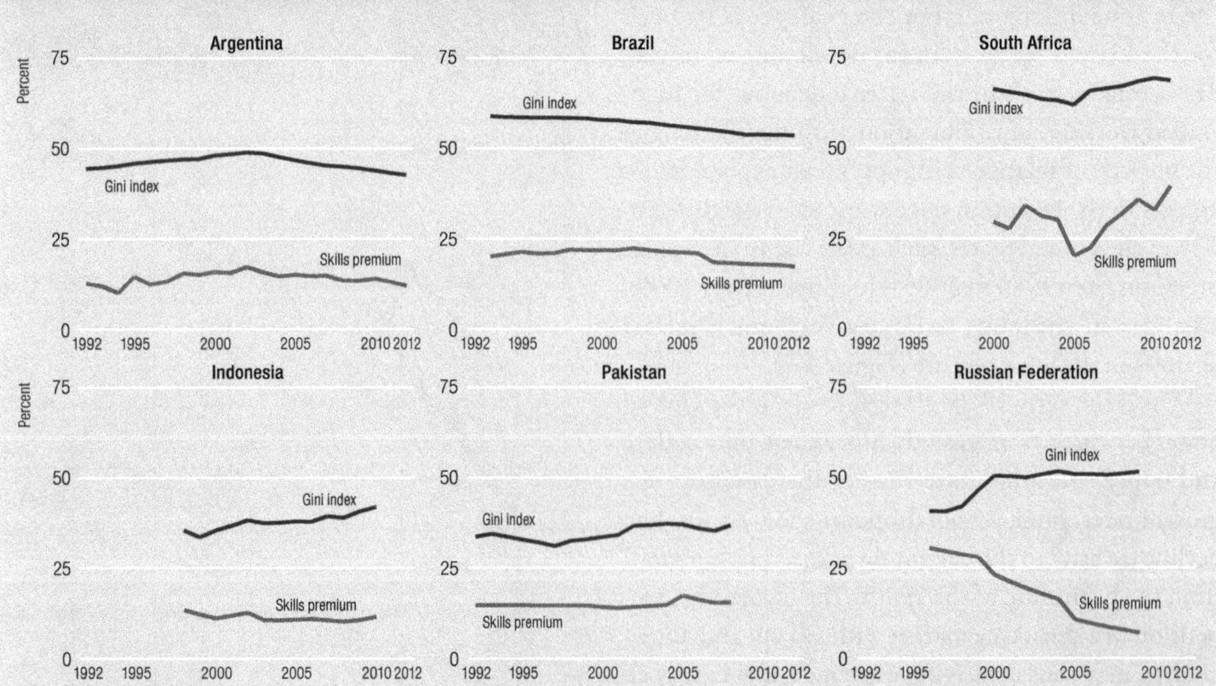

Note: Gini index of inequality in equalized (square root scale) household market (pre-tax, pre-transfer) income.
Source: UNIDO elaboration based on the Standarized World Income Inequality Database (Solt 2009, 2014) and Montenegro and Patrinos (2014). Missing values extrapolated.

> **The decline in the skills premium in many developing countries reflects improvements in higher education**

decline in the skills premium in many developing countries reflects improvements in higher education. A greater volume of higher education, however, makes it in turn more profitable for firms to invest in and adopt new technologies that require more skills, raising the demand for tertiary education and setting in motion a self-reinforcing cycle with wage gaps the outcome of this "race between technology and education" if education supply cannot keep up (Acemoglu 2003; Bénabou 2005; Mishel, Shierholz and Schmitt 2013). So far, however, many developing countries seem to have avoided this, either because of success in expanding higher education or because of a decrease in the demand for skilled labour (reflecting perhaps natural resource–driven growth in some countries), or both.

Fragmentation of production

As has been pointed out in Chapters 2 and 3, GVCs generally provide opportunities for developing countries to promote manufacturing and obtain access to foreign technology. Developing countries could start manufacturing by joining an existing supply chain, without having to build an entire one for themselves (Baldwin 2011). This may enable developing countries to develop manufacturing capabilities in certain areas before (or even without) building up broader manufacturing capabilities—they can learn by starting small (Milberg, Jiang and Gereffi 2014). Over time, the benefits of GVCs can be extended if countries upgrade the position of their manufacturing firms in the GVC—by shifting their structure of production from lower to higher value-added parts of the GVC (Jiang and Milberg 2012).

Despite the opportunity to import technology, between-country inequality may persist because not all developing countries adopt new technologies at the same intensity (Figure 4.8). The gap in the number of workers or capital units that benefit from the

Figure 4.8
Average difference in penetration of significant technological innovations between developed and developing countries, 1780s to the present

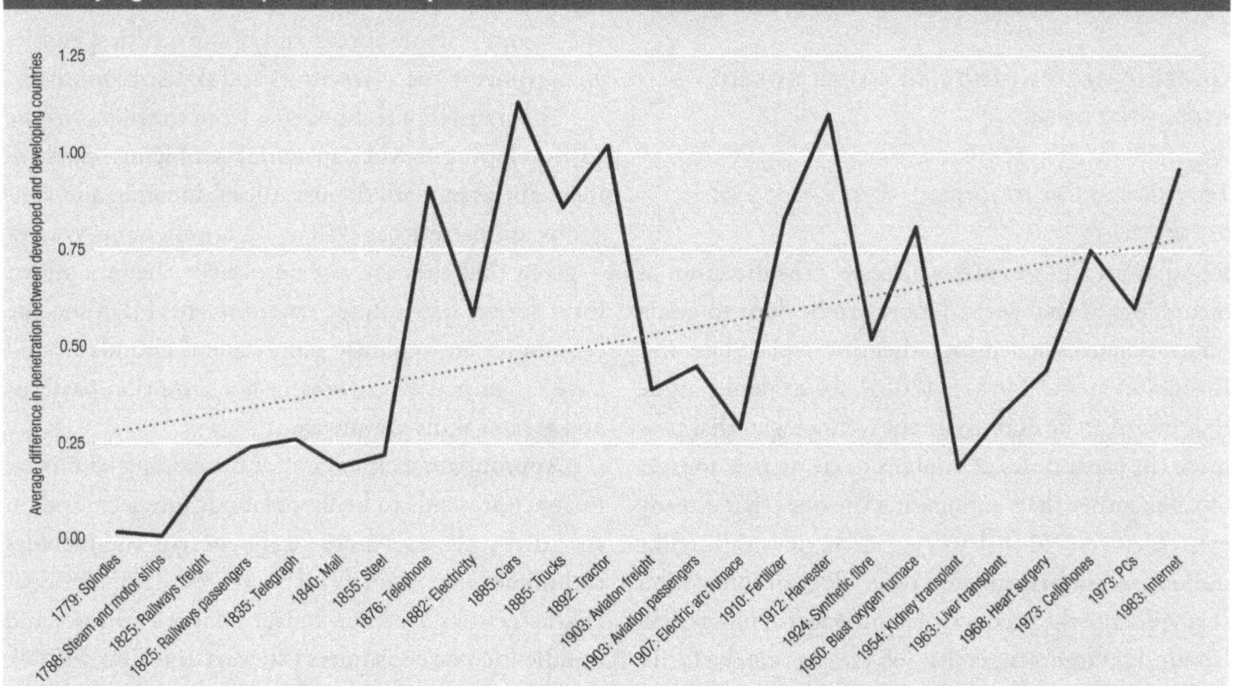

Note: Penetration is defined on the basis of the intensive margin of adoption for each new technology. See Comin and Mestieri (2013) for details.
Source: UNIDO elaboration based on Comin and Mestieri (2013).

> **More can be done to invent technologies that promote the capabilities of humans to create new sources of value**

technology (the "intensive margin" of technology adoption) has increased over time and can explain up to 80 percent of between-country inequality (Comin and Mestieri 2013).[14]

Much research and policy advice has been concerned with the increasing adoption of foreign technologies in developing countries, especially as domestic capabilities are required to absorb technology. A fundamental obstacle is not always the lack of skills but also the lack of market size, as manufacturing innovation is subject to fixed costs and thus gains from economies of scale require increasing specialization.

A country's development level also endogenously limits the penetration of technologies and national R&D efforts through the abundance of low-wage and low-skilled labour. With low wages, it is not always profitable to adopt and apply new technology, whether through a country's own R&D efforts or imported machinery (Allen 2012). In contrast, higher wages are an incentive for technological innovation in rich countries. This takes us to the conditions and techniques needed in developing countries to combat the social disparities induced by technological change.

Getting technology to drive social inclusiveness

Technological trajectory and choice of techniques

Regulations and incentives help steer the direction of technological change, and more can be done to guide innovation to complement rather than replace humans. Brynjolfsson and McAfee (2014), for example, argue that more can be done to invent technologies that promote the capabilities of humans to create new sources of value rather than automating the ones that already exist. Cowen (2014) shares a similar view, calling for more directed or regulated technological innovation that will raise the productivity and wages of low-skilled labour. His argument is that job creation can be facilitated if access to and use of information and communications technology (ICT) becomes easier, for instance if robots and machines are "easier to operate."

It may also be necessary to support technological innovations with organizational change, helping to flatten hierarchies and decentralize management responsibilities. Marsh (2012), for example, states that the industrial revolution started out "centred on new technology *plus new methods of organization* originating in a small group of countries" [our emphasis] (p. 241). Brynjolfsson and McAfee (2012a) consider many of the most recent ICT innovations as offering opportunities for certain human skills that are "more valuable than ever." Exploiting these opportunities, however, requires new forms of organizations aimed at complementing new technologies with human skills to deliver new products and services.

For developing countries, this concern has been addressed by an extensive literature on the so-called appropriate technology (Jéquier 1976; Kaplinsky 1990; Schumacher 1973). In its simplest definition, appropriate technology is "the technology which best makes use of a country's resources to achieve its development objectives" (Stewart and Ranis 1990, p. 4). This refers both to product and process innovations. Appropriate technology depends on the nature of the country, its resources and opportunities, and its changes over time as resources and skills accumulate.

The transfer of technologies from the high-income to developing world can entail problems—such as inequalities in work opportunities, incomes and consumption patterns as well as high underemployment —given that they are not necessarily the best suited for a developing-country environment. High-income economies are typically more capital abundant, and have larger markets, different consumption patterns and greater skills prevalence.

Appropriate technology for developing countries, in contrast, tends to be more labour intensive and on a smaller scale, uses more local materials and includes techniques and products for rural production. Product characteristics are then better adapted to low- and middle-income consumers (Stewart and Ranis 1990). Countries should try to use technologies that are better suited for their characteristics, reflecting their factors, skills and endowments (Lin and Zhang 2009). In

> **Innovation and industrial policies are fundamental in shifting the innovation path towards a more inclusive trajectory**

recent years, this view has been gaining ground in the development agenda (Box 4.2).

Given the "path-dependent" nature of technological knowledge and innovation, the choice of techniques is crucial for the long-run development of a country (Chataway, Hanlin and Kaplinsky 2014). The different orientations of agriculture in East Asia and Latin America exemplify this. The former relied heavily on small, relatively unmechanized agriculture and was more inclusive than the latter, which was oriented towards large-scale, capital-intensive agriculture. Similarly, specialization in capital and scale-intensive extractive industries (base metals, oil and gas) involves technologies that lead to more non-inclusive innovations, as seen in South Africa where high exclusion and unemployment are linked to the large capital-intensive minerals–energy complex (Chataway, Hanlin and Kaplinsky 2014).

Innovation and industrial policies are therefore fundamental in shifting the innovation path towards a more inclusive trajectory—determining the structure of prices, factor costs, infrastructure and the availability of alternative technologies (and the knowledge that firms have about these technologies)—all affecting firms' investment decisions on their choice of technology (Stewart and Ranis 1990).

Box 4.2
Examples of appropriate technology in developing countries

From the end of the 2000s, development agencies in the Republic of Korea have supported appropriate technology (KOICA 2014). The emphasis is on existing technology and resources and encouraging productive capacity among the local community. The areas include agriculture (organic fertilizer production, small-scale irrigation, crop improvement), energy (bio-energy, smart-grid systems), water and sanitation (rainwater collection tanks and purifiers, water treatment) and small-scale manufacturing (waste material recycling, heating tents, leather coating and dyeing). The Korean Intellectual Property Office (KIPO) also runs cooperation programs on charcoal manufacturing (Chad), soil brick manufacturing (Nepal) and simple water purifiers (Cambodia) (KIPO 2015).

Some global bodies foster inclusive innovations in developing countries, including the UN Office for South-South Cooperation (UNOSSC). In 2008 it launched the South-South Global Assets and Technology Exchange, a virtual and physical platform allowing entrepreneurs in developing countries to obtain technology, assets and finance (UNOSSC 2015).

Compensation mechanisms for labour-saving innovations[15]

Even when the direction of technological change leads to labour-saving process innovations, compensation mechanisms can mitigate the impact on employment creation. But their effectiveness depends heavily on the country situation.

The effectiveness of the price mechanism, for example, relies on competition—an oligopolistic regime weakens this compensation mechanism as cost savings are not always translated into lower prices. The effectiveness of the investment mechanism depends on the investment behavior of firms, which may fail to translate into new investments if they have pessimistic expectations (even drawing on their accumulated profits obtained by innovation). By the same token, the effectiveness of the wage mechanism depends on the degree of substitutability between capital and labour.

A strong competition policy can favour the employment adjustments to counterbalance the initial job losses due to process innovation, while expansionary policies can accelerate compensation by counterbalancing the initial decrease in aggregate demand associated with the dismissed workers and encouraging investment. Conversely, as downward wage flexibility is often ineffective where path-dependent technological change implies very low labour–capital substitution elasticity, an economic policy favoring wage moderation may be useless as a means of compensating for technological unemployment.

Finally, the ultimate effect of technological change on growth and employment is always mediated by a country's absorptive capacity (Chapter 2). In other words, only developing countries that enjoy a sufficient level of endogenous R&D and innovation capabilities

“Policies to establish a better framework for career development are critical

can fully develop the growth and employment potential of new technologies.

Skills and factor endowments

As seen in the previous chapters, education, skill formation and fostering of innovative abilities (such as conducting R&D) are all advocated in the literature as complements to attract foreign technology and to generate the absorptive capacity for technology penetration.

When technological progress is skill biased, a skill mismatch can offset the positive effect of technology-driven structural change and lead to a net increase in inequality. So expanding education and training programs, especially in ICT and related areas, is important, as are the right education policies to match labour-market needs. Indeed, the Economic Commission for Africa (UNECA 2013) concludes that non-inclusive, jobless growth has characterized Africa's recent economic performance and is based on "the misalignment of the educational curricula with the needs of the labour market." Many other economies also struggle on this. And matching the curriculum with jobs yet to be invented is even harder (Box 4.3).

Given that education is expensive and subject to fixed costs, gains in financial-market efficiency for funding it are important for workers to access the opportunities opened by new technologies. In low-income countries, underdeveloped financial markets and the lack of credit in poor rural areas stymie such access. Credit market imperfections can constrain the occupational choices and labour market mobility of unskilled workers, entrenching higher income inequality. Inequalities can even increase under moderate rates of technological innovation, if workers with low skills are unable to access costly education (Canidio 2013).

Beyond the formal education system, aligning skills with the needs of industry is important, including on-the-job training and predictable career ladders, but these often are underappreciated conditions for innovation-led industrial development. Lazonick and others (2014), for example, argue that "collective and cumulative learning" matters for technological innovation, underlining that it is embedded within organizations. In their view, collective and cumulative careers in which the individuals simultaneously develop and utilize their skills are essential foundations for prosperity with a broad-based middle class. It follows that in both advanced and developing countries, policies to establish a better framework for career development are critical.

Box 4.3
Education for jobs that do not yet exist

Frey (2011) averred that "60 percent of the jobs ten years from now have not been invented yet." At the same time, the demand for jobs will likely grow for tasks where computerization is less likely, such as jobs requiring social and creative intelligence as well as in jobs related to art and entertainment (Brynjolfsson and McAfee 2012a, 2012b; Autor and Dorn 2013). Others predict that in the future human labour will almost exclusively complement robots and machines —a switch from practice to date.

Whatever the direction of the demand for labour it will be either necessary to develop new curricula for emerging skills and occupations or to develop education and support of technologies that will make robots and machines "easier to operate" (Cowen 2014).

Entrepreneurship on hand?

The availability of many medium-skilled workers and cheaper technology offers unique opportunities for entrepreneurs to provide "unmet needs of human populations across the world" (Brynjolfsson and McAfee 2012a). They call for policies to support entrepreneurship (Box 4.4).

Entrepreneurial activity is crucial to discover a country's comparative advantage. Even though the social benefits of entrepreneurial activities may be large, local conditions may lead to an undersupply of entrepreneurs (Hausmann and Rodrik 2003; Naudé 2010). This is particularly important in the current context of increasing fragmentation of global production. Entrepreneurs are responsible for matching the output profile of the firm with the requirements

> **"Entrepreneurial support should consider the level of industrialization**

Box 4.4
Basic conditions remain as important as ever

As high-income economies and countries like China increasingly replace low- and medium-skilled routine tasks in manufacturing, the prospects for countries abundant in low-skilled labour to industrialize through assembly-type manufacturing become less viable. But the same technologies replacing routine tasks may open up new opportunities for entrepreneurs, including network production and additive manufacturing.

In South Africa—the world's second-largest supplier of titanium ore—entrepreneurs with government and research institutions have been developing 3D printing systems to accelerate the additive manufacturing of titanium metal parts, including hip joints. Production times are up to eight times faster than with older technologies (Wild 2014).

But to use these opportunities, the entrepreneurial ecosystem needs to have access to high-quality infrastructure, transport and logistical services, and ICT. Nothing has changed here.

defined by the international network, helping to disseminate imitative learning, which is more difficult than outsourced nodes governed by a foreign headquarters (Szirmai, Naudé, and Goedhuys 2011). Such learning improves the chances for sustainable transfers of technology—if it can be mastered by local firms and entrepreneurs—through spillovers into the local economy.

Entrepreneurial support should consider the level of industrialization, since without careful government interventions and policies, the market alone will underinvest in knowledge and innovation (Ács and Naudé 2011). As seen before, the better the innovation system, the more a developing country can tap into global technology, the more rapidly will knowledge circulate within the domestic economy and the more rapidly the economy may embark on the process of technological upgrading. This is no mean task: a complete overhaul of existing institutions could be required.

Redistribution and social protection

Good social protection is key for inclusive structural transformation as it can generate substantial efficiency effects in reallocating labour from less to more productive sectors. As Atkinson (2013) outlines, social protection is vital for "facilitating economic change while promoting social inclusion." (p. 5) Social protection also provides cyclical stabilizers of aggregate demand during economic downswings.

Redistributive tax policies can smooth income and wealth inequalities. Piketty, Saez and Stantcheva (2011) among others propose higher taxation, inheritance taxes and a global wealth tax to shift the balance in favor of labour. Based on simulations using data from the Unites States, Piketty, Saez and Stantcheva (2011) suggest that the top marginal tax rate there could be increased to 83 percent without creating disincentives. Baker (2012) makes a case for a tax on financial speculation, arguing that the financialization of the U.S. economy has been one of the most important determinants of income and wealth inequalities there. Higher marginal taxation and transfers have a huge impact on narrowing income inequalities in many countries (Figure 4.9).

Figure 4.9
Gini index, before and after taxes and transfers, 2009

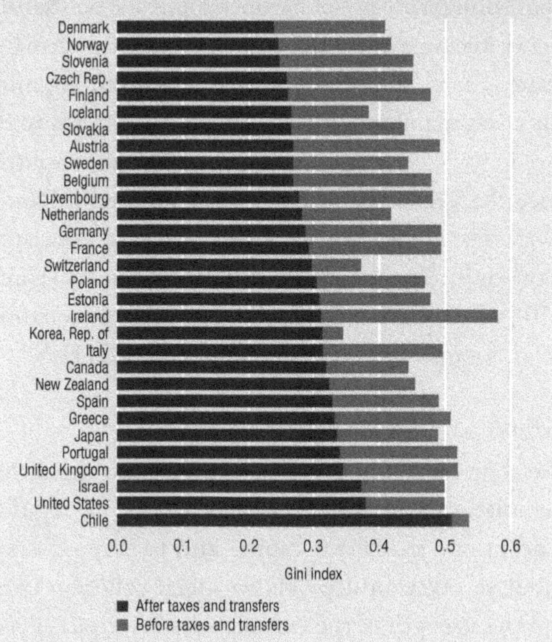

Source: UNIDO elaboration based on OECD Social and Welfare Statistics Database (OECD 2014).

> **"Redistributive tax policies can smooth income and wealth inequalities**

In many countries, and particularly developing countries, distributive tax policies are not easy to pursue. Taxing (or retaxing) richer groups is difficult, because they oppose these policy changes through political, administrative and legal means (Tanzi 2014). It can be even harder in developing countries, given the entrenched powers of local elites and the ease of moving capital offshore. Preferences for redistribution also have structural, cultural and historical causes, complicating the adoption of redistributive tax policies.

Finally, labour market regulations that ensure the regulation of the labour practices of MNEs can be important—to ensure decent labour practices, such as work and health protection and bans on child labour. Consumer-oriented "seals of approval" can help reinforce this approach, underlining "fairness" and "non-exploitation" characterizing a desirable product for many consumers.

Rising income inequalities

Piketty (2014) ascribes rising inequality to the struggle between capital and labour, with labour receiving an ever-smaller share of national income, partly due to a systemic weakening of its bargaining power, stemming from falling trade union membership, trade policies and higher unemployment due to lack of sufficient demand. There is some disputed evidence that declining minimum wages and union membership contributed to the rises in wage inequality in some high-income countries since the 1980s (Bénabou 2005; Card and DiNardo 2002). The United States has a strong inverse association between income inequality and union density (Figure 4.10), while selected developing countries show similar trends for total union membership (Figure 4.11).[16]

Figure 4.10
Income inequality and trade union density, United States, 1960–2011

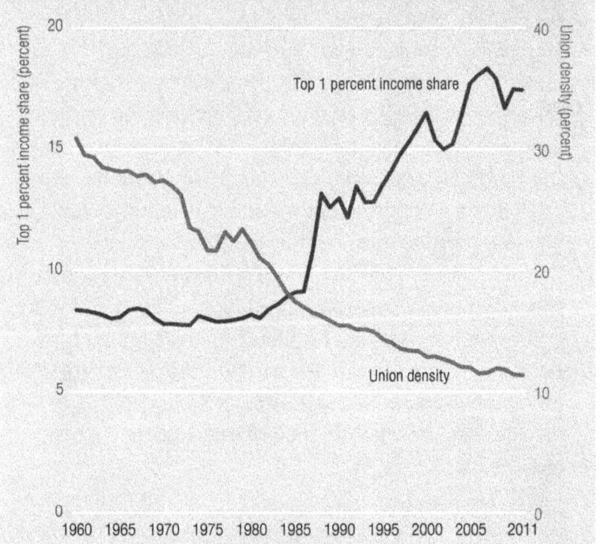

Note: Union density is defined as net union membership as a share of total employed wage and salary earners.
Source: UNIDO elaboration based on Database on Institutional Characteristics of Trade Unions, Wage Setting, State Intervention and Social Pacts (Visser 2013) and World Top Income Database (Alvaredo and others 2014).

Establish or reset the social contract

The ability of countries to enact policies to strengthen the relative bargaining power of labour, to regulate innovations that favor capital and to impose redistributive taxation depends on social cohesion itself, and the strength of the existing social contract. Posen (2014) argues that the social contract in Nordic countries explains their relatively low levels of inequality. This is possibly the most basic precondition of all, and it is doubtful that much progress will be made towards, for example, social justice[17] in societies that do not already have such tight social cohesiveness.

Notes

1. See for example McMillan, Rodrik and Verduzco-Gallo (2014); Szirmai (2012b) and Verspagen (2004).
2. Box 1.1 uses a wider definition, but here we focus only on technological innovation.
3. This is also applicable to some middle-income countries in Latin America and to South Africa, where import penetration of manufactures from relatively low unit labour cost producers, such as China, have contributed to deindustrialization (see Chapter 1).
4. Specifically the poverty measure used is the share of the population that lives on less than 2 dollars PPP per day. This information has been taken from the World Bank WDI Database.

Figure 4.11
Income inequality and unionization, selected developing countries, 2000–2011

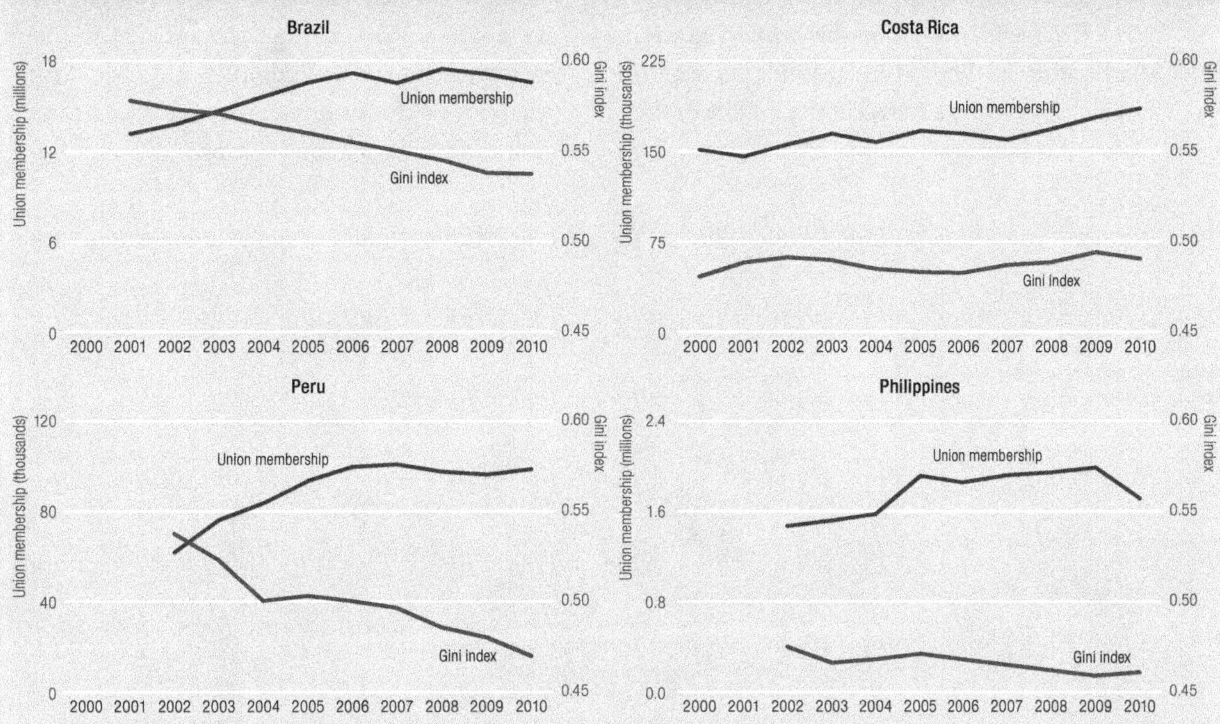

Note: Gini index of inequality in equalized (square root scale) household market (pre-tax, pre-transfer) income.
Source: UNIDO elaboration based on Trade Union Membership Statistics Database (Visser, Hayter and Gammarano 2015) and the Standarized World Income Inequality Database (Solt 2009, 2014).

5. The main data source used to calculate this index is the Standardized World Income Inequality Database (SWIID) 5.0, complemented with data from van Zanden and others (2014) to fill some gaps. The Gini index refers to inequality in equalized (square root scale) household market (pre-tax, pre-transfer) income.
6. See Lavopa (2015) for the details on the definitions of the variables and methods used. All explanatory variables are introduced in log form because the original variables showed a highly skewed distribution.
7. See Lavopa (2015) for details.
8. See Vivarelli (2013) and (2015) for a recent review of this literature.
9. Bogliacino, Piva and Vivarelli (2012) using a panel of European firm data find a positive employment effect of R&D in these sectors but not in technologically more mature manufacturing sectors.
10. his section draws on Gries and others (2015).
11. See Katz and Autor (1999) for a survey.
12. For the United States see Autor and others (2003); Autor and others (2006, 2008); Autor and Dorn (2013); for Germany: Spitz-Oener (2006); Dustmann (2009); for the United Kingdom: Goos and Manning (2007); for other Western countries: Goos, Manning and Salomons (2014), Michaels and others (2014); and Van Reenen (2011).
13. See Timmer and others (2015) for a description on how this database has been calculated. Skills in this database are defined on the basis of educational attainment levels.
14. The extensive margin of technology adoption refers to the timing of a country's adoption of a new technology. The intensive margin refers to how many units of the good embodying the new technology are adopted (for a given size economy). See Comin and Mestieri (2010) for details.

15. This section draws on Vivarelli (2013).
16. In a linked study, Card, Lemieux and Riddell (2004) find evidence from the United States, Canada and the United Kingdom that union membership is "concentrated in the middle of the skill distribution."
17. Including equal access to opportunities (for example, through education), ease of social mobility (for example, through entrepreneurship and inclusive finance) and compensating benefits for the poorest (for example, through social protection and labour market policies).

Chapter 5

Moving towards greener structural transformation

Under what conditions will countries promote enough technological change to induce environment-friendly growth? Today's countries had to pay a heavy price in pollution to pursue industrialization and economic growth. The question here is whether (low- and middle-income) countries industrializing in the 21st century will leapfrog and decouple their growth from pollution and the use of natural resources. Many firms have invested in technological change to increase their energy and resource efficiency, maximizing profits and also reducing pollution. But that type of technological change is not sufficient to overcome the scale effect of increasing world production. Additional environment-friendly technologies, whose introduction is not fully justified by an economic rationale, must be adopted—such as pollution abatement technologies, carbon capture and storage systems, and more costly but more environment-friendly inputs, such as renewable energy. Establishing those technologies often can be economically viable only through economic instruments such as subsidies or carbon taxes.

Historically, growth has taken a heavy toll on the environment. Growth requires the use of inputs, but nearly all are finite and will be depleted. Moreover, growth that produces economic goods for consumption also produces "bads," such as greenhouse gases or waste. According to DARA and the Climate Vulnerable Forum (2012), climate change caused global losses of 1 percent gross domestic product (GDP) in 2010. Least developed countries (LDCs) were hit hardest, losing more than 7 percent of GDP.

LDCs and all the other industrializing countries face a conundrum. They need to industrialize to spur growth and improve their living standards, but industrializing increases pollution and puts great pressure on resources, expressed as domestic material consumption.[1] Over 1970–2010 the correlations were strong between manufacturing value added (MVA) and carbon dioxide (CO_2) emissions per capita, and between GDP per capita and domestic material consumption (Figures 5.1 and 5.2).[2]

Three forces govern the pollution path. Pollution is *increased* by growth, which requires a higher scale of production, usually involving polluting production processes. It can be *reduced* through technological change. It can also be reduced through structural change by increasing the share of less polluting sectors in economies (such as services) as income per capita increases.[3]

Over time, the scale effect is countered by technological and structural change. Technological change mitigates the scale effect at all stages of development, without any significant break over time. UNIDO's *Industrial Development Report 2011* shows that countries with the highest GDP per capita are those with the lowest energy intensity (as an emissions-to-GDP ratio). Over 1960–2011, world GDP per capita increased monotonically, whereas emissions intensity decreased (Figure 5.3). Although the period saw environment-friendly technological change, that change was not enough to decouple pollution from economic growth. Market pull forces stimulated environmental improvements, but that pulling effect was not enough to reduce or even to stabilize emissions or general environmental pollution.

Technological change for environmental sustainability operates mainly through two channels—the production process and the production structure—involving environmental, economic and social trade-offs (Figure 5.4).

Change in the production process

Technological change drives change in the production process through two channels: profit driven (reflecting market pull factors) and non–profit driven (regulations and international agreements).

A profit-driven channel

This channel boosts environmental sustainability through greater efficiency in the use of resources. The

> **Technological change for environmental sustainability operates mainly through the production process and the production structure**

Figure 5.1

Manufacturing CO_2 emissions and real manufacturing value added per capita, by country income, 1970–2010

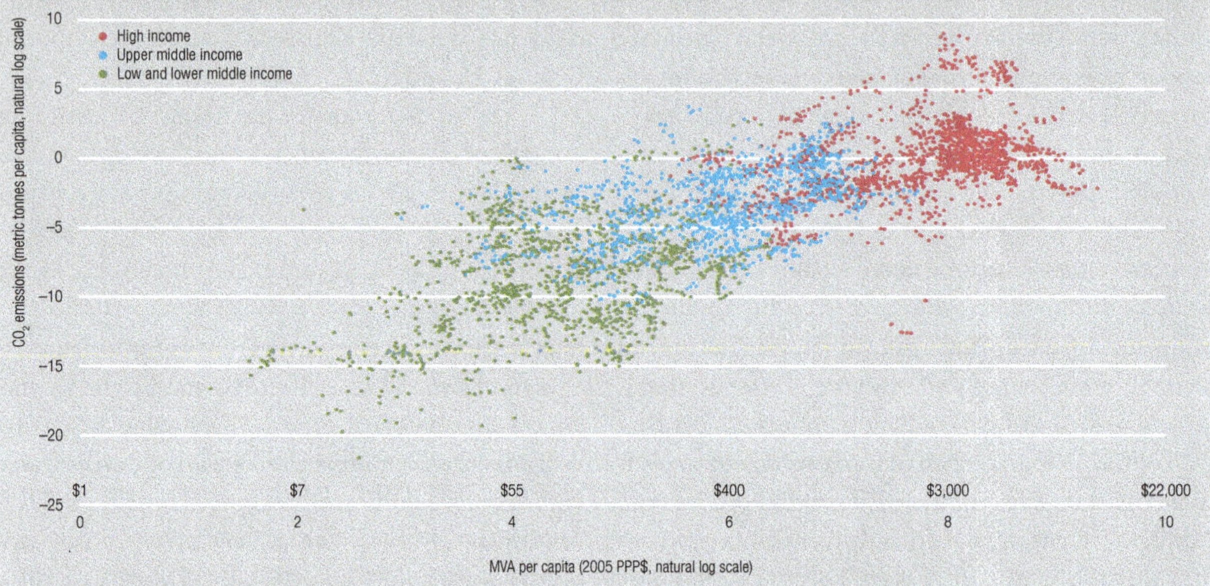

Note: PPP is purchasing power parity; MVA is manufacturing value added. Sample of 70 countries. Income classification based on Annex A1, Table A1.1.
Source: UNIDO elaboration based on Fuel Combustion Statistics (IEA 2015b), World Development Indicators (World Bank 2015a) and Manufacturing Value Added Database (UNIDO 2014d).

Figure 5.2

Domestic materials consumption and GDP, per capita and by income group, 1980–2009

Note: PPP is purchasing power parity; MVA is manufacturing value added. Sample of 70 countries. Negative values for domestic materials consumption exist when domestic materials consumption (expressed as tons per capita) lies in the range [0, 1] and is transformed in logarithms. Income classification based on Annex A1, Table A1.1.
Source: UNIDO elaboration based on Material Flows Database (SERI and WU Vienna 2015) and World Development Indicators (World Bank 2015a).

> **"The production process changes through a more efficient use of natural resources**

Figure 5.3
CO_2 emission intensity and GDP per capita, worldwide, 1960–2011

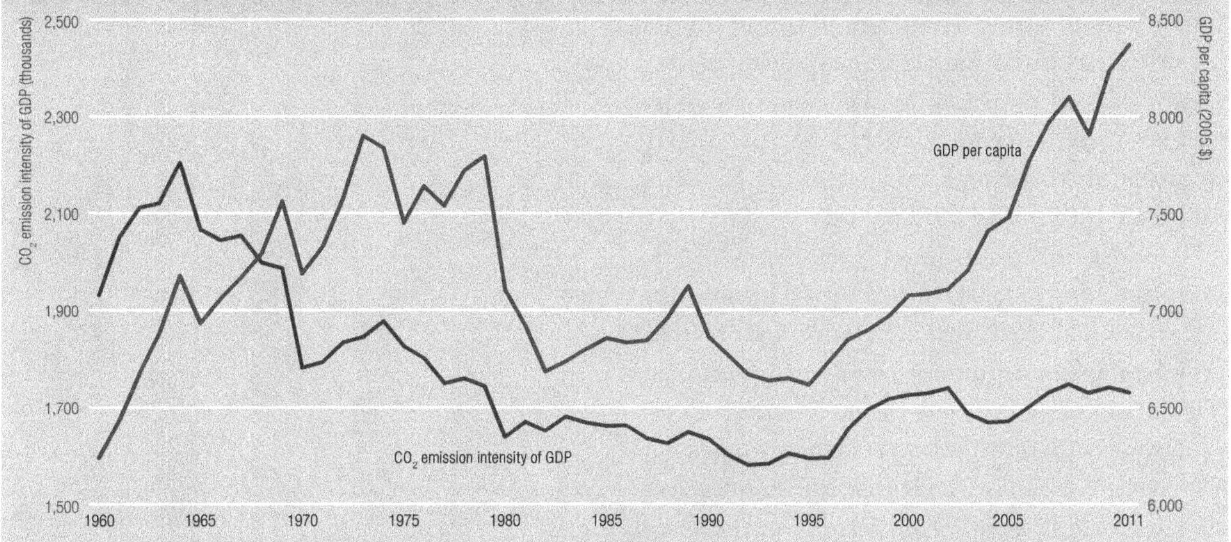

Note: GDP is gross domestic product. Carbon dioxide emission intensity is calculated as emissions over GDP.
Source: Adapted from Mazzanti and others (2015) based on World Development Indicators (World Bank 2015a).

Figure 5.4
Conceptual framework: Technological change for environmental sustainability

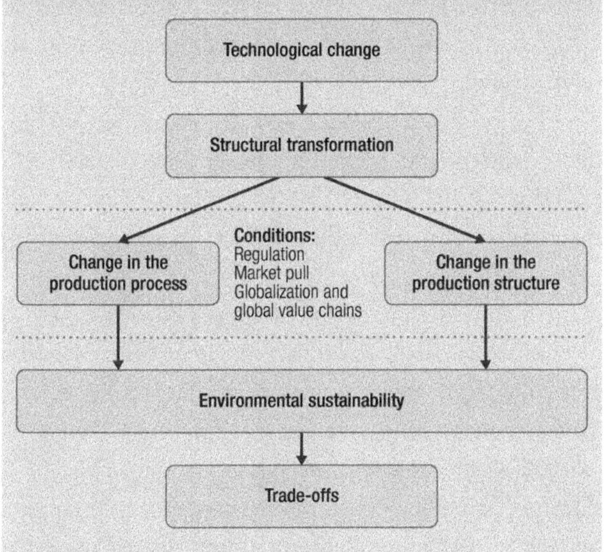

Source: UNIDO elaboration.

argument is that profit-seeking firms naturally tend to change the production process by maximizing the output with a minimum use of inputs.

The production process changes through a more efficient use of natural resources, such as non-renewable energy and materials, helping firms be more cost competitive. Under ideal conditions, the costs of renewable inputs are comparable to fossil fuel energy. Some pollution abatement technologies are done at low cost, and production processes are re-engineered to minimize resource use. Waste, normally a "bad" outcome of the production process, becomes a key input to be re-used directly through materials recovery or waste-to-energy technologies. Such transformations are possible, however, only if environmental technologies exist and conditions, including the relative prices facing producers, are right to change the production process in an environmentally positive direction. Some transformations —such as a global transition to the use of renewable energy or a drastic reduction of costs for pollution abatement technologies—are still far from materializing, but firms tend to use energy inputs more efficiently even if not driven by policies.

Increasing energy prices is an important vehicle of environment-friendly innovation in the medium to long term because energy costs stimulate firms to

> **Global emissions increased only 16 percent in manufacturing, compared with 29 percent for the economy as a whole**

invest in energy efficient technologies (Porter and van der Linde 1995). Firms naturally tend to maximize output by minimizing input costs. The more innovative sectors, such as manufacturing, are more exposed to profit-driven measures. In the short term, however, an increase in energy prices reduces real GDP, especially for energy-importing countries (Cantore, Antimiani and Anciaes 2012).

Global emissions increased over 1995–2009 by 29 percent but by only 16 percent in manufacturing, which had the lowest increase on four metrics, including emissions intensity and energy intensity (Figure 5.5). Manufacturing energy consumption can be decomposed into five factors: energy efficiency, intermediate input use, trade structure in intermediate input use, share of manufacturing in the global economy, and changes in final demand (Figure 5.6). The two most important driving forces are energy efficiency (leading economies to reduce total energy consumption) and increase in final demand (leading economies to increase total energy consumption). The former effect is outweighed by the latter over 1995–2009, but energy efficiency contributes the most in *reducing* energy consumption.

Energy efficiency can also be explained by the usual tendency of firms to replace depreciated capital (Diaz and Puch 2013). When firms replace old machinery, they typically purchase more technologically advanced equipment, which usually is more productive and doesn't add to the energy burden.

Efficiency also pushes firms to invest in technologies that allow the recycling of waste and materials. There is increasing awareness of how technologies can help firms re-use materials as inputs in the production process. Sharp price increases in primary materials in the past decade communicated the scarcity of resources and the need to manage them more sustainably. Recycling, in many cases, becomes more economically viable than discharging materials and waste, and production is transformed into a circular process, whereby formerly economic "bads" acquire value (Box 5.1).

Figure 5.6
Decomposition of manufacturing energy consumption, 1995–2009

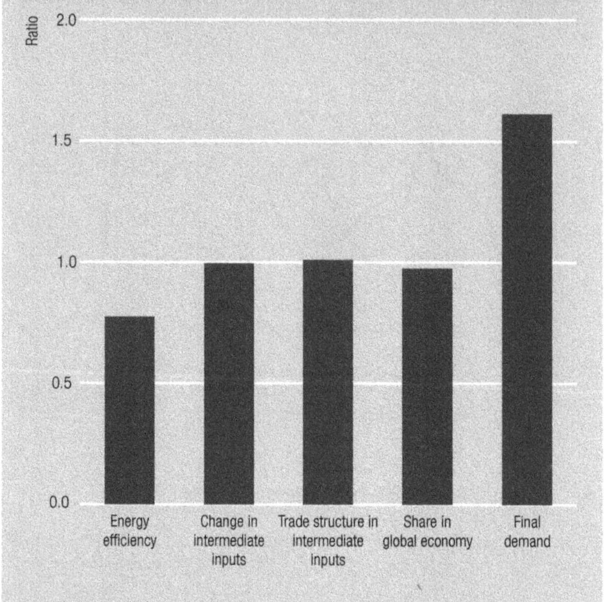

Note: See Figure 5.5.
Source: UNIDO elaboration based on Zhong (2015).

Figure 5.5
Change of global energy consumption, energy intensity, total emissions and emissions intensity, by sector, 1995 and 2009

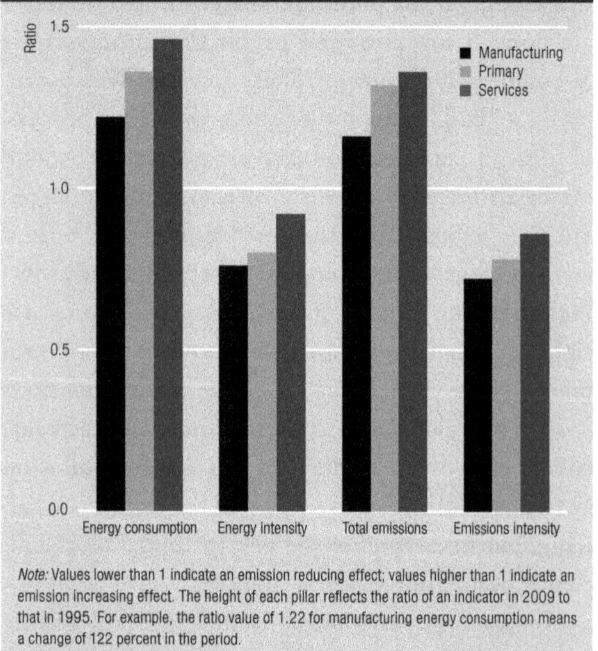

Note: Values lower than 1 indicate an emission reducing effect; values higher than 1 indicate an emission increasing effect. The height of each pillar reflects the ratio of an indicator in 2009 to that in 1995. For example, the ratio value of 1.22 for manufacturing energy consumption means a change of 122 percent in the period.
Source: UNIDO elaboration based on Zhong (2015).

> **"More efficient use of materials can translate into important cost savings**

Box 5.1
Zero liquid discharge system and chromium recovery in the tanning industry in India

Tanning is an old (and important) employment generator in India, but some plants have adopted modern mitigation approaches. The zero liquid discharge system recycles and reuses effluent water from tannery discharge. The high operating costs are justified by a very high recovery of water (90–95 percent), more rational plant management and possible use of sewage for industrial and municipal use (Hussain 2014).

The chrome used can be recovered directly (direct reuse of exhaust chrome liquor after simple screening as tanning liquor for the next batch) or recovered and reused indirectly (through precipitation as hydroxide using an alkali; the precipitated chrome slurry is then dissolved in sulphuric acid, and that solution can then be used as a tanning liquor).

For 130 tanneries of Vaniambadi, total chromium in wastewater is less than 2 mg/litre, compared with 165 mg/litre before treatment. The chrome recycling technology has an important effect on the health and safety of India's workers and the population living near the factories (Environmental Compliance Assistance Centre 2015).

More efficient use of materials can translate into important cost savings for the companies by reducing material and waste-management costs and by identifying new areas of revenues or business tied to using resources better, using underused resources and selling by-products (Box 5.2).

Process innovation in the chemical sector is also moving apace. Steam cracking is the mainstream process for converting hydrocarbon feedstock into olefins —around 95 percent of ethylene and 70–75 percent of propylene is produced that way in Western Europe (Ren, Patel and Blok 2006). Olefin is the main feedstock for such chemical processes as plastics, fibres and other chemicals. The demand for olefins is estimated to be higher than for any other chemical. Olefin production through steam cracking is very energy intensive, despite substantial efficiency improvements in recent decades. The main feedstocks for steam cracking are those from crude oil, such as naphtha, and those from natural gas.

A technology roadmap highlights the potential of catalytic technologies to replace steam cracking, resulting in savings of 143 megatonnes of carbon dioxide-equivalent globally—in an optimistic scenario that assumes high deployment rates (IEA, ICCA and DECHEMA 2013). Olefin production by catalytic cracking of naphtha and methanol has an energy-saving potential of some 10–20 percent (Ren, Patel

Box 5.2
Some process innovations and technologies

- Sensor technologies—condition monitoring of material and structure, mobile electronic control and feedback control techniques, autonomous distributed microsystems.
- Surface technologies—surface refinement, surface functionalization with nanotechnologies, optimization of tribological systems, new coating technologies.
- Process technologies—vibration cleaning techniques, drying technologies: simulation methods, new transformation technologies for steel, waste-free processes.
- Process intensification techniques—microreactor and processing techniques, new catalysis techniques, combination of conventional process techniques with biological process techniques.
- Water management—membrane technologies for special applications, process water circulation, decentralized water management.
- Recycling infrastructures and technologies (for example, recycling of complex products, such as ships; separation processes for complex material composites).
- Material technologies—material with high functional integration, use of secondary raw materials for earth moving, use of material diversity for light construction.
- Technologies for use of renewable raw materials—plants as production platform and raw material suppliers, especially algae and bioplastics.
- Technologies for energy supply—energy-saving technologies and storage media, heating and cooling techniques, organic photovoltaic, renewable energies, such as offshore wind plants.

Source: Adapted from Lang-Koetz, Pastewski and Rohn (2010).

> **Countries should stimulate firms to use fossil fuel inputs more efficiently or incentivize the adoption of renewable energy**

and Blok 2006). A state-of-the-art facility in China that uses a catalytic process for naphtha cracking has reduced energy consumption by around 20 percent (IEA, ICCA and DECHEMA 2013).

Other alternative technologies for producing olefins from conventional or heavy feedstock include gas stream technologies, ethane oxidative de-hydrogenation, propane oxidative dehydrogenation and hydropyrolysis of naphtha. They, too, have potential for cutting energy consumption (Ren, Patel and Blok 2006).

A non-profit-driven channel: Renewable energy

Renewable energy, not yet cost competitive, will require a steep fall in the cost of generation to make it cost competitive. Even when energy efficiency is profitable, market failures—particularly a lack of information or incomplete pricing of inputs—may affect the adoption of energy-efficient technologies. In those cases, countries should stimulate firms to use fossil fuel inputs more efficiently or incentivize the adoption of renewable energy. For example, countries setting a carbon tax on emissions would lead firms to pay a higher bill for using polluting inputs, such as coal and oil. Firms could react to a carbon tax by investing in technologies that minimize the use of such polluting sources of energy. A very ambitious carbon tax of $300 per ton over 2010–2030 would increase energy prices and reduce world energy demand (Table 5.1). Countries would also have lower energy intensity and higher value added, especially in innovative sectors such as manufacturing and information and communications technology (ICT).

With renewable energy more expensive than traditional fossil-fuel sources, policy-makers can subsidize the purchase of renewable energy. From an economic point of view, subsidies are justified as a correction of market externalities (Box 5.3). Clean air is an underprovided public good, as firms pollute the atmosphere with carbon sources of energy. Public interventions to boost the demand of renewable energy also generate intertemporal positive externalities. Bosetti, Carraro and Galeotti (2006), for example, show that increasing experience reduces production costs of renewable energy and stimulates an increasing penetration of carbon-free inputs.

Indeed, although the jury would seem to still be out, work by the International Renewable Energy Agency (IRENA 2014) claims that biomass, hydropower, geothermal (Box 5.4) and wind can now provide electricity competitively. In particular, the levelized cost of solar photovoltaics has fallen by 50 percent over 2010–2014. Ultimately, renewable energy, such as solar photovoltaics (Box 5.5), should become much more frequently a profit-driven, environment-friendly technology.

As reported by the International Energy Agency (IEA 2013), if fossil fuel sources remain substantial in countries' energy portfolios, pollution abatement technologies will need to become more of a priority to accomplish the ambitious agreed-on target limit of 2 degrees temperature increase. The IEA estimates that technology—carbon capture and storage—will have to contribute one sixth of total reductions in CO_2 emissions required by 2050 and 14 percent of the cumulative emission reductions through 2050 (against a business-as-usual scenario). Unfortunately, carbon capture and storage is not a mature technology and has yet to be demonstrated on a large scale. The development and evolution of that technology has an

Table 5.1
Impact of a global $300 per ton carbon tax versus baseline, 2010–2030 (percent change)

Sector/industry value added						World energy		
Agriculture	Energy	Materials	Manufactures	Services	ICT	Prices	Demand	Intensity
0.16	–9.10	0.04	0.23	–0.91	1.29	22.55	–8.54	–7.70

Note: ICT is information and communications technology.
Source: UNIDO simulation with the International Futures model (see http://pardee.du.edu).

“Pollution abatement technologies will need to become more of a priority

Box 5.3
The importance of resource policies

Resource policies are crucial in tackling negative externalities on unsustainable patterns of resource use, while promoting opportunities through innovations at the product, process and system levels, as outlined by Bleischwitz (2009).

Guidelines for resource policy

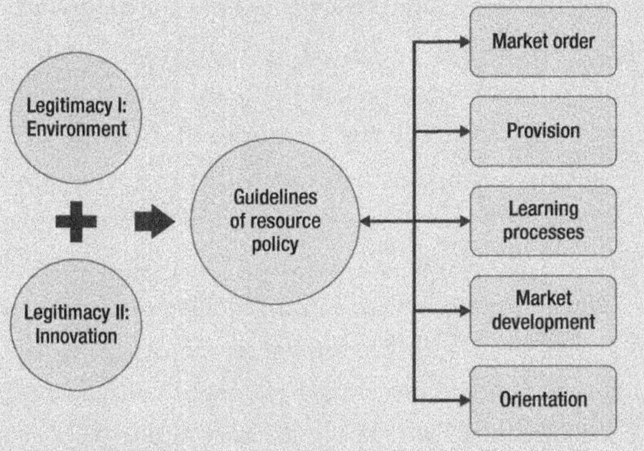

Source: Bleischwitz (2012).

Market order sets the framework conditions for efficient use of resources. This requires internalizing externalities through a regulatory regime based on the polluter-pays principle and, at an international scale, on precautionary principles. It also requires identifying instruments that foster eco-friendly design of products and producer responsibility throughout the whole life cycle of the product or service.

Provision refers to the need to tackle information and data deficits that prevent a more efficient use of resources and a better understanding of the environmental impacts and health hazards linked to unsustainable patterns of resource management.

Learning processes expand knowledge on sustainable practices of resource use and facilitate local knowledge transfer and capability building. Without it, data sources are insufficient. Learning is facilitated by formalized processes, such as improved education systems and inclusion of sustainability and resource management issues in the curricula of technical studies. But learning is also supported by collaborative activities at the level of firms and industries through, for example, benchmarking, reporting guidelines, audit tools or business platforms. Such activities are especially relevant for low- and middle-income countries.

Market development through industrial policies fosters sustainable manufacturing and the uptake of radical innovations, and it fosters the transition towards a circular economy.

Orientation refers to the long-term vision of sustainable and resource-efficient growth at a global and regional level that provides guidance for policy design and target setting by providing some policy certainty and guidance.

Box 5.4
Policies for geothermal production in Kenya

Over 2010–2014, Kenya saw sustained economic growth —but lower than the 10 percent a year assumed in its Vision 2030—while the manufacturing share fell from 12.5 percent to 11 percent. Manufacturers blame the price of electricity, especially relative to South African and Egyptian companies (Njoroge 2014).

The three main sources of electricity in Kenya are hydropower, fossil fuels and geothermal plants. Hydro is an unreliable source because of droughts and erratic rainfall. Fossil fuels impose a heavy toll in terms of imports. For those reasons, the government introduced policies to boost the production of geothermal energy, including feed-in tariffs, aimed to stimulate new investment.

Results so far have been impressive. From 2010 to 2015, Kenya increased geothermal electricity production from 202 to 504 gigawatt hours and the installed capacity from 1,430 to 2,848 megawatts. Kenya is in the world's top 10 for geothermal installed capacity (5 percent) and production (3 percent; Bertani 2015).

Further development of geothermal promises boosting growth and industrialization in the country. The increase of production can stimulate the reduction of costs for consumers and manufacturing producers by up to 30 percent and generate savings up to $24 million per month (Richardson 2015). Technological change is helping reduce production costs. The use of wellhead generators, which allow early generation of power before a conventional plant is built, promotes production with a more aggressive timescale and lower risk for investors.

"Eco-innovations can be incremental or radical and disruptive

Box 5.5
Solar photovoltaic in China

China now leads the global solar photovoltaic sector. Seven of the top 10 global solar panel manufacturers in 2013 were based in China. In 2013, a record 13 gigawatts (GW) of capacity were added to a total of 20 GW existing capacity, and the target capacity for 2017 is 70 GW (IRENA 2014b). Projections indicate that by 2050, China will account for about 37 percent of total photovoltaic capacity globally, maintaining its position as world market leader (IEA 2014b).

Solar photovoltaic prices decreased from around \$4 in 2008 to \$0.8 per watt in 2012. They are forecast to drop to \$0.40 per watt by 2035, assuming that big efforts are still made to increase solar photovoltaic capacity. (IEA 2014b). In the Chinese solar photovoltaic value chain, 1.6 million people were employed in 2013, up from 0.3–0.5 million in 2011 (IRENA 2014c).

economic rationale only if the medium- to long-term economic impact is factored into prices.

Change in the production structure

Eco-innovations can be incremental or radical and disruptive, according to the degree of change they promote. Incremental innovations improve products, processes or organizational practices without changing the parameters of the manufacturing system. Over time, the accumulation of incremental changes can lead to substantial changes that may require the adaptation or redefinition of the whole production system. Radical innovations point to green innovations that promote paradigm shifts and system disruption, which could mean creating a new manufacturing sector, reconfiguring the whole system, introducing new products or services and profoundly changing technological systems (Eco-Innovation Observatory 2013).

A "natural" tendency of the economic process

Technological change reduces pollution because it helps change the production process, and firms change the production technique to produce more output by minimizing inputs. But the economic structure also must change at the macro level.

Countries have a natural tendency to industrialize by transitioning towards more emissions-reducing high-tech sectors.[5] Low-income countries generally show the highest share of value added in low-tech sectors, but since the 1970s, that share has been decreasing. Medium-income countries have the highest shares of medium-tech sectors, and high-income countries, -of high-tech sectors. And the share of high tech tends to rise across all income categories (Table 5.2).

This natural tendency to shift from low- to high-tech sectors also brings in its wake a natural tendency towards pollution. The lowest environmental productivity (expressed as the value added-to-pollution ratio) is associated with medium-tech sectors (Figure 5.7). The medium-tech sector also shows the highest pollution intensity for other pollutants besides CO_2 emissions, such as particulates, sulphur dioxide (SO_2) and nitrogen dioxide (NO_2), although with lower abatement costs than other sectors.

Low- and high-tech sectors have higher environmental productivity—in other words, they generate fewer emissions when producing \$1 of value added. Sectoral specialization towards high-tech sectors reduces emissions intensity.[6] In short, a natural economic tendency contributes to inclusive and sustainable industrial development (ISID).

But environmental protection improvements from the low- to high-tech transition may not be enough to decouple economic growth from pollution. Countries need to enforce actions to curtail environmental harm, even if those actions are not strictly related to the production process (as for pollution abatement technologies). But that non-profit-driven technological change can be expensive (Figure 5.8).

The steep cost of abatement is one of the main factors deterring firms from intervening massively for pollution reduction well beyond any "natural tendency" and countries from adopting emissions caps. Low- and middle-income countries, especially, are reluctant to adopt environment-friendly technologies because adoption costs could hamper growth. But

Countries have a natural tendency to industrialize by transitioning towards more emissions-reducing high-tech sectors

Table 5.2
Distribution of manufacturing value added across technological and income groups, selected years, 1970–2010 (percent)

		Low income	Lower middle income	Upper middle income	High income
1970	Low tech	79.4	53.0	45.1	33.4
	Medium tech	5.8	17.4	23.0	19.8
	High tech	14.8	29.7	31.9	46.8
1980	Low tech	64.6	47.7	38.5	31.9
	Medium tech	10.4	17.0	25.0	19.9
	High tech	25.0	35.3	36.5	48.2
1990	Low tech	63.9	45.1	28.8	30.2
	Medium tech	10.7	20.3	22.7	19.7
	High tech	25.4	34.6	48.5	50.1
2000	Low tech	68.4	42.9	35.4	26.6
	Medium tech	8.0	18.4	21.7	19.8
	High tech	23.6	38.8	42.9	53.6
2010	Low tech	59.6	32.8	28.5	23.0
	Medium tech	19.3	22.8	25.8	21.2
	High tech	21.2	44.4	45.7	55.9

Note: See Annex A1, Table A1.1 for income classification and endnote 5 for tech categorization.
Source: Mazzanti and others (2015).

Figure 5.7
Environmental productivity in the manufacturing sector (value added/CO_2 emissions), worldwide, by technology class, 1970–2010

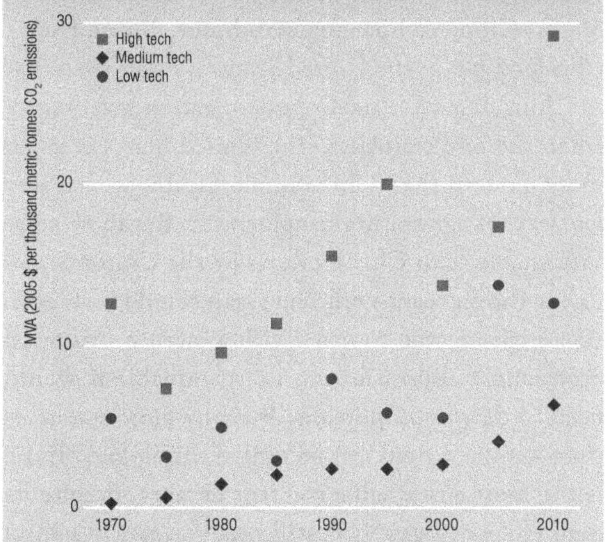

Note: MVA is manufacturing value added; CO_2 is carbon dioxide. See endnote 5 for tech categorization.
Source: Mazzanti and others (2015).

changing the way emissions are measured changes the picture relative to the contribution of different country income groups to global emissions (Box 5.6).

The problem is particularly complex for global pollutants, such as greenhouse gas (GHG) emissions, which require massive and urgent action at the international level, where the change required goes far beyond what the market can induce through profit-maximizing firms. Industrializing countries have not committed to reducing atmospheric carbon concentrations, which were mainly generated by high-income countries. Moreover, emissions-abatement efforts through the adoption of new environment-friendly technologies are asymmetrical across countries. Countries that already committed to emissions-reduction policies under the Kyoto Protocol have already used the low-cost emissions-reduction options, and further emissions-abatement actions would be much more expensive. The problem of equality and responsibility now deters countries from reaching a global agreement for emissions reduction. Thus, every effort at pollution abatement should be tailored to a country's stage of structural change.

"The steep cost of abatement is one of the main factors deterring firms from intervening massively for pollution reduction

Figure 5.8
Pollution intensity versus abatement costs for different technological categories of industrial sectors

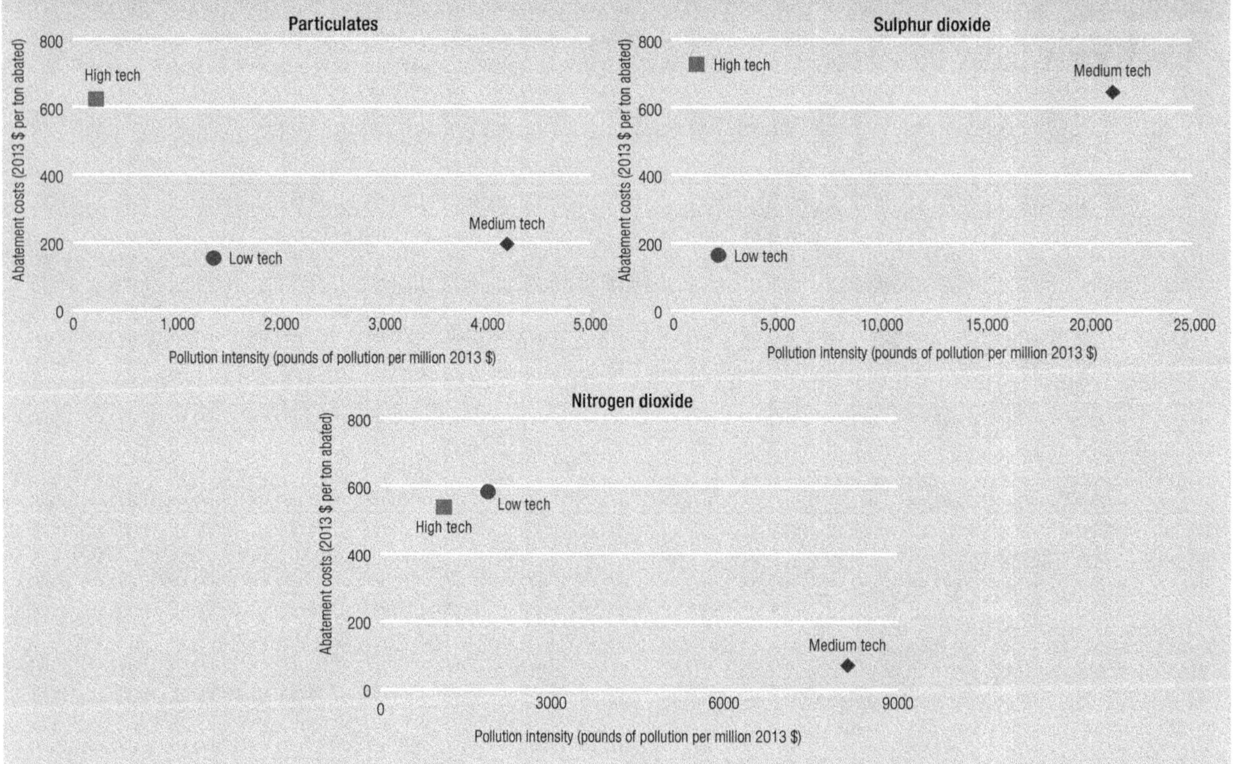

Note: These graphs are based on the ISIC Rev. 2 classification for sectors 31–39 (2, 3 and 4 digits) from Hatzichronoglou (1997) This classification can be converted to ISIC Rev. 3; see UNSD (2015). Monetary values related to pollution intensity (1987 $) and abatement costs (1994 $) are inflated with a standard deflator technique. 15 out of the 81 observations referring to industrial sectors used to elaborate these graphs are missing and filled with data of sectors belonging to the same 3-digit category.
Source: UNIDO elaboration based on the World Bank's IIPS Pollution Intensity and Abatement Cost Datasets (Hettige and others 1995).

Apportioning pollution abatement costs

To what extent is a country efficient at generating emissions, given a certain level of GDP and consumption of inputs? The countries with the highest technical inefficiency over 1995–2009 are China, the United States, India and the Russian Federation (Figure 5.9), together accounting for 55 percent of global emissions in 2013.

Although they are most responsible for global CO_2 emissions, their contribution for reducing those emissions would ideally take into account their stage of development, not just the amount of emissions they generate (Box 5.7).

The United States, the Russian Federation, China and India are the four countries with the largest share of global emissions, but their sustainable modernization levels are very different (Figure 5.10). The United States–China comparison stands out. The emissions efficiency gap in the United States is lower than that in China, but sustainable modernization is very similar in the two countries. The United States is much more modern than China economically, with a very low level of agricultural employment. But to be more sustainable than China, efforts by the United States to cut the emissions efficiency gap should be bigger. As a high-income country with a mature structural economic composition, to be sustainable it would need to devote proportionally many more resources than China to environmental technologies. India is the most sustainable country because, despite its high efficiency gap, it is still predominantly a rural economy. Russia is the least sustainable, showing a very high efficiency gap despite high, non-agricultural employment.

“Countries at a lower level of structural change should be apportioned relatively less in emissions-reduction burden costs

Box 5.6

Apportioning the abatement costs in international agreements for emissions reduction: does it matter what countries produce or what they consume?

As high-income countries tend to practice “environmental dumping”—relocating the dirtiest production activities to lower income countries—production-based rather than consumption-based indicators wrongly show that high-income countries are becoming greener when they are just moving the problem to other countries.

Mazzanti and others (2015) use an input–output approach to analyze the ratio between the footprint of domestic consumption of manufacturing goods and the domestic direct production emissions of manufacturing sectors. That ratio increases from high- to low-income countries because of the greater development of manufacturing in the former. Yet the low-income countries and the high-income countries' ratios tend to converge progressively, as the former tend to decrease the ratio because of increasing industrial production and the latter tend to increase the ratio because of relocation and offshoring (box table).

Emission ratios, consumption- versus production-based

	Consumption perspective/ production perspective					Share of global direct CO_2 emissions (production perspective) in manufacturing (percent)				
Income group	1970s	1980s	1990s	2000s	1970–2009	1970s	1980s	1990s	2000s	1970–2009
Low income	14.9	11.5	9.4	10.0	11.1	1.0	1.0	1.0	1.0	1.0
Lower middle income	6.2	4.4	2.7	2.3	3.1	3.0	5.0	10.0	12.0	8.0
Low and lower middle income	8.1	5.7	3.4	2.8	4.0	4.0	6.0	11.0	13.0	9.0
Upper middle income	2.0	1.7	1.5	1.3	1.5	21.0	27.0	35.0	42.0	32.0
High income	1.4	1.6	2.0	2.3	1.8	75.0	67.0	54.0	46.0	59.0
Upper middle and high income	1.6	1.6	1.8	1.8	1.7	96.0	94.0	89.0	87.0	91.0
Total	1.8	1.9	2.0	1.9	1.9	100.0	100.0	100.0	100.0	100.0

Note: CO_2 is carbon dioxide. Consumption-based emissions is total environmental pressure corresponding to the final demand for manufactured goods. Production-based emissions is environmental pressure directly exerted by the production of manufactured goods. Income classification based on Annex A1, Table A1.1.
Source: Mazzanti and others (2015).

The difference between the adjusted sustainable modernization index (ASMI) and the sustainable modernization index (SMI) is substantial, especially in comparing between the United States and China. The United States is now the least sustainable country, followed by Russia, China and India (Figure 5.11). The biggest difference is the efficiency gap of the United States, which is now very inefficient in generating income per capita from emissions. With one unit of emissions, China produces much more welfare per capita than the United States. India remains the most sustainable from a structural change perspective because of its rural employment profile, whereas Russia remains highly unsustainable from a structural change perspective because it is a highly polluting country with a small return in terms of GDP per capita.

Countries at a lower level of structural change, such as India and China, should thus be apportioned relatively less in emissions-reduction burden costs than the United States and the Russian Federation in international agreements ahead of Paris 21. On both the SMI and the ASMI, China and India have the same emissions-efficiency gap as the Russian Federation. But they are still transitioning from a rural to a modern economy and should have the opportunity to devote a higher proportion of resources towards growth and development. The low emissions intensity of the United States is counterbalanced (compared with China) or more than counterbalanced (compared with India) by a higher level of modernity, which should induce the United States to be keen to accept the biggest abatement cost burden. Reinforcing this finding is the welfare indicator in the ASMI formula.

> **“Abatement measures present new markets, jobs and trade opportunities**

Figure 5.9
CO_2 emissions inefficiency across selected countries, 1995–2009

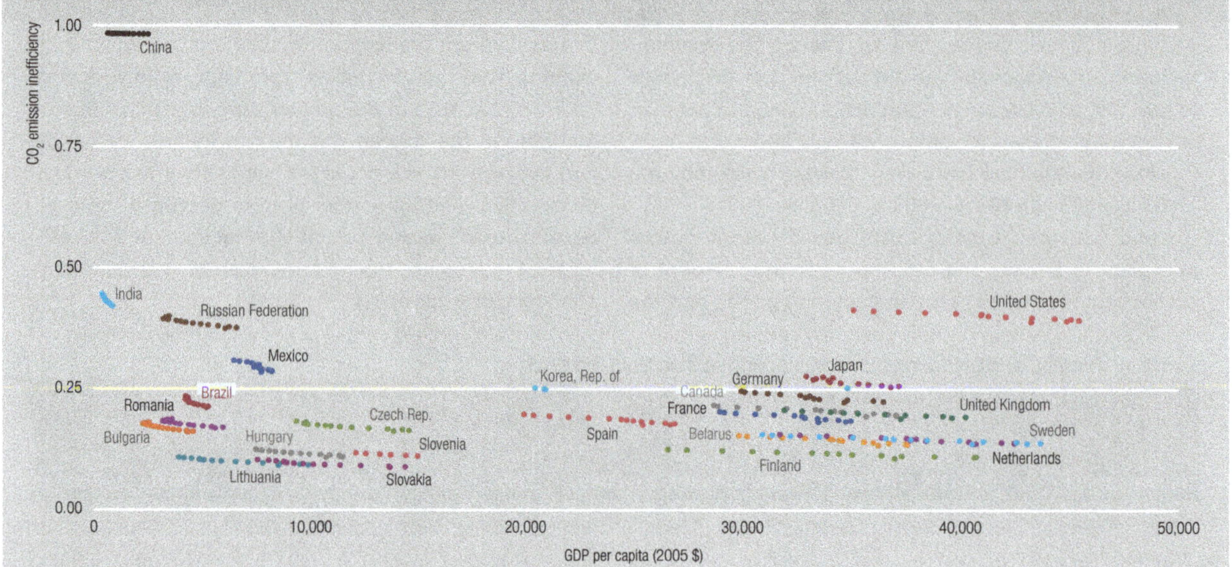

Note: GDP is gross domestic product. For details of stochastic frontier approach used, please see Annex A4.
Source: UNIDO elaboration based on Fuel Combustion Statistics (IEA 2015b), World Development Indicators (World Bank 2015a) and World Input-Output Database (Timmer and others 2015).

Table 5.3
Leading carbon dioxide emitter countries from fossil fuels, 2013

Country	Tons per capita	Kilotons
United States	16.6	5,300,000
Korea, Rep. of	12.7	630,000
Russian Federation	12.6	1,800,000
Japan	10.7	1,360,000
Germany	10.2	840,000
China	7.4	10,280,000
India	1.7	2,070,000
World	4.9	35,270,000

Note: CO_2 is carbon dioxide.
Source: UNIDO elaboration based on Emission Database for Global Atmospheric Research (European Commission, Joint Research Centre and Netherlands Environmental Assessment Agency 2014).

Exploiting opportunities from abatement measures

Over the medium to long term, expensive abatement measures present new markets, jobs and trade opportunities. By breaking past trends, low-income countries should be able to leapfrog and to exploit business opportunities in emerging environmental goods markets.

UNIDO’s *Industrial Development Report 2011* showed that environmental interventions in firms, such as adopting energy efficiency technologies, may be profitable for an individual manufacturing firm. But such business-driven interventions alone will not contribute significantly to inclusive and sustainable industrial development through structural change. On the production and export sides, demand for environmental goods has not yet been great enough to generate relevant structural change. Steenblick (2005) defined a methodology to classify “green goods,” those that define a green growth path.[7] The share of environmental goods exports across the globe is still tiny and heavily concentrated in high-income countries (Table 5.4). But that share could increase with more aggressive environmental regulations across the globe.

On the distribution of exports, countries with higher GDP per capita are currently more active in exporting environmental goods (Figure 5.12) to, usually, high-income destinations (Figure 5.13). Leapfrogging has a long way to go to catch up.

" Demand for environmental goods has not yet been great enough to generate relevant structural change

Box 5.7
The sustainable modernization indicator

Consider a sustainable modernization indicator (SMI), defined as follows:

$$SMI_{it} = \left(1 - \frac{\dfrac{GDP_{it}}{CO_{2it}}}{\dfrac{GDP_{it}\,bench}{CO_{2it}\,bench}}\right) {}^{*}NONAGREMP_{it} \qquad \text{E1}$$

It has two parts. The first represents the emissions efficiency gap—the distance of a specific country from a benchmark country showing the best performance on GDP per unit of emissions. The second represents structural change—the share of employees in the non-agricultural sector.[7]

Countries with the same SMI are placed in the same isoquant as with countries A and B in the box figure. The further outward to the right the isoquant, the less sustainable[8] the position of country because it has a bigger emissions efficiency gap and because more modernity also implies more production, use of inputs and pollution. For example, country A has a higher emissions efficiency gap than does country B. But A and B show the same SMI and sustainable structural change. Country B is a more modern and industrialized country than country A, but it can sustain that level of modernization with a very low emissions efficiency gap.

Along the same isoquant, the lower the emissions efficiency gap, the more a country can enjoy higher levels of modernization by keeping the same sustainability level. Countries C, D and E are less sustainable than country A. Country D is more modern because the share of industry and services employment is higher than in country A, but it has the same emissions efficiency gap, whereas given its more modern structure, it requires more efficiency to remain on a sustainable path. So, even though it is more modern than A, D did not use that modernity to improve its emissions intensity technology relative to A. C is evidently less sustainable than A because it is more modern and industrialized and even less efficient from an environmental point of view than A. Country E is more efficient than A in emissions efficiency but has much a higher modernization level, and its emissions efficiency level is not high enough to sustain that level of energy demand.

Interpretation of the structural change sustainability index according to the isoquants

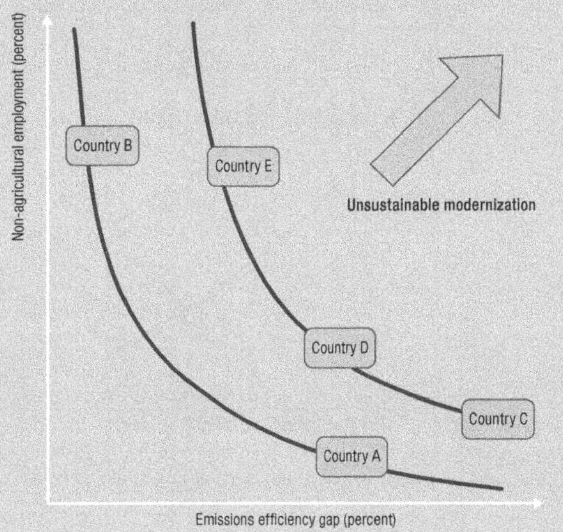

Source: UNIDO elaboration.

The adjusted sustainable modernization index (ASMI) is an alternative measure:

$$ASMI_{it} = \left(1 - \frac{\dfrac{GDPpercap_{it}}{CO_{2it}}}{\dfrac{GDPpercap_{it}\,bench}{CO_{2it}\,bench}}\right) {}^{*}NONAGREMP_{it} \qquad \text{E2}$$

The difference from the SMI is that the ASMI is the ratio between GDP per capita and emissions rather than a simple ratio of emissions and GDP. The ASMI represents the welfare impact of each ton of emissions, the SMI the GDP impact of each unit of emission. And whereas the former represents a measure of efficiency gap in terms of the capacity of each unit of emission to generate wealth, the latter represents a measure of the productivity of emissions.

Low- and middle-income countries need to pursue green industrialization, and green technologies have to be spread around the world so that poor countries can better grasp business opportunities in emerging environmental sectors.

Adopting eco-innovations

Eco-innovations. These are products that reduce their overall life-cycle environmental impacts by favoring reparability, disassembly, recyclability and recoverability. Environmental standards improve products'

> **“Environmental standards improve products’ attractiveness on the international market**

Figure 5.10
Sustainable modernization indicator analysis for the United States, the Russian Federation, China and India, 2009

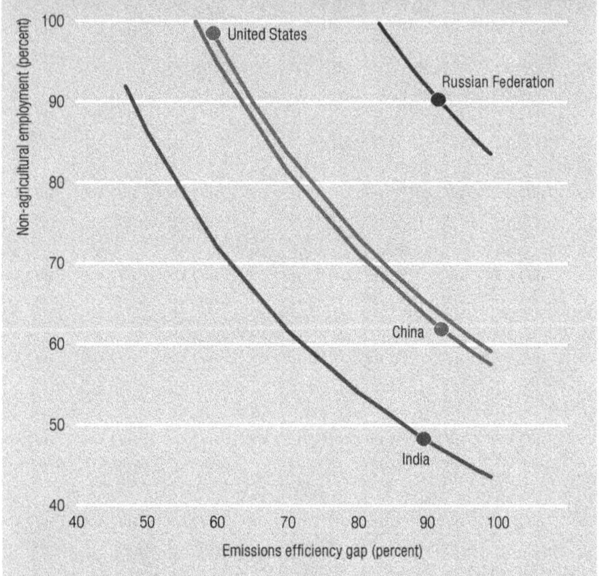

Note: See Box 5.7.
Source: UNIDO elaboration based on World Development Indicators (World Bank 2015a).

Figure 5.11
Adjusted sustainable modernization indicator analysis for the United States, the Russian Federation, China and India, 2009

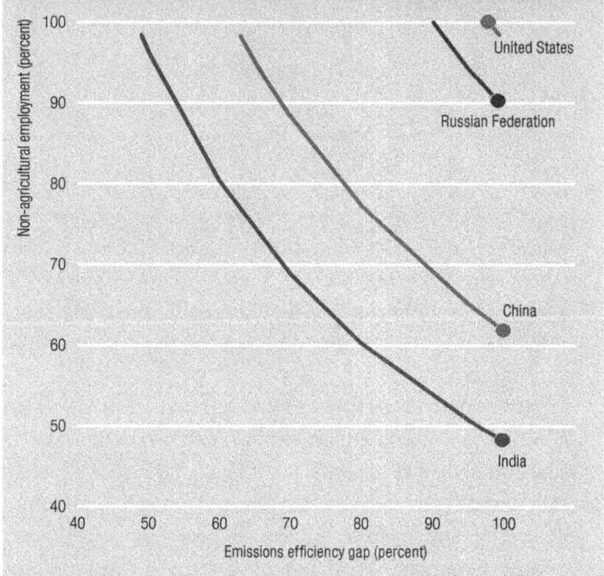

Note: See Box 5.7.
Source: UNIDO elaboration based on World Development Indicators (World Bank 2015a).

Figure 5.12
Relationship between GDP per capita and exports of environmental goods, 2013

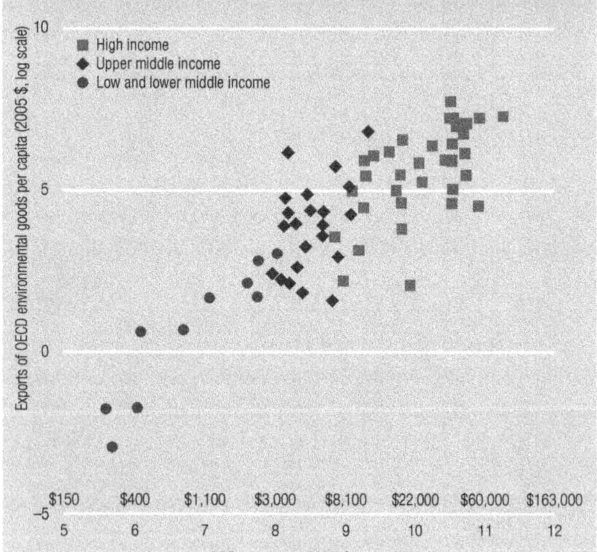

Note: GDP is gross domestic product; OECD is Organisation for Economic Co-operation and Development. Income classification based on Annex A1, Table A1.1.
Source: UNIDO elaboration based on World Development Indicators (World Bank 2015a) and United Nations Comtrade Database (UNSD 2015a).

Table 5.4
Share of green goods exports at different levels of GDP per capita, 2003 and 2013 (percent)

	Low income	Lower middle income	Upper middle income	High income
2003	0.2	1.2	2.8	4.6
2013	0.1	1.9	3.9	4.5

Note: GDP is gross domestic product. Because of missing values for certain categories of products, we consider for each income category only those countries with full data for every environmental good category. A sample of 70 countries was considered. Income classification based on Annex 1, Table A1.1.
Source: UNIDO elaboration based on United Nations Comtrade Database (UNSD 2015a).

attractiveness on the international market. In many cases, those new goods are more productive inputs in the production process of other manufacturing industries, increasing their competitiveness.

Steel and chemical. One innovation to increase resource efficiency is using less metal to produce steel beams, turning out “lightweight” steel. The new products save an average of 30 percent in metal, without any change in performance (Allwood and others 2013). Another

“ Waste can be part of the production process as an input, presenting interesting new markets

Figure 5.13
Main destinations of environmental goods exports from high-income countries, by income group, 2013

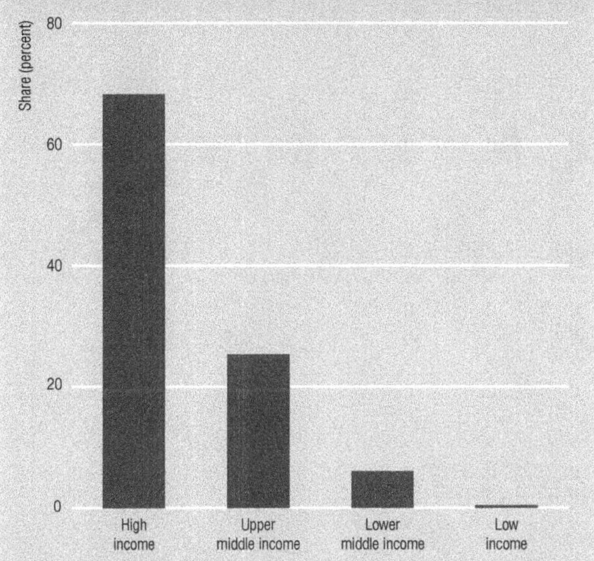

Note: Income classification based on Annex A1, Table A1.1.
Source: UNIDO elaboration based on United Nations Comtrade Database (UNSD 2015a).

is to reuse blast furnace slag from steelmaking in the cement industry, in plastering and in environmental technologies (Das and others 2007). The use of slag in cement production reduces the release of potentially harmful substances to the environment and yields economic benefits. For steelmakers, it helps cut costs tied to waste management and generates additional revenue. And for cement manufacturers, it reduces energy requirements (blast furnace slag cement requires 75 percent less energy than Portland cement) and lowers production costs sharply (Yüksek 2012).

Chemical innovation has improved energy efficiency in housing through new insulating materials. Transport has also benefited, with lighter and tougher materials and the production of alternative fuels. Food waste can be converted into usable, high-value chemicals. Waste orange peel can produce D-limonene, pectin and mesoporous cellulose using a single hydrothermal, low-temperature microwave process (Pfaltzgraff and others 2013). That pilot application also estimated that associated direct costs could be in the region of more than $7.2 million for producing 1,232 metric tonnes of products that could be sold for more than $18.4 million (Pfaltzgraff and others 2013).

Lastly, fast pyrolysis, followed by chemical characterization with mass spectrometry, can convert wheat straw and extract fine chemicals, such as potassium oxide (K_2O) and silicon dioxide (SiO_2), which have applications in formulating bio-derived adhesives (Schnitzer, Monreal and Powell 2014). The examples illustrate the opportunities for low- and middle-income countries to produce chemicals more efficiently.

Recycling waste

Waste disposal was traditionally a typical pollution-abatement activity. But waste can be part of the production process as an input, presenting interesting new markets.

Some eco-innovations not only improve the production process. They also create new product lines that change the structure of manufacturing—as in the waste recycling industry. In 2012, the global value of waste exports was $150 million, or 1 percent of global trade, up from 0.5 percent in 1993 (Figure 5.14). Worth around $1 trillion, global waste management is projected to double by 2020 (Bank of America 2013).

Recycling extracts valuable resources from waste streams so that they remain in the productive system and generate large economic and resource savings. Recycling covers a wide range of technologies to treat specific waste streams, with the main ones metals, electrical and electronic equipment, paper, glass and plastics.

Metal recycling has important environmental and economic benefits. It reduces the volume of virgin ores that have to be mined and refined, substantially reducing the environmental impacts and costs associated with those activities (Bigum, Brogaard and Christensen 2012). It also reduces the energy requirements for metal production by as much as a factor of 10 or 20, depending on the metal (Rankin 2011, cited in Reck and Graedel 2012).

Metals used in large quantities in modern-day industrial processes are referred to as “critical metals.”[8]

“ Metal recycling has environmental advantages

Figure 5.14
Cumulative average growth rate of waste trade, 1993–2012

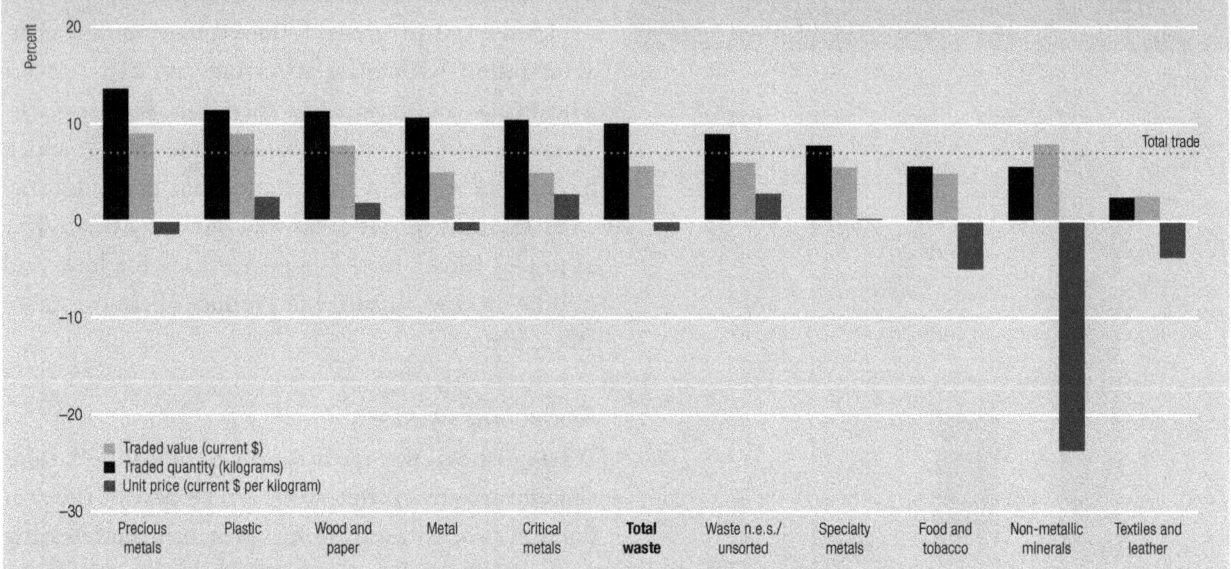

Note: n.e.s. is not elsewhere specified.
Source: UNIDO elaboration based on United Nations Comtrade database (UNSD 2015a).

They are mined in a purer form, have a straightforward recycling process and can be easily recycled.

The number of specialty metals produced in small quantities, often as a by-product in large mines, is also growing. They consist of 35 elements, 16 that can be grouped together as rare earth elements. Although recycling rates for widely used metals such as steel and aluminum can be in the region of 85–90 percent and more than 50 percent for end-of-life scrap (postconsumer scrap), the recycling percentages of specialty and precious metals are uncertain (Reck and Graedel 2012). Recycling of metals generally includes the phases of collection, pre-processing and end-processing. Emerging technologies have been developed in recent years, although mature technologies such as shredding and mechanical separation still prevail for mainly economic reasons (Reck and Graedel 2012).

Besides its economic profitability and resource-efficiency benefits, metal recycling has environmental advantages. State-of-the-art recycling facilities can help cut GHG emissions, given that primary production (mining) has a huge CO_2 impact because the concentration of metals in the ores is very low. For instance, electrical and electronic equipment's annual demand for gold is around 300 metric tonnes. Given that around 17,000 tonnes of CO_2 are released for each tonne of gold mined, 5.1 million tonnes of emissions are generated by gold for electrical and electronic equipment use alone. Copper leads to 15.3 million tonnes of CO_2 emissions (3.4 tons of CO_2 per tonne) as a result of the high annual total demand for copper in electrical and electronic equipment. The cumulative values of the e-waste metals account for 23.4 megatonnes of CO_2 emissions, or nearly 1/1,000 of global CO_2 emissions.

Primary production of metals also uses considerable amounts of land and creates wastewater and sulphur dioxide. For instance, the production of 1 kilogram aluminum by recycling consumes only one tenth of the energy required for mining and avoids the formation of 1.3 kilogram of bauxite residue and 0.011 kilogram of sulphur dioxide emissions. For end-processing materials, the most environmentally friendly production method is hydrometallurgy because it avoids the release of dangerous sulphurous gases and other greenhouse effects (Liew 2008). It also allows

> **Policy-makers need to correct biases to create the right market environment**

some treating of hazardous gases and wastewater with common established technologies that require little capital investment.

Conditions facilitating the adoption of environment-friendly technologies

Market conditions and the way markets are organized play a role in driving—or deterring—eco-innovation. The demand for new products and the progressive incorporation of environmental features in existing products have driven the adoption and diffusion of eco-innovations (Rehfeld, Rennings and Ziegler 2007). Market demand has also been shaped by developments in the policy agenda that define what consumers expect from the environmental impact of products and services. Rising prices of, for example, metal products have created incentives to reuse metal elements in buildings.

Firms may be interested in polluting reduction actions simply because they are profitable, but market externalities may prevent them from exploiting market opportunities. In those cases, policy-makers need to correct biases to create the right market environment. But the commitment of top management can matter more than policy measures in increasing a firm's likelihood of investing in energy efficiency (Cantore 2011). Environmental regulation is effective only when it is flexible enough to allow plants to experiment with alternative solutions (Luken, Van Rompaey and Zigová 2008). And firms are more likely to invest in energy efficiency measures when they have already done it (Cantore 2011).

Regulations

Regulations are key factors in eco-innovation, with external market conditions and internal socio-technical factors as determinants (Ghisetti, Marzucchi and Montresor 2013). National regulations have been a major driving force for innovation decisions in the United States, Japan and Germany (Popp 2006). Regulatory pressure and stringency rather than the presence or absence of specific types of regulations are often motivating factors for eco-innovation decisions (Arimura, Hibiki and Johnstone 2007; Frondel, Horbach and Rennings 2007). Regulations and regulatory pressure are generally understood not just as the current regulatory framework but also as perceptions of the future direction of regulations (Khanna, Deltas and Harrington 2009).

Different types of regulatory approaches may trigger different types of innovations. Whereas regulatory standards may trigger pollution abatement solutions, environmental management systems or integrated regulatory systems can incentivize cleaner and more resource-efficient technologies (Frondel, Horbach and Rennings 2007). And for resource-efficient eco-innovations and cleaner technologies, both regulatory pressure and cost savings seem to be pivotal (Horbach, Rammer and Rennings 2012).

Innovation effects of regulations can also vary according to the environmental area targeted. Whereas standards may set minimums for recycled or recyclable content in products, packaging and other eco-design considerations, economic instruments tackle market failures such as externalities of environmental impacts linked to resource use. Landfill taxes have proven very successful in diverting waste from landfills to reuse and recycling. Some countries in Europe, such as the United Kingdom and the Czech Republic, have introduced aggregated taxes to set incentives for reducing consumption of materials and to create incentives for the formation of secondary markets.

Low- and middle-income countries generally lack a comprehensive regulatory framework for environmental issues and are most focused on localized pollution problems. In recent years, they have started to develop green growth strategies and regulatory frameworks to incentivize better use of resources. Green regulations have been shaped by the additional pressure of increased risks linked to local pollution, such as air pollution in China.

International agreements

For global pollutants in a post-Kyoto world, the main problem is to reach a coordinated agreement for cutting emissions globally. Even mild emission-reduction

"Fairness counts overwhelmingly in negotiations

agreements such as the Kyoto Protocol represent a cost for all signatory countries. For Europe, the cost is estimated to be 0.31–1.50 percent of GDP, and for the United States 0.42–1.96 percent (UNEP and GRID-Arendal 2005). Even flexible efficient mechanisms, such as the emission permit market, do not completely eliminate costs. That market allows sellers of carbon credits to gain money from the sale of those permits and purchasing countries to abate emissions by purchasing emission allowances that minimize total expenditures, reducing costs for Europe to 0.13–0.81 percent and for the United States to 0.24–0.91 percent.

In the European Union, the biggest such market, the price of permits is persistently low at around €5/tonne of CO_2 (Knopf and others 2014), too low to stimulate innovation in abatement technology. Demand-side fundamentals accounted for only 10 percent of the price variation, whereas other factors such as regulatory and political announcements were more important (Box 5.8). The market for futures shows that the expected price by 2020 could be less than €5/tonne of CO_2, despite announcements by the European Commission to put in place a market stability reserve through the strategic release or withdrawal of permits (Figure 5.15).[9]

Standard environmental-economics literature stresses difficulties in reaching global environmental agreements when the gains from free riding are very high. Barrett (1994) showed that self-enforcing international environmental agreements can sustain a large number of signatories but only when the difference in net benefits between the non-cooperative and fully cooperative outcomes is very small. That happens when the benefit from additional expanded participation is marginal. When benefits are high, countries have stronger incentives to abandon coalitions to enjoy gains from pollution reduction without bearing the costs. The difficulties in generating participation in environmental agreements could be overcome with appropriate systems of money transfers, which could induce reluctant countries to join global environmental agreements (Carraro, Eyckmans and Finus 2006).

Fairness is another issue that counts overwhelmingly in negotiations. The reluctance of poor countries to join international agreements is prompted by the historical responsibility of rich regions in generating atmospheric carbon concentration. Likewise, rich

Figure 5.15
Prices of carbon permits, 2011–2014

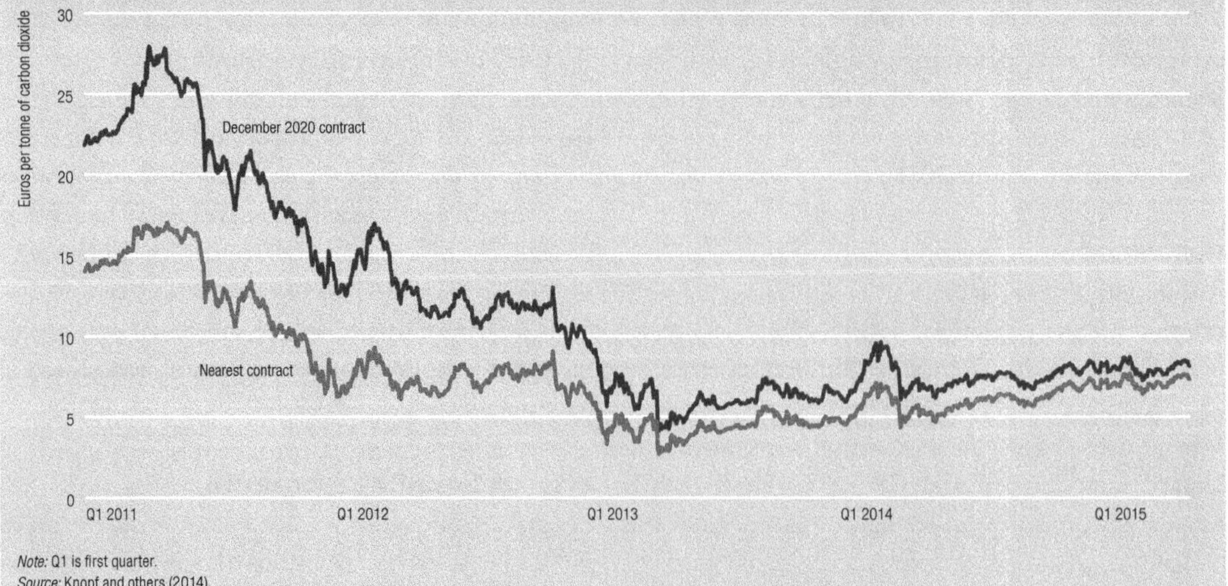

Note: Q1 is first quarter.
Source: Knopf and others (2014).

“ If the new technologies are not too complicated, firms can absorb them

countries claim that emissions-stabilizing policies will be effective only when developing countries contribute fully to reducing emissions. Rich countries are reluctant to sign heavy agreements, aware that developing countries will generate future emissions. Only a simultaneous convergence of industrial technology gaps and environment-friendly technology gaps between developed and developing countries can guarantee the joint target of emission-reduction and emission-per capita convergence (Cantore and Padilla 2010). The previous analysis of SMI and ASMI indicators showed that if countries eliminated the emissions efficiency gap, inequality in modernization levels would remain an issue in apportioning the emissions-reduction cost burden.

A third element that affects climate change negotiations (beyond compliance costs and fairness) is the asymmetric previous engagements of countries in adopting environment-friendly technology, which has an impact on marginal costs. High-income countries involved in Kyoto have already exploited the low-hanging fruit and are likely to pay higher bills for further improvements, whereas in other regions, low-cost opportunities are more available.

Technology transfer

Eco-innovation in resource efficiency requires knowledge of material flows at different system levels. Although descriptive and analytical tools such as material flow analysis and life-cycle assessment could provide detailed descriptions of how materials are used through the economy, they require data that are currently difficult to access and interpret, especially by businesses. Also, firms must be able to access and benefit from those tools. But if the new technologies are not too complicated, firms can absorb them—especially in low-income countries.

In Germany, a network of resource efficiency agencies supports small and medium-size enterprises (SMEs) in identifying opportunities to increase resource efficiency through environmental audits and sharing best practices and technologies. The eco-innovation literature has looked at how knowledge and information flows are generally channeled as spillovers between clusters, regions and sectors (Mazzanti and Zoboli 2005) but also as an emerging culture of firms' cooperation and collaboration (De Marchi and Grandinetti 2013).

Surprisingly, neither patenting nor spillovers have much effect on emission intensity on almost all continents, and spillovers are relevant and significant only in Asia, but only when they are used as consumption-based rather than production-based indicators of emissions intensity (Mazzanti and others 2015). So there must be strong barriers across continents in diffusing technology and in countries using technology to improve their environmental performance.

Greater distance is associated with a lower probability of knowledge-technology flows (Verdolini and Galeotti 2011). And the more similar any two countries, the more likely the flows. Flows are also more likely among lead innovators than among followers and the closer the two countries are to the innovation frontier. Linguistic similarities and trade relations between sending and receiving countries also play a role (Box 5.8).

Greening through global value chains, environmental standards and consumer preferences

GVCs. The greening of global value chains (GVCs) can create opportunities for collaborative approaches to eco-innovation that permeate and benefit all actors involved. More companies are committed to stricter and more stringent ways to identify material sources and to certification schemes that ensure the sustainable supply of different materials.[10] Regional and national support systems that provide access to specific knowledge and that help companies (especially smaller ones) in introducing, adopting, or even developing new technologies may be particularly important.

The performance of vertically related firms generates green knowledge along the value chain, especially where environmental regulation is weak (Ghisetti and Quatraro 2013). In well-integrated value chains, downstream firms demand green innovation from

"Defining standards is important to enhance competitiveness on the supply side

Box 5.8

South–South offshoring of environmental services: An opportunity for Costa Rica

Costa Rica is recognized worldwide for its unique approach to environmental protection. Due to conservation incentives put in place in the 1980s, today tropical forest covers more than half the country. Illegal farming is down to just 15 percent, and farmers are paid to manage and protect their natural surroundings (Conservation International 2012). To date, however, this know-how has been used principally to support domestic priorities. Experts work for national non-governmental organizations and foundations, and the country has not yet seized the opportunity to commercialize the significant expertise it has built over many years.

Because of its critical mass of qualified human capital to sustain this niche, Costa Rica is in an excellent position to export high-demand environmental services, such as natural resources management, environmental impact studies, threatened and endangered species assessments, protected areas evaluations, and environmental education and training (Chassot 2012, Rodriguez 2012).

More than 18 other countries, including China, have consulted Costa Rica about its conservation policies (Conservation International 2012). As with many developing countries, however, limited knowledge of potential markets and undeveloped entrepreneurship skills undermine the potential for translating those consulting opportunities into profitable service exports (Chassot 2012). Promotion of this industry will require the internationalization of local firms and the attraction of foreign environmental firms to use Costa Rica as a platform to export environmental services. Linking the two types of firms will be critical for developing this niche activity.

Source: Gereffi 2015.

upstream firms because of corporate social responsibility strategies and because of the need to anticipate future environmental regulations.

The transfer of environmental knowledge along value chains is induced by lead firms shaping governance mechanisms to improve the environmental performance of the firms in the network (De Marchi, Di Maria and Ponte 2013). Suppliers may lack environmental knowledge, whereas big players on the purchasing side have deep knowledge of how to address environmental problems. They support suppliers with knowledge of the product, process and organization—and sometimes with finance.

Environmental standards. Large companies are twice as likely to have environmental performance monitoring systems (UNEP 2014).[11] But there is a huge mismatch between commitments, declarations, targets, setting and actions. Large companies are not always in a position to set environmental standards for SMEs along the chain, and even if they are, they can have problems monitoring implementation.

Defining standards is important to enhance competitiveness on the supply side. A recent example of this approach and its benefits is the work done in the certification of cocoa. Unilever, as part of its wider Sustainable Living Plan, committed to source all cocoa sustainably. That has led to certification schemes to ensure that producers engage in agriculturally sustainable practices. It has also sparked new collaborations among local traders, non-governmental organizations, international wholesalers and manufacturing companies, increasing farm productivity, reducing or controlling pesticide use and leading to new forms of governance and knowledge transfer (Afrane and others 2013).

One straightforward advantage for firms to adopt environment-friendly technology is to optimize resources along the supply chain. ISO 14001–registered plants have higher system investments for the adoption of waste reduction and cost efficiency technologies and higher returns on investment than do nonregistered plants (Curkovic and Sroufe 2011).

Consumer preferences. Managers and executive employees identify brand building and reputation as the most important factors in competitiveness leverage, followed by cost savings and supply chain cost optimizations (Berns and others 2009). The credibility of the institutions issuing certifications is crucial. Unfortunately, the scientific tools to analyze

“Consumer demand for green goods is still not high enough to generate a big upswing in environmental pressure

the economic, environmental and social impacts of supply chains—representing the basis for any accurate branding policy—are still limited (Seuring 2013).

In the organic agri-food market, market power mechanisms are important not only to determine the distribution of the margins between producers and retailers along the supply chain but also to shape consumer preferences (Richards, Acharya and Molina 2010). The reputation of retailers' or manufacturers' brands and retailers' willingness to promote organic products are key factors in attracting consumers (Llorens, Puelles and Manzano 2011). But consumer demand for green goods is still not high enough to generate a big upswing in environmental pressure.

Environment-friendly technology change is driven by firms seeking efficiency in the use of inputs to maximize profits (production process technological change) and a "natural tendency" of countries to transition to more environment-friendly high-tech manufacturing sectors (production structure technological change). Also observed are the creation and expansion of new world markets of environmental goods (especially in developed countries) and a radical shift in the conception of waste from a "bad" to an input to the production process. Waste-recycled products are now a booming market at the international level—and one of the few examples of win-win-win success stories of technology. These changes cannot yet guarantee that the world is on a sustainable path, but policy interventions to strengthen the conditions to promote inclusive and sustainable industrial development are keys for developing countries to leapfrog. Regulations and global agreements to control or limit GHGs are important, as are improving the markets to correct externalities and shifting consumer preferences towards environmental goods, matched by well-organized GVCs.

Notes

1. Such consumption is defined as all solid, liquid and gaseous materials (including waste) that enter the economy for further use in production and consumption processes.
2. CO_2 emissions per capita and domestic material consumption are used as a proxy for a broader range of indicators of environmental degradation and threat.
3. As will be discussed later in this chapter in more detail, in some cases less polluting sectors gain shares only because the most polluting sectors are located in other countries.
4. Stern (2006) estimates the optimal carbon tax internalizing externalities at more than $300 per ton, even though from a practical point of view, this number may seem unrealistic.
5. The OECD classification (also adopted for Chapters 3 and 4) allows us to identify low-, medium- and high-tech manufacturing sectors. The International Standard Industrial Classification (ISIC) Rev. 3–based classification adopted in this chapter is slightly different from the OECD standard classification, for consistency with the IEA classification on emissions. Across ISIC sectors 15–37, what we define as the high-technology macrosector also includes sector 28 ("Manufacture of fabricated metal products"), which is classified as a medium-technology sector, whereas what we define as the medium-technology macrosector also includes sector 33 ("Manufacture of medical," high-technology), sector 36 ("Manufacture of furniture," low-technology) and sector 37 ("Recycling," low-technology).
6. This finding is complementary with findings explained in Chapter 3, which points out that high-tech sectors are those delivering a growth premium, and in Chapter 4, which emphasizes that specialization in high-tech sectors delivers gains in social inclusiveness.
7. This methodology is borrowed from Lavopa and Szirmai (2015) and adapted to the environmental context.
8. In this analysis, we discuss environmental sustainability just by analysing the path of emissions, but we acknowledge that the sustainability covers a wider spectrum of pollutants.

9. These are the OECD classifications: *Pollution management*: air pollution management; waste-water management; solid waste management; remediation and clean-up; noise and vibration management; environmental monitoring, analysis and assessment. *Cleaner technologies and products:* cleaner/resource-efficient technologies and processes; cleaner/resource-efficient products. *Resources management group*: indoor air pollution control; water supply (Steenblik 2005).
10. Such metals include iron ore, aluminum, copper, zinc, lead and tin, titanium, chromium, manganese, nickel and molybdenum. Uranium is also considered a critical metal because it contributes to producing 14 percent of the world's electrical supply.
11. Reuters (2015) questions this expectation. Prices could rise to around 20 euros by 2020 with the help of the reform.
12. The European Environment Agency (2015) reports that over the period 2003–2010, the number of organizations adopting environmental management systems increased by 50 percent.
13. Based on a survey of 1,712 respondents in 113 countries.

Chapter 6

Designing and implementing inclusive and sustainable industrial development policies

Industrial policy-makers face huge challenges in designing and implementing policy frameworks adapted to their own country's resource endowments, economic and political systems, new products and processes and the changing international geography of production and consumption. Previous chapters discussed the importance of technology and innovation as critical drivers of structural transformation—and as enablers of growth that is not only long-lasting but also mindful of environmental boundaries and the imperatives of social inclusion. So, this chapter dwells on a very specific yet highly promising subset of the policy framework: policies to foster technological progress and innovation in industrializing countries.

The chapter explores industrial policy instruments offering the easiest route to structural transformation: technology policies, innovation policies and competitiveness policies. Competitiveness policies are of key importance for absorbing innovation capabilities from global value chains (GVCs) and maximizing the expected spillovers across the whole economy.

Some words of caution on the possible trade-offs between the economic, social and environmental dimensions may alert policy-makers to the hurdles facing them. Some technological solutions, especially those characterized as win-win-win, can reconcile the three dimensions of inclusive and sustainable industrial development (ISID). But the chapter does not provide guidance on how to prioritize the economic, social and environmental dimensions. Strategic decisions about how best to define goals and balance priorities are left to policy-makers, since this depends heavily on the country context.

Innovation is crucial at all stages of development. First, it must be a focus of governments in developing and emerging countries and of international donors. Second, the process of innovation is different at different stages of development, so policy measures should reflect this. Third, governments should engage in a learning process to build the institutions and competencies needed at their current stage. To this end, innovation policy should stimulate the type of innovation appropriate to a country's stage of development. It should result from a public–private dialogue, be evaluated for performance, and be coordinated with other industrial policy measures.

Policy-making can be effective only if it follows accepted good practices. These include ensuring transparency in implementation; measuring, monitoring and evaluating policy outcomes; and fostering strong coordination and collaboration among different national and global stakeholders. The crucial role of technology and innovation in the Sustainable Development Goals is a welcome step in that direction.

Managing trade-offs

One trade-off is that the productivity growth associated with rapid upgrading tends to reduce the demand for labour (Massa 2015), but this reduction is not inevitable at lower levels of per capita income, when manufacturing tends to be more labour intensive. If productivity growth goes hand in hand with accelerated growth of output, the net effects on employment can be positive. Moreover, if structural change and industrialization promote economy-wide rapid growth through linkages and spillovers, this can increase employment and the absorption of labour. Where synergies between sustained growth and inclusive development are most prominent is in reducing poverty. Countries that accelerate growth and catch up, such as China, India, Mauritius and Vietnam, can realize huge reductions in poverty (Naudé, Szirmai and Haraguchi 2015; Ramani and Szirmai 2014).

Economic pros versus social and environmental cons

Trade-offs between sustained growth and income inequality can be very pronounced. In almost all countries experiencing sustained growth and catch-up, there have been increases in inequality as measured

> **Trade-offs between sustained growth and income inequality can be very pronounced**

by the Gini index. This has to do with the balance between the supply of skilled labour and the demand for it. Where technological change is skill-biased and the labour supply fails to keep up with the demand for skilled labour, inequality will tend to increase. This is not an inevitable outcome, but it characterizes most growth experiences in past decades.

The final trade-off is between sustained growth and environmental sustainability. Here the record so far has been disappointing, and the negative environmental impacts of growth on carbon dioxide emissions and global warming have been larger than the positive impacts of technological advance.

Social pros versus environmental cons

Potential trade-offs can emerge between the social and environmental dimensions when new technologies are introduced, but synergies can also emerge (Box 6.1).

Biotechnology. In developing economies biotechnology is a good example of technological innovations bringing social benefits but environmental harm. Biotech crops can alleviate poverty among small farmers by increasing their incomes, but the adoption of genetically modified crops may also have adverse impacts on the environment. First, the presence of living modified organisms may pose serious challenges to biodiversity (Kaphengst and Smith 2013). Second, transgenic crops may negatively affect the soil and soil organisms (Kaphengst and Smith 2013). Third, the development of resistance to pesticides and herbicides targeted to biotech crops may lead to an even higher use of pesticides. For example, Wang and others (2009) argue that in China the use of biotech cotton and the associated lower level of insecticide spraying have led to secondary insect infestations and therefore to an increased use of pesticides.

Biofuel production. Similarly, biofuel production can lead to rural employment, though the magnitude of this effect depends on the type of feedstock grown as well as on the degree of agricultural mechanization (Diop and others 2013). The replacement of fossil fuels with biofuels may also lead to important public health benefits by improving air quality (USAID 2009). Still, biofuels may lead to a series of adverse environmental impacts, as reported by Timilsina and Shrestha (2010). The conversion of natural landscapes into biofuel plantations and processing plants may have severe effects on biodiversity. In Indonesia and Malaysia, palm oil plantations have replaced natural forests (Koh and Wilcove 2008). In Brazil, parts of

Box 6.1

Synergies between social and environmental gains: Solar drying of fruit and vegetables in Uganda

Low-cost solar food-drying technology is enhancing women's empowerment in Uganda, not only by adding value to the production of fresh fruits and vegetables, but also through significant productivity and quality increases over traditional sun drying methods.

With traditional drying methods, such as open sun-drying, fruit and vegetables losses are estimated at around 30–40 percent of production. The quantity and quality of the dried product suffers contamination by rain, dirt and dust and infestation by insects, rodents and other animals. Furthermore, traditional methods are very time consuming, require large areas for drying, and cannot be expected to result in standard or uniform products.

Well-designed solar drying systems can greatly improve quality, quantity and productivity. They provide faster drying rates, reduce risks of spoilage and contamination and can be constructed from locally available materials at low capital costs and with no fuel costs.

In Uganda, traditional drying is still widely practiced, but the major exporters of dried fruits already use solar dryers. Uganda's biggest exporter of dried fruits uses 110 solar driers at 85 sites in Mbarara and Kayunga (Ribbink, Nyabuntu and Kumar 2005). Its 139 producer groups—70 percent female—typically employ between one and five labourers to help with the fruit preparation and are supplied by around 930 farmers. Most producers use simple, timber-framed cabinets to dry 20–30 kilograms of fresh fruit in each cabinet down to 5–12 kilograms of dried fruit.

> **Trade-offs can emerge between the social and environmental dimensions when new technologies are introduced, but synergies can also emerge**

the Mata Atlantica region (a biodiversity hotspot) and the Cerrado (the world's most biodiverse savannah) are being converted into sugarcane and soybean plantations (Timilsina and Shrestha 2010).

Biotechnology innovation. This can also increase the vulnerability of the poorest smallholder farmers, who are encouraged to switch from growing a wide variety of crops to monocultures of biotech crops, thus increasing the risk to their already precarious socioeconomic situation in the event of a failed harvest. This was the case in South Africa, where the introduction of biotech cotton has contributed to the vulnerability of poor farmers as well as to socioeconomic inequality (Witt, Patel and Schnurr 2006). Nevertheless, biotechnology may contribute to a better environment, as it allows a reduction in pesticide use. There is evidence, for example, that in Argentina, China, and India the introduction of biotech cotton has led to decreases of up to 75 percent in the amount of insecticides applied to cotton fields (Carpenter 2011).

Renewable energy. The noise from wind turbines can cause headaches, sleep disturbances and hearing loss —and even damage the vestibular system humans use for balance (Pedersen 2011; Punch, James and Pabst 2010). But in another trade-off, renewable energy industries can reduce greenhouse gas emissions and boost employment (Panwar, Kaushik and Kothari 2011).

Environmental protection and growth often conflict, although "natural" environment-friendly technological change is a reality and "artificial" or policy-induced technological change can become a business opportunity.

Environmental pros versus economic cons

Biofuel technologies. As with the social and environmental impacts discussed in Chapter 4, biofuel technologies exemplify environmental and economic trade-offs. Biofuels can yield significant reductions in greenhouse gas emissions against fossil fuels—up to 90 percent relative to petrol (OECD 2008). But their production often exerts an upward pressure on food prices (FAO and others 2011).

Textiles and clothing. These two industries are huge exporters and employers in some developing economies but are linked to serious environmental issues, including the use of harmful chemicals; high consumption of water and energy; generation of large quantities of solid, liquid, and gaseous wastes; emissions; and animal exploitation. Huge volumes of water and energy are consumed not only in textile production but also in subsequent laundering by consumers (Sherburne 2009).

Steel. The steel industry supplies basic products to other industries and can be an important sector in the middle stage of development. Yet its production technologies have considerable adverse environmental impacts, such as massive quantities of wastewater and emissions from blast, open-hearth or basic oxygen furnaces. Direct-reduction furnaces and electric arc furnaces are less polluting but still produce substantial emissions of dust and carbon monoxide and are highly electricity intensive.

Reconciling the three dimensions

Carbon capture and storage.[1] The innovative technologies used in the iron and steel industry offer an example of reconciling environmental and economic impacts by aiming to cut carbon dioxide emissions generated from iron-making.

Renewable electricity. Environmental, economic and social goals can be reconciled. Cantore, Nussbaumer and Kammen (2015) investigated costs and employment generation of a higher penetration of renewables in electricity production in Africa. They analyse a reference scenario assuming implementation of current policies and two more aggressive climate change mitigation scenarios for energy efficiency and renewable energy penetration. The more environment-friendly scenario targets an atmospheric concentration

> **Technology can deliver win-win-win solutions, simultaneously balancing growth, environmental and social concerns**

of carbon dioxide of 450 parts per million and sees an economy–environment trade-off between reduced emissions and higher generation costs. But from a social point of view, scenarios with the highest energy efficiency and share of renewable energy assume the highest levels of job creation and the lowest generation cost per job created (as energy efficiency and renewable energy increase, employment grows more than generation costs).

Policy framework and taxonomy

Again, the purpose of this chapter is not to provide guidance to policy-makers on how to address the trade-offs detailed above but to discuss the challenges of managing them. In many circumstances, technology can deliver win-win-win solutions, simultaneously balancing growth, environmental and social concerns. When it cannot, there can be no substitute for a societal agreement on how to prioritize the three dimensions and decide which type of technology to foster, always mindful of possible negative impacts, at least in the short term.

Technology and industrial policies for innovation need to be complemented by macroeconomic, business-enabling, trade and investment, industry representation and infrastructure policies to support a country's competitiveness (Figure 6.1). These policies are prerequisites for integrating into GVCs, but should be complemented with a more radical macroeconomic approach and more strategic investment policies. Complementary policies are also needed to address trade-offs and ensure a balance between economic, environmental and social objectives.

Technology policies—early, middle, and late stages

Technology as defined by the Organisation for Economic Co-operation and Development (OECD) refers to the state of knowledge about ways of converting resources into outputs (OECD 2011). Taking the indirect measurable output of technology patents into account, we restrict the OECD definition to "the state of knowledge concerning ways of converting resources into outputs *that could be subject to patent protection*" [our emphasis].

Technology policy implemented by government facilitates the development of technological capabilities and infrastructure for private and public firms. It pursues three aims: expanding and accelerating the rate of technological change to raise productivity and hence social welfare; meeting social needs such as defence, education, health or the environment; and improving the process of technology generation, diffusion and utilization (Steinmueller 2010).

Technology policy varies by an economy's development stage. Weiss's (2015) proposed taxonomy corresponds to three development stages of industrialization (or World Bank income levels) along a continuum: early, middle and late stage. Each stage is characterized by some regularity in factors such as the complexity of market structures, technological content, productivity and degrees of specialization and skill level of the labour force. Within each stage there is a choice between general horizontal measures available to all firms and selective vertical ones applied selectively to priority targets, whether subsectors or specific firms. In addition, there can be market-based interventions and public inputs. The former affect prices and taxes and thus operate through pricing links. The latter reflect the provision of goods or services that firms themselves would not supply adequately, either because they cannot be marketed or because significant external benefits are involved.

The early stage includes countries at relatively low incomes experiencing the first structural change (the transfer of low-skilled workers from agriculture into more labour-intensive activities) as well as the use of simple technologies. In the transition from agriculture into manufacturing, the acquisition of technological know-how, usually through imitation, is essential. The technology policy therefore is likely to focus mainly on ensuring that firms import technology, which can mainly be achieved by two approaches: through foreign direct investment (FDI), where local producers use the technology and product design of the parent firm; or through licensing, where technology

> **Technology policy implemented by government facilitates the development of technological capabilities and infrastructure for private and public firms**

Figure 6.1
Policies targeting inclusive and sustainable industrial development

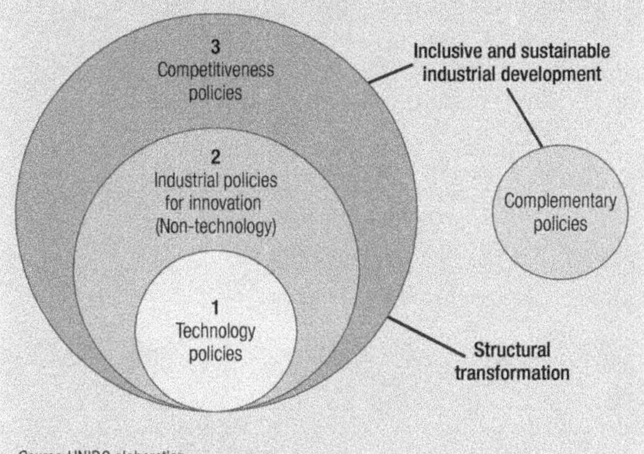

Source: UNIDO elaboration.

is embodied in equipment, and staff of the supplier give advice on its use. Governmental policy support is therefore essential to facilitate one or both of these mechanisms.

Once a firm has reached a minimum level of competence, policy-makers should focus more on encouraging and facilitating the mastery of technology and the ability to adapt, modify and improve foreign technologies. As market-based interventions will be premature at this stage, there might be more focus on public intervention through supporting investment agreements and technology licensing contracts with foreign firms, with public investment promotion agencies assisting in the initial search for partners or in the subsequent negotiations. Furthermore, some technology extension programs through a national technology institute may provide training and advisory services for the application of known technologies, particularly to subject matter experts.

The middle stage includes economies that have more sophisticated production capabilities and higher real wages. Industrialization at this stage is characterized by a relative decline in the roles of labour-intensive and resource-based manufacture and a shift into medium-tech activities or labour-intensive segments of high-tech goods. The promotion of higher-value-added medium- and high-tech products—as local adoption of foreign technology develops—is essential, as are public investment in research and development (R&D), research infrastructure and human capital, and incentives for private R&D.

Further policy coordination between different agencies of the government and a form of public–private council might be fostered to guarantee a strategic view of the overall direction of technology policy. Consequently, the main market-based measures to encourage private investment refer either to direct subsidy payments to innovating firms or to tax relief on R&D expenditure. Public good inputs include direct funding for research in universities, public–private research collaboration (both seeking to create or adapt knowledge) and technology extension programs to reduce the cost of searching for information on existing technologies.

The late development stage is characterized more by the support of development activities working with frontier technologies and education and science-based infrastructure. The support of restructuring for "sunset" activities (which are no longer competitive), as well as the pursuit of catch-up policies, are important at this stage. The same instruments can be used as in the first two stages, with specific measures to support innovation, including state funding for research as well as credits for higher-risk innovative investment. Some of the less wealthy European Union states have used foreign investment to introduce best practice technology and management. The current model in high-income economies is generally based on the premise that growth in economies at or close to the technology frontier must be driven by innovation, which is essential for long-term competitiveness (Weiss 2015).

Industrial policies for innovation

Industrial policies for innovation are a broad concept for combining technology and non-technological policies for innovation, for which we offer a taxonomy (Table 6.1) and a framework for different kinds of innovations at different stages of development (Table 6.2). For analytical purpose, this taxonomy

> **Effective policy-making calls for integration of different policy instruments into a coherent policy mix**

Table 6.1
Taxonomy for innovation policy (including technology and non-technological industrial policies)

Policy domain	Market-based	Public goods/direct provision
Technology market	R&D subsidies, grants	Technology transfer support, technology extension programme, public-private research consortia, public research institutes
Product market	Tax exemptions for innovation investments, attraction of foreign direct investment, R&D tax incentives, import tariffs, duty drawbacks, tax credits, investment/foreign direct investment incentives	Use of public procurement for innovation, protection of intellectual property rights, procurement policy, export market information/ trade fairs, linkage programmes, foreign direct investment country marketing, one-stop shops, investment promotion agencies
Labour market	Wage tax credits/ subsidies, training grants	Training institutes, skills council
Capital market	Subsided credit for innovative firms, directed credit, interest rate subsidies	Loan guarantees, skills council
Land market	Subsidized rental	Promotion of technology and production clusters, creation of technology parks, establishment of special economic zones, export processing zones, factory shells, infrastructure, legislative change, incubator programmes

Source: Adapted from Weiss (2015) and Warwick (2013).

Table 6.2
Innovation for different stages of development for developing and emerging countries

Country category	Objective of innovation	Type/source of innovation	Main agents involved
Early stage	Improve productivity and process technology	Incremental innovation based on adoption of foreign innovations and technologies; Innovation needs to respond to specific "local" conditions for outcomes	Universities and research institutes, private businesses, especially those with exposure to foreign markets
	Favour the generation of inclusive innovation to improve welfare and access to business opportunities	Incremental innovation based on combination of foreign technology and/or local, traditional knowledge	Nongovernmental organizations, GOs, small firms, public and private associations engaged in disseminating knowledge via networks
Middle stage	Build up innovation capacities to reach the world technological frontier	Incremental and radical innovation capacity to compete with leading world innovators	Private firms, universities and research institutes, public institutions
	Build-up niche competencies	Incremental innovations based on applying foreign innovations and technologies strategically to support industrial development	Public institutions to address coordination challenges, private sector initiative including foreign companies
Middle and late stage	Climb the value ladder in global value chains	Incremental and radical innovation capacity to differentiate contributions	Private sectors with support from public agents, intermediaries, diasporas, large firms
	Keep competitiveness in frontier industries	Innovation is identical to that in developed countries exposed to the global market	Private sector in interaction with public research institutions, universities and large firms

Source: Adapted from OECD (2012b).

distinguishes between five conceptually different market domains. This distinction should not be seen as an indication that policies belonging to different domains cannot complement (as opposed to substitute for) each other. Effective policy-making calls for integration of different policy instruments into a coherent policy mix solving the variety of market failures that can hinder innovation and technological change. Some examples of such integration follow in this section.

"The first step is to understand the type of innovation that has to be targeted

Innovation (see Box 1.1 for definitions) is a highly heterogeneous phenomenon whose characteristics depend on a host of various conditions at various levels (firm, industry, region, country or world). Possible determinants of innovation at the firm level include level of R&D expenditure, degree of sectoral innovation opportunities, ease of access to finance, availability of skilled workers, market conditions (such as degree of competition and demand conditions), regulation of intellectual property rights (IPR), degree of knowledge spillovers and so on.

One crucial element determining the emergence, development and expansion of innovation activities is government intervention. Governments in developed and developing countries increasingly make innovation a key issue, recognizing its potential to promote economic growth and address social and environmental challenges:

> "Since the late 1990s, countries started to design and implement national innovation strategies in order to boost the potential of their economies to produce a stream of commercially successful innovations. They invest in innovation with the hope that it drives economic growth and competition and as a consequence will lead to high and sustainable levels of economic and employment growth, increasingly depending on the strength of their innovation ecosystem" (Atkinson and Ezell 2012, p. 133f.)

The main argument for government support is that a market economy cannot generate by itself the optimal levels of investment in innovation because of the existence of market failures: partial appropriability due to spillovers and information asymmetries that lead to serious funding gaps. These market failures mean private firms under-invest in innovation activities, thus depriving the economy of one of the key levers of sustained growth.

To counter this, governments provide different forms of support to firms' investment in innovation, often through (sometimes overlapping) policy instruments (see Table 6.1). Table 6.2 lists the objectives, sources and agents of support by stage of development.

To identify the optimal intervention, the first step is to understand the type of innovation that has to be targeted because, for instance, product and process innovation differ in their impacts on firm or economy-wide performance. Objectives such as introducing new products or increasing the range of exported goods are more likely to require technological innovations than non-technological innovations. (Non-technological innovations refer to product, labour, capital and market at Table 6.1) Innovation policy traditionally tends to favour technological innovation, yet evidence suggests that success often also depends on accompanying non-technological innovation. Policy-making should therefore be broadened to take into account non-technological innovation.

Finally, policy-makers should consider that the same measure may affect the various types of innovation differently. For example, measures to increase demand for innovation (such as vendor-supplier cooperation) are more likely to favour generation of incremental rather than radical innovation, which often stems from large public-funded projects and supply-push policies (Nemet 2009; OECD 2009).

The barriers to innovation also differ by type and stage of innovation. For instance, cost factors can be relevant for all types of innovations, while market factors, such as uncertain demand for innovative goods or the weakness of property rights, may affect mainly product innovation, not process innovation. In contrast, weak engineering and technical skills are often associated with lack of process innovation, especially in developing countries (Hirsch-Kreinsen 2008). Proactive and comprehensive government policies are a prerequisite to establish an overall innovation policy framework as well as the need for interaction among the actors and government institutions involved, especially at local level, as innovation primarily takes place in local milieus with a concentration of knowledge, talents and entrepreneurs (World Bank 2010).

> **The instruments for labour market failures complement those related to failures in the technology market**

Middle-income countries catching up

Two case studies of middle-income countries catching up—Brazil moving into the aerospace industry and Malaysia transforming from agriculture to palm oil production—feature government interventions in five of the policy domains proposed in the taxonomy (see Table 6.1). As indicated above, the distinction between different policy instruments is deliberate and aimed at facilitating a fuller understanding of the policy decisions taken to achieve the two objectives. The complementarity among some of these policies instruments is highlighted when relevant.

Brazil—moving into aerospace industry

Brazil's experience was characterized by the need to develop the manufacturing capacities and technological competencies to join the exclusive ranks of countries that possess an aircraft assembly industry (Box 6.2). Brazil had to learn everything about aircraft production before finding an opportunity to specialize in design and system assembly. To avoid technological dependency, the country followed the unusual strategy of limiting the use of technology licensing to acquire the competencies required to locally assemble planes—state support was critical to facilitate acquisition of any missing technological and organizational capabilities. Brazil's Empresa Brasileira Aeronáutica S.A. (EMBRAER) is now the world's fourth largest aircraft manufacturer. Export-oriented catching-up strategies favoured the development of more flexible innovation systems in Brazil; strong education and research systems provided relevant assets to facilitate smooth adjustment of the emerging aerospace industry in the face of changes in global environment. The choice of segment specialization was also relevant, as Brazil centred on the commercialization of commuter aircraft. Embraer found a niche in small commercial and military aircraft.

From a technology market perspective (see Table 6.1), the rise of the local aerospace industry was made possible thanks to co-production and licensing arrangements with foreign partners, as such agreements helped to boost market penetration without technological dependence. For the product market, traditional strategies to develop an indigenous aerospace industry were used, including massive public funding and assembling aircraft under licensing and public procurement contracts (often related to the military), which served to boost demand for the aircraft. The Brazilian government's threat to impose or increase import duties was used to force foreign producers of commercial aircraft into collaboration agreements; foreign firms were compelled to provide the technical and organizational know-how to assemble the final product. These kind of agreements helped the country to become the sole source of some parts for some of the major global aircraft manufacturers (Goldstein 2002).

The instruments for labour market failures complement those related to failures in the technology market and relate to a balance between education policies targeting long-term core competencies and the ability to respond to short-term skills requirements. Catch-up required the creation of a strong knowledge base in natural sciences and engineering for excellence in aeronautics, material science, electronics, information

Box 6.2
Key elements to creating an aerospace system of innovation in Brazil

- Finding a market niche (in this case, commuter aircraft capable of serving airports with poor infrastructure).
- Channelling finance and design efforts to develop a new product for this niche.
- Establishing a company to ensure commercialization of innovations.
- Creating new linkages to provide capital (government launch support, government commissioning and acquisition of the bulk of new planes, and a corporate tax incentive scheme channelling private capital).
- Creating linkages to access technology (through exclusive co-production contracts, licensing agreements and support for R&D in aerospace and connected activities).

Source: Vértesy (2011).

❝ Public interventions have played a catalytic role promoting the growth and participation of private firms and research communities

technologies and even management (Vértesy 2011). Facilitating close collaboration and information exchange between universities, specialized training institutes and industry was necessary from the early stages of development of the industry, as it allowed the education system to incorporate and adapt early to the industry's needs. In the capital market, public funding was (and remains) necessary, though on its own it is not enough to overcome the technological and capital barriers in the aerospace industry. The government acted both as an investor and as a customer. A complex set of funding mechanisms have been made available for the aerospace industry through various Brazilian government agencies, including the establishment of a development bank, financing for R&D, short- and long-term financing and a preferential income tax system, among others. (Goldstein 2002; Pritchard 2012)

Additionally, in the land market, aerospace companies tend to concentrate in well-identified clusters where knowledge-producing institutions have a prominent presence, accelerating the knowledge spillovers and technological change (Martínez 2011). Accordingly, the government competed to attract firms in the supply chain, by establishing aerospace business parks, providing tax breaks and supporting R&D (Santiago 2015). See Box 6.3 for UNIDO recommendations for catching up in the palm oil industry.

Box 6.3
Recommendations for catching up in the palm oil industry

- Good governance: comprehensive, multi-level, multi-ministry, multi-year plans.
- Incorporate legislation, research, education policy, financing and investment, legal frameworks, taxation schemes, infrastructure, rural development and land distribution.
- Increased collaboration with funding bodies, non-governmental organizations and universities to foster research.
- Fiscal policies to support the development of the domestic market.
- Pro-active strategy to promote export orientation and as the industry consolidates, export diversification.
- Pro-active promotion of technological diversification.
- Environmental controls to ensure sustainability in the long term.

Source: UNIDO's elaboration based on Craven (2011) and Rasiah and Azmi (2006).

Malaysia—transforming from palm oil production

The second case study concerns a natural resource–driven middle-income country, Malaysia, that has positioned itself at the technological frontier as the world's dominant palm oil producer and exporter; this local industry is not only the leading innovator but also controls the industry's value chain (Rasiah and Azmi 2006).

Public interventions have played a catalytic role promoting the growth and participation of private firms and research communities. The interaction among these and other agents has sustained the long-term transformation of the industry from an agricultural activity into a buoyant processing segment. Malaysia has industrial master plans (IMPs) to promote industrialization, technological change and upgrading of palm oil activities within a broader framework of industrialization and the alleviation of poverty and inequality in the country (Rasiah and Azmi 2006).

In the technology market, the government controls the value chain from raw materials to final consumer goods and is the engine for new product development. The local industry has developed the capacity to become internationally competitive based on value-added product development and is an example of successful incremental innovation. Many IMPs include generous incentives to manufacturing R&D, including tax credits of 50 percent on qualifying R&D expenditures over a 10-year period as well as many expenses (Rasiah and Azmi 2006). A research fund for universities and research institutes was established.

Furthermore, in the product market, the penetration of new export markets and the capture of more value added in the product chain have been reinforced by promotional strategies, including the creation of a palm oil promotion council and the state-led IMPs. The first two plans identified palm oil as a priority

> **Governments need information on the implementation of their innovation strategies**

industry, setting goals for the development of the different segments of the value chain (Rasiah and Azmi 2006). Specific targets included the provision of institutional support to improve refining technology, stimulate palm oil R&D, and develop complementary domestic industries, including biofuels (Abdullah and others 2009). Later, the emphasis was on productivity growth and the search for input suppliers, because local processing capacity surpassed local input production. Incentives were granted to labour-intensive and agro-processing firms to enhance downstream activities with an impact on value added (Rasiah and Azmi 2006). The later IMPs were oriented to "new agriculture," which involved large-scale commercial farming, the wider application of modern technologies, production of high-quality and value-added products, biotechnology, and the participation of entrepreneurial farmers and a skilled workforce (Malaysia Ministry of Human Resources 2008). It also looked to curtail the rising costs of production and dependence on foreign labour in upstream activities through foreign investment in oleochemical-based products, bulking facilities and R&D and through centralized procurement of agricultural inputs such as fertilizers and pesticides to lower input costs (EPU 2010).

Complementing the technology and product market, in the labour market, the Malaysian government intervened by creating a human resource development fund (Rasiah and Azmi 2006). The fund upskills workers through financial incentives for employers, who can claim financial assistance for up to 100 percent of training costs. In addition, employees with no formal education but with ample practical experience and knowledge can have their experience certified based on their competency levels under a "recognition of prior learning" scheme. The Palm Oil Institute of Malaysia was established and has been the key public and privately coordinated institution for advanced training in the sector (Rasiah and Azmi 2006). The former policies were strengthened through policies in the capital market. Financial support was given through the tax system, research funding and other mechanisms depending on the stage of development of the industry. In the land market, there was some progress in better defining land ownership, including ample access to land by the government. The government capitalized on some earlier investments in infrastructure, the organization of large-scale production activities and public investments in irrigation (Santiago 2015). See Box 6.3 for UNIDO recommendations for catching up in the palm oil industry.

Because of its mandate, UNIDO has been over the years significantly engaged in supporting its Member States, particularly low-income countries, in fostering technological absorption and the transfer of green technologies towards low-, middle- and high-income countries. Further examples of the instruments deployed for the dissemination of innovative technologies are available in publications such as UNIDO (2011b, 2015b) and UNIDO and UNU-MERIT (2014).

Measuring, monitoring and evaluation: Technology, innovation and international cooperation

Governments seeking to foster innovation need surveys to obtain information on the implementation of their innovation strategies and to understand how these contribute to fostering the competitiveness of particular enterprises and to enhancing economic and social development more generally (World Bank 2010). Such measuring, monitoring and evaluation (MME) is indispensable to policy-making for all countries, both to learn from the past and more instrumentally to justify continuing those policies to a sometimes skeptical audience (Georghiou and Roessner 2000).[2]

Measuring technology

Because the technological content of products and processes is difficult to measure directly, Keller (2010) suggests three ways of measuring technology indirectly: inputs (measuring R&D data), output (measuring patents) and impact (measuring productivity gains). Patents traditionally have been the most

> **International innovation policy-making must be subject to measuring, monitoring and evaluation to keep international cooperation on the right track**

commonly used proxy for technological content, especially in low- and middle-income countries, than R&D, expenditures and productivity gains. Data on patents are normally recorded more substantively and for longer periods of time, but their limitations should not be understated, since a minority of patents account for most of the value of all patents and many innovations have never been patented (Keller 2010).

Further methods and corresponding indicators have been proposed (Georghiou and Roessner 2000):

- Evaluation of publicly supported research carried out in universities and public research organizations, including R&D spending, patents, licenses, contractor reports of inventions, citation counts, and changes in consumer/producer surpluses.
- Evaluation with a focus on linkages, including citations to scientific papers in patent applications.
- Evaluation of R&D support programs specifically aimed at industrial collaboration.
- Evaluation of diffusion and extension programs.

Measuring innovation policy

Innovation policy can be measured in various ways with different data, indicators and interpretations. Measuring different types of innovation (that is, process or product) calls for different types of indicators (Table 6.3 and Table 6.4). Further ways to improve measurement are discussed in Box 6.4.

Measuring international cooperation on innovation

Similarly, international innovation policy-making must be subject to MME to keep international cooperation on the right track. Ex-ante and ex-post evaluations are essential, as is the highest possible transparency (European Commission 2012b). Such cooperation can be measured by the indicators in Table 6.5.

International action is needed, particularly in improving the international evaluation possibilities: development of innovation metrics that can be linked to aggregate measures of economic performance,

Table 6.3
Input indicators and measurement

Government support to private R&D	Total amount of R&D subsidies and tax incentives
Foreign technology transfer	Number of licensing agreements, imports of capital goods and number of joint ventures with foreign firms
Quality and quantity of technical tertiary education	Share of public expenditure on tertiary education, share of engineers in total tertiary government disbursements for training programmes and public support to universities and research centres
Number of domestic research institutes	Number of domestic research institutes
Innovation human capital	Employed personnel with science and technology qualifications
Innovation infrastructure	Support activities for R&D, capital and investment in technology-based equipment
Labour market	Employment in creative sectors as a share of employment
Diffusion of new technologies	Business use of mobile internet, 3G (and higher) coverage, e-Intensity Index

Source: UNIDO elaboration based on Amsden (1989), Guadagno (2013), Holbrook and Godin (2011) and Hall and Jaffe (2012).

Table 6.4
Output indicators and measurement

Number of new products
Number of new processes introduced
Number of patents
Share of innovative firms in the economy
Number of firms entering new and high-tech sectors and share of high-tech export of the country

Source: UNIDO elaboration based on Amsden (1989), Guadagno (2013), Holbrook and Godin (2011) and Hall and Jaffe (2012).

> **The main constraint is poor competitiveness policies**

Box 6.4
Options to improve measuring, monitoring and evaluation of innovation in developing countries

Due to the long-term nature of innovation policies, the lack of a standard set of metrics, lack of funding and the relative weakness of statistical systems, measuring and evaluating innovation policy is often challenging for developing countries (Rood 2013; OECD and Eurostat 2005).

National statistics offices must be involved in innovation surveys. Personal interviews by adequately trained staff are recommended as they improve data quality, response rate and the quality of results (OECD and Eurostat 2005). The World Bank also emphasizes that qualitative and quantitative approaches are needed (World Bank 2010).

The OECD emphasizes that the measurement exercises of developing countries should focus on the innovation process rather than its outputs. Innovation activities and capabilities of firms and organizations are therefore at least as important as the resulting innovations. Additional emphasis should be placed on other aspects as human resources, information and communications technologies and their incorporation and use, as well as expenditure on innovation activities and organizational change. The OECD also recommends undertaking innovation surveys every three to four years in the developing country context, preferably at the same time as other major international innovation surveys are undertaken, to exploit benchmarking possibilities (OECD and Eurostat 2005).

Table 6.5
Measuring international cooperation

Number of international co-authorship of scientific publications
Number of researchers, workshops and conferences, share of researcher recruitment from abroad, mobility of PhDs
Number of scientific and technical cooperation agreements
Number of international joint programmes, open programmes
Number of provision of incoming fellowships and outgoing fellowships
Number of measures aimed to raise the attraction of domestic universities and research institutes
Shared research infrastructure
Number of bilateral agreements, foreign reviewers and panellists
Number of research offices located abroad; recruitment from abroad
Budget and share of total budget from non-national resources

Source: UNIDO elaboration based on CREST Working Group (2007), Edler (2010) and European Science Foundation (2012).

investment in a high quality and comprehensive statistical infrastructure to analyse innovation at firm level, promotion of metrics of innovation in the public sector and for public policy evaluation and promotion of measurement of innovation for social goals and social impacts of innovation (OECD 2010).

Competitiveness policies and global value chain integration

Extending the traditional competitiveness policy toolbox

The Aid for Trade report of the OECD and WTO (2013) stresses that the main constraint on the growth of developing countries and the expansion of their private sectors is poor competitiveness policies. This weakness limits the integration and upgrading of firms' processes and products in GVCs. Since the orthodox toolbox is not always a key to success, governments need to extend traditional competitiveness policies with the technology and innovation toolbox to achieve structural transformation (see Figure 6.1) and to enable easier integration with GVCs and enhance upgrading possibilities. A sound policy mix of technology, innovation and competitiveness policies is crucial, because competitiveness policies alone are not sufficient.

The OECD and WTO list developing countries' main constraints as poor infrastructure, lack of access

" Strategic investment policies need to work alongside the traditional competitiveness toolbox

to finance and weak standards compliance. Private sector firms also list five more constraints: customs procedures, lack of skilled labour, licensing requirements, transportation costs, and unsupportive regulatory and business environments. OECD and WTO recommend a traditional competitiveness package of promoting trade and trade facilitation programs; building human, institutional and infrastructural capacities; improving institutions; making policy and regulations more trade friendly; and creating a supportive business environment (particularly with stable macroeconomic policies) and an attractive investment climate.

Presenting alternatives to this conventional view is, for example, Rodrik (2014), who confirms that a poor business climate featuring the constraints mentioned above is a challenge for investors in developing countries (particularly in Africa), as all these hurdles raise the costs of doing business for an investor starting or expanding a manufacturing operation. Nevertheless, he argues that if such costs act as a tax on tradable industries, the problem could be solved with the exchange rate. For example, a real exchange rate depreciation of 20 percent is effectively a 20 percent subsidy on all tradable industries. This could be a way of undoing the costs imposed by the business environment in a relatively easy and quick manner and could foster a country's competitiveness both externally and in the domestic market. Crucial to this is the implementation of a monetary and fiscal policy framework to sustain a competitive and undervalued real exchange rate. Once an economy shows higher growth rates, Rodrik recommends fixing individual problems of the countries associated with their poor business climate and dealing with these problems over time, reducing the reliance on the real exchange rate. Although this approach could be an alternative to standard competitiveness tools, the appropriate monetary and fiscal policy framework is very difficult to achieve for developing countries.

Farole and Winkler (2014) recommend focusing on long-term investments and attracting investors in sectors with potential for higher rents, as they are more likely to engage in activities to support spillovers. Regional investors are advantageous in that they tend to be most integrated with domestic markets. Nevertheless, the authors point out that a more favourable investment climate for local firms depends on access to finance and imported inputs, enforcement of contracts, reliable regulatory standards, adequate power and other infrastructure support, adequate competition in domestic economies and support for building capacity and competitiveness. Strategic investment policies therefore need to work alongside the traditional competitiveness toolbox. The authors suggest further that domestic investors should not be unfairly disadvantaged by incentive regimes and should have the same access to infrastructure and services as foreign investors. They argue that investment incentives to promote action and measurement systems should be introduced along with capacity building to support spillovers.

Based on UNIDO's own decades-long experience in promoting the competitiveness of developing countries, the centrality of domestic technological capabilities cannot be overemphasized. The most significant spillover of FDI on productivity gains is often through the transfer and diffusion of technology—the simplest linkages involving, for instance, the contractual supply of goods and services, result in local firms' experiencing learning effects (UNIDO 2007).

In the course of its activities in the field of trade capacity building, UNIDO has catalogued an impressive range of policy instruments related to promoting competitiveness, meeting standards and ensuring access to markets. These are presented in detail in the Trade Capacity Building Resource Guide (UN 2015c).

Table 6.6 links the factors in the competitiveness of developing countries with GVC integration, based on OECD, WTO and World Bank's (2014) analyses. The table shows that not only the characteristics of firms but also public policies affect a country's competitiveness. The institutional context in which firms operate affects their capacity to meet the requirements

> **The centrality of domestic technological capabilities cannot be overemphasized**

Table 6.6
Factors affecting developing country competitiveness in global value chains

Factor		Description
Business environment and macroeconomic management	Macroeconomic stability and public governance	• Macroeconomic stability exists when key economic relationships are in balance. Exchange rate volatility affects costs paid for inputs and price netted for exports. • Governance includes traditions and institutions by which authority is exercised (rule of law, corruption, government effectiveness). Volatility can affect the timely delivery of goods and raise risk of inventory theft.
	Ease of opening a business and permitting/licensing	• The procedures, time and cost for a new business to start up and operate formally and the process to obtain construction permits, water and mineral extraction permits, and so on. • Comparatively lengthy procedures can deter foreign direct investment due to other potential country alternatives, while undermining the development of domestic firms.
	Standards and certification	• Codified public and private product and process requirements used to standardize supply across multiple suppliers. • Standards can drive upgrading by disseminating information on improving quality and productivity; yet, developing country firms often lack the capital and expertise to master multiple certification requirements.
Infrastructure and services	Transportation, information and communications technology (ICT), energy and water	• Impact of the cost and quality of these factors is compounded as fragmented production means inputs and intermediate goods must be transported between multiple locations. • ICT facilitates the transmission of codified design specifications between actors in product-based chains and is the main medium for participation in cross-border service exports. Energy drives cost competitiveness in capital-intensive assembly and processing segments of the chain.
Industry institutionalization	Industry maturity and coordination	• Experience of firms in participating in GVCs. Presence of key chain actors such as input and service providers and the establishment, influence and representativeness of an industry association to reduce transaction costs for meeting requirements.
	Public–private coordination	• Linkages and cooperation among private sector, government, educational institutions and other industry stakeholders. • Essential to rapidly identify and overcome challenges to GVC participation.
Trade and investment policy	Market access	• Extent of tariffs and import restrictions in potential target markets affect potential to engage with different end-markets. Tariff escalation is particularly damaging as GVC trade takes place in similar tariff lines.
	Import tariffs	• Tariffs charged on imported components, services and capital equipment required for the production or provision of exports become taxes on exports in GVCs.
	Export-import procedures	• Complexity of and time taken to complete customs procedures managing imports and exports of products and services reduces reliability and timeliness of delivery.
	Border transit times	• Time taken to move products and services through border crossings. Inefficient border crossings affect timeliness of product delivery to next stage of GVC or end-market.
	Industry-specific policies	• Investment and export promotion policies designed to support specific industry participation and upgrading in specific segments of different value chains.

Note: GVC is global value chain.
Source: Adapted from OECD, WTO and World Bank (2014).

of GVCs. Governments can enact policies that either promote or reduce firms' capacity to enhance their competitiveness, attract investment and insert themselves into GVCs. Good governance in general is important, because it signals to prospective investors and traders that a country is a good place to invest their capital.

There should be a distinction between policies that aim to foster competitiveness economy-wide (such as through building infrastructure) and sector-specific interventions that seek to alter production patterns directly. The second approach involves more precise and targeted policies, such as tariffs and other trade restrictions.

“Global value chains can be important assets to achieve structural transformation in the long run

The World Trade Report 2014 confirms the significance of a competitiveness pillar but also emphasizes the orthodox approach:

> “Countries that have a more favourable domestic business environment have been found to be more integrated into global value chains. Trade policy also plays a role in facilitating supply chain participation. Obstacles to GVC integration include infrastructure and customs barriers. Trade facilitation addresses these obstacles and helps to reduce trading times and improve the predictability of trade, which have been found to be significant determinants of trade in general and within value chains in particular.” (WTO 2014, p. 122)

GVCs can be important assets to achieve structural transformation in the long run with the right policy mix and have gained great importance in industrial policies. Unbundled production enables countries to specialize on certain tasks within an industry sector instead of on products or industries as a whole. This has been fostered by globalization, the information and communications technology revolution, changing lead firm strategies, and a policy shift in developing countries that made capacities available for export production (WTO 2014). Joining GVCs became an objective in itself for some industrial policy planning, not a means to achieve structural change; this planning was often characterized by standalone traditional competitiveness policies, which are not enough. The “wider” approach incorporating technology and innovation policy that is recommended above can help an economy go beyond simple engagement with GVCs and diminish the risk of getting trapped in low-value stages or sectors as well as improve knowledge spillovers.

Integrating and upgrading in global value chains

The literature shows a widely shared view of the importance of GVCs in industrial policy. Nevertheless, several proposed GVC policy frameworks tend to be too simple, with too strong a focus on the standard competitiveness policies. The exact nature of GVCs’ impact on the learning and innovative processes of firms in developing countries is still controversial and understudied.

The literature also shows a general expectation that firms coordinating the GVC—the lead firms—produce a positive impact on suppliers by transferring to them valuable knowledge with which to compete in global end-markets.[3] Indeed, for small firms in developing countries, participation in GVCs is probably one of the few opportunities to both obtain information about the type and quality of products demanded by consumers in global markets and to actually gain access to those markets.

Lead firms play a fundamental role in innovation processes, as do domestic technological capabilities at the firm-, industrial cluster, regional, and local innovation system levels (Pietrobelli and Rabellotti 2007, 2011). Upgrading in GVCs can be an important channel for innovation, industrialization and structural change, whether through moving into higher value-added functions within the same chain or by jumping into related but more technologically sophisticated value chains (Humphrey and Schmitz 2001, 2002). But such upgrading is not automatic and requires government policy (Nixson 2014). It is advisable for policy-makers and firms to examine and understand the governance structures and the control and upgrading possibilities of distinct value chains in different sectors, because these characteristics determine strongly the extent of firm entry and development as well as the chances for local firms to innovate and upgrade within the global industry (Gereffi and Fernandez-Stark 2005, 2011).

The heterogeneity among local suppliers in developing countries should be considered, as they differ widely in the capacity to absorb, master and change the knowledge and capabilities that lead firms in GVCs may transfer to them. Factors affecting their upgrading include changing global dynamics and power distribution along the chain and limitations within developing countries (Bamber and others 2014). The relative importance of these factors differs by product and industry.

> **Upgrading can be addressed in a social and environmental way**

To upgrade to higher-value segments, investment in innovation and knowledge-based capital, as well as economic competencies such as organizational know-how and branding, must be assured (OECD, WTO and World Bank 2014). The highest proportion of value creation is often found in some upstream activities (new concept development, R&D, manufacturing of key parts and components) and in certain downstream activities (marketing, branding and customer service). These activities require tacit, non-codified knowledge in original design, the creation and management of cutting-edge technology and complex systems, and management know-how.

Staritz (2012) highlights further ways of upgrading in a GVC: "Market entry and upgrading is achieved by assisting supplier firms and producers to access information and resources, develop linkages with other firms and producers, comply with lead firm requirements and standards, increase productivity, acquire new skills, competencies and capabilities, and take on new functions associated with higher value added activities" (p. 11). International organizations can help firms meet the requirements and standards of lead firms by facilitating benchmarking and international comparisons (UNIDO 2015f; UNIDO and ICSHP 2013). Pietrobelli and Rabellotti (2011) argue that adoption of international standards plays a crucial role in learning.

Upgrading can also be addressed in a social and environmental way, not just in economics terms, and requires more environment-friendly production as well as job quality improvements (Barrientos, Gereffi and Rossi 2011; De Marchi, Di Maria and Micelli 2013; Giuliani and Macchi 2014). Here, too, is a flip-side risk—economic *and* social and environmental downgrading, which should be addressed with technological and non-technological innovation and environmental, competitiveness and complementary policies (UNIDO 2010a).

It is important to know which type of policy support is needed in individual GVCs in terms of the framework within which learning takes place. Pietrobelli and Rabellotti (2011) identified three learning mechanisms:

- Mutual learning from face-to-face interactions typically appears where firms deal with complex and not easily codified transactions in GVCs and therefore need highly complementary skills. Maintaining and strengthening production and linkage capabilities to interact with lead firms in the GVC is a prerequisite for suppliers to interact with the lead firms. As learning efforts have costs and are time consuming, parties are more bound to continue their engagement.
- A diffused learning mechanism is characterized by direct, formal or informal training of the local workforce undertaken by the lead firm when it takes direct ownership of some operations in the chain. This can lead to fruitful local–lead firm collaboration.
- Knowledge transfer from GVC lead firms, confined to a narrow range of tasks, occurs mainly when suppliers lack specific skills. Lead companies provide them with support to avoid the risk of non-compliance. If the support is limited to a narrow range of tasks, local suppliers risk getting trapped in low-value segments, acquiring certain technologies and skills but remaining unable to develop other strategic capabilities (Pietrobelli and Rabellotti 2011).

GVC lead firms might require their local suppliers to adopt international standards if they are skilled and fully competent or the product is a standardized commodity. Lead firms can also demand that they adjust to specific technical and quality standards and take full responsibility for the process technology.

Lead firms do not always become directly involved in the learning process, but impose pressure on their suppliers for innovating and keeping abreast of technological advancements. Worse, lead firms do not always enrich local firms with knowledge transfer or support upgrading processes and could in fact prevent upgrading. Therefore it is crucial to understand the structure of value chains, the processes of structural change and the power asymmetries between firms

"Understanding the trade-offs is a precondition for developing the right complementary policies

Table 6.7
Learning mechanisms within a global value chain

Governance type	Complexity of transactions	Codification of transactions	Competence of suppliers	Learning mechanisms within GVCs
Market	Low	High	High	• Knowledge spillovers • Imitation
Modular	High	High	High	• Learning through pressure to accomplish international standards • Transfer of knowledge embodied in standards, codes, technical definitions
Relational	High	Low	High	• Mutual learning from face-to-face interactions
Captive	High	High	Low	• Learning via deliberate knowledge transfer from lead firms • Confined to a narrow range of tasks (such as simple assembly)
Hierarchy	High	Low	Low	• Imitation • Turnover of skilled managers and workers • Training by foreign leader/owner • Knowledge spillovers

Source: Adapted from Gereffi, Humphrey and Sturgeon (2005) and Pietrobelli and Rabellotti (2010).

that determine how entry barriers are created and how gains and risks are distributed (Staritz 2012). Table 6.7 summarizes some of the mechanisms by which firms learn in GVCs (Pietrobelli and Rabellotti 2010).

Complementary policies

Technological change can lead to enormous advantages for economy and society, but, as it is often the case in the manufacturing sector, can also result in three awkward trade-offs as mentioned at the start of this chapter: economic versus social, social versus environmental and environmental versus economic. Understanding these trade-offs is a precondition for developing the right complementary policies—that is, policies that can enhance the speed of innovation while at the same time providing targeted support to the people who stand to suffer the most from accelerated innovation.

Integrative policy approaches considering the full range of positive and negative consequences of innovation and promoting interactions among all actors and sectors of the economy are needed to achieve gains on all three dimensions. Another important action is to provide incentives to innovate and diffuse technologies.

So far governments have been unable to develop integrative approaches to the full range of consequences of technological change, partly because of knowledge and implementation gaps. These may stem from several factors, such as few incentives for a country to take unilateral action; concerns about consequences of policies for household income, employment, or the competitiveness of firms and sectors; or governments ill-equipped to engage with the cross-cutting and long-term nature of the three dimensions. But governments can take these actions towards a coherent policy framework (OECD 2001):

- Provide permanent incentives to innovate and diffuse technologies that support sustainable development objectives by expanding the use of market-based approaches in environmental policy. When market-based instruments are inappropriate, performance standards might be used in preference to measures that prescribe and support specific technologies.
- Support long-term basic research through funding and efforts to build capacity (for example, development of centres of excellence). Increase research on ecosystems, the value of the services they provide, the long-term impact of human activity on the environment, and the employment effects of new technologies.
- Address unintended environmental and social consequences of technology.

A one-size-fits-all approach to economic policy is unlikely to bring structural changes

- Support applied research activities when they are clearly in the public interest (such as protection of public health and the environment) and unlikely to be provided by the private sector:
  - Cooperate with the private sector to develop and diffuse new technologies.
  - Facilitate public-private and inter-firm collaboration with the innovators of cleaner technologies and practices.
  - Seek out opportunities for greater international collaboration on research, especially on issues critical for sustainable development.
  - Allow competition among technologies that can meet the same policy objective and equal access to learning opportunities (such as protected niche markets and similar schemes) for both foreign and domestic investors.

Successful complementary policy measures were adopted by Singapore (Box 6.5) and India (Box 6.6), countries at different income levels (see the taxonomy in Table 6.1).

Matus, Timmer and Appleby (2013) emphasize the challenge in overcoming an industrial paradigm that does not account for ecological and social externalities. They argue that manufacturing must be defined in the context of a fair and socially just economic system that meets the needs of all people while maintaining what they called the Earth's "life support" systems. In innovation, they focus on interconnected elements like invention, selection, production, initial adoption, widespread or sustained use, adaptation and redesign, and withdrawal. They propose an industrial symbiosis that involves establishing relationships between organizations to manage resources better. It attempts to move industry towards a circular or non-linear economy, where waste-to-input linkages and goal-oriented inter-firm relationships substitute for the current manufacturing structure. An industrial symbiosis network requires the sharing of natural resources (materials, energy, water, waste by-products), as well as social technologies (use of assets, logistics, contracting, expertise) within an industrial cluster or along a supply chain (or both). This can bring benefits in the form of better resource and energy efficient processes and sustainable livelihoods for workers (Chertow and Ehrenfeld 2012).[4]

Good practices in formulating policy

Having examined the aims of policy-making, we now turn to its good practice, which is of course crucial if policies are to achieve results. UNIDO's *Industrial Development Report 2013* highlighted that a one-size-fits-all approach to economic policy has not succeeded in past decades and is unlikely to bring structural changes in the future—especially because country heterogeneity demands a flexible approach to policy design. Evidence-based and realistic, country-tailored industrial policy conducted in a consensual way is key for policy effectiveness and requires the following preconditions.

Use—do not fight—the political system. A fact of political life is that no policy will be underwritten unless those in power agree to it (UNIDO 2013a): "It is not sufficient to just propose good economic policies; one must propose a way in which they will be endogenously chosen by those with the political power to do so" (Robinson 2009, p. 27).

Strengthen political leadership. This will set a national transformation agenda that aims, in low-income countries, to create and nurture productive activities or, in middle-income countries, to advance technologically (UNIDO 2013a). Political leadership at the top is crucial for raising the profile of industrial policies and for ensuring the required coordination, oversight and monitoring (Rodrik 2004).

Encourage public–private dialogue. Governments should join forces with their industrial private sectors to design interventions based on their combined expert knowledge and to ensure that decisions are supported by key stakeholders. Especially in developing countries with low public sector capacity, private sector input can make a large contribution to successful policies (Altenburg 2010). The new industrial policy needs

“ Evidence-based and realistic, country-tailored industrial policy is key

Box 6.5

Establishing Singapore as a biotech hub to catalyse transitions towards a sustainable, knowledge-based economy

Faced with the prospect of losing global market shares in technology-intensive products and processes to other economies, Singapore became an aggressive suitor for new niche industries. In 2000 the government defined life sciences—medicine, pharmaceuticals and biotechnology—to be one of its four industrial pillars (alongside electronics, engineering and chemicals), hoping to ensure that the country would emerge as a knowledge economy once the traditional industries were gone (Arnold 2003). The pressure of providing Singapore's ageing and gradually more affluent population with modern healthcare also provided impetus for this policy change.

In previous structural stages, policies were formulated to build a scientific infrastructure and attract foreign companies, particularly large multinational enterprises (MNEs) that could provide the country with knowledge-intensive subsidiaries to learn from. This FDI policy worked well: in 2010, around 75 percent of Singapore's manufacturing output was contributed by MNEs and "foreign capital provided nearly two-thirds of the equity capital of its manufacturing firms" (Siddiqui 2010).

In 2000 the Biomedical Sciences Initiative (BMSI) was launched to make Singapore a leading biotech hub, creating mainly export-based growth and development through higher-skilled and better-paying jobs (Pereira 2006). During the first 10 years of the BMSI, the government invested more than $3.14 billion in a mixed set of public goods inputs to promote innovation and market-based policy instruments providing incentives to the private sector (Chong, n.d.; Haseltine 2013).

The first BMSI phase (2000–2005) included policy instruments focused on technology, such as the expansion of two public biotech research institutes; in the land market, construction of an R&D biotech hub, Biopolis, close to the National University of Singapore, the Singapore Science Parks and the National University Hospital; in the product market by offering world-class intellectual property protection; and in the labour market, where the A*STAR Graduate Academy supported students with scholarships to pursue PhDs in biomedical sciences locally and overseas, provided they work in Singapore for up to eight years after graduating (Van Epps 2006).

Within five years of the program's start, Singapore was among the largest recipients of FDI directed at biotechnology in Asia (Pereira 2006). Today it is a global leader in the industry with more than 30 of the world's leading biotech companies—many of them at Biopolis, which has become Singapore's main biocluster and the fastest growing biocluster in Asia (EDB 2014). According to a global biotech innovation report by Scientific American, Singapore has become the third most desirable location for biotech firms, not just because of the percentage of biotech patents granted but also due to the strength of such patents internationally (Grindrod 2015).

Policies have made the biotech industry economically viable: between 2000 and 2006, biomedical industrial output expanded almost 400 percent from $4.4 billion to $16.6 billion, employed over 5,000 private and public researchers and created some 10,000 high-value manufacturing jobs (A*STAR 2013; Watch 2010).

The second BMSI phase (2006–2010) continued developing infrastructure and capabilities for the biotech industry but focused more on clinical research capabilities. A*STAR established another five institutes and Biopolis was expanded. A one-stop coordinating office, the Singapore Biomedical Sciences Industry Partnership Office, was set up to facilitate collaboration between industry and research institutes, and A*STAR established a technology transfer program to identify technologies with potential for commercialization and help them quickly to the market by facilitating licensing deals and spin-offs with industry (A*STAR 2013).

Between 2006 and 2011, the number of biotechnology, pharmaceutical and medical technology patent applications increased dramatically (OECD 2015). In the course of a decade, Singapore moved from a technological specialization one degree below the world average to the third highest position after Denmark and Belgium (OECD 2013). Manufacturing output increased to $17 billion in 2010 (8.4 percent of gross domestic product) and value added to $7.9 billion (3.9 percent of gross domestic product; EDB 2012).

The expanding biotech sector has had positive implications on other sectors in the economy: manufacturing sees a continued demand for more than 300,000 jobs in related industries such as chemicals, electronics and engineering (BFTA 2014) to support the pharmaceutical and biotechnology industry.

> **There is no single, correct recipe**

Box 6.6

Market-based industrial energy efficiency policy in India

India has adopted an emissions permits market for economic and environmental reasons: its energy demand is expected to climb from 700 million tons of oil equivalent (MTOE) in 2010 to 1,500 MTOE by 2030. Dependence on energy imports is also expected to increase from 30 percent to over 50 percent in the same time period (McKinsey & Company 2014). From an environmental point of view India, as one of the big players in the climate change arena, should act decisively to decrease its emissions.

In 2012 India introduced the Perform Achieve Trade (PAT) scheme. Over 2012–2015 PAT has targeted 478 companies and eight sectors: thermal power plants, iron and steel, cement, fertilizer, aluminum, textile, pulp and paper, and chlor-alkali. The companies account for 164 MTOE of energy consumption (54 percent of India's total). Each company is subject to an energy consumption reduction target calculated using production and annual energy consumption data over five years (2006–2010) submitted by designated consumers, who receive tradable, certified energy savings credits if they hit efficiency gains beyond their targets.

In parallel, the central government promotes complementary actions aimed at correcting market failures in the adoption of environment-friendly technologies. In particular it subsidizes manufacturing firms by a linked capital subsidy scheme to mitigate the high up-front costs of new technologies that aim to replace old machines with less energy-intensive technologies. These subsidies reduce the payback period for energy saving investments, which would otherwise average about three years.

Even before the end of the first PAT pilot period, results were encouraging. In 2013, an industry survey by the Confederation of Indian Industries of 55 Indian companies producing 10 percent of India's total emissions and 45 percent of its industrial emissions revealed that 93 percent of these companies were implementing emission-reduction initiatives. These companies together were cutting 2.5 million metric tons of carbon dioxide-equivalent annually (Clough 2015).

to be based on such dialogue and not on top-down planning. UNIDO's *Industrial Development Report 2013* emphasized that fruitful dialogue requires private sector organizations that allow companies to articulate their needs and give input to the policy process. But in some low-income countries and industries, producers may be overshadowed by more powerful trading companies in Chambers of Commerce. Consequently, governments might consider strengthening manufacturers' representation in national chambers, because retailers and wholesalers have policy concerns that are different from those of manufacturers (on, for instance, tariffs and import quotas). Rent seeking—an inherent risk in public–private dialogue—can be mitigated by governance mechanisms that avoid focusing policy outcomes too narrowly on meeting the interests of certain groups (te Velde and Leftwich 2010). Consensual decision-making is needed (UNIDO 2013a).

Boost industrial policy management capabilities. UNIDO's *Industrial Development Report 2013* emphasizes the importance of fostering these capabilities, ideally through learning by doing and especially in developing countries with these capacity gaps. Each step of the policy cycle requires strong analytical and implementation capacities. Special emphasis (again, often in low- and lower-middle-income countries) is needed in defining priorities and building a broad consensus, establishing clear rules for market-based competition conducted transparently and efficiently, delivering services effectively and avoiding political capture (Altenburg 2011). Management capabilities should also be built, ideally through learning by doing, as industrial policy may fail anywhere and should thus be seen as experimental (UNIDO 2013a). Developing countries should therefore initiate their own national industrial policy experimenting and learning processes, including MME, to identify high-impact solutions.

There is no single, correct recipe, nor can all governments privatize, stabilize and liberalize in similar ways. Industrial policy-makers, especially in developing countries, might gradually shift their attention from investigating and imitating international best

> **International cooperation and network building in technology and innovation are complicated**

practices to identifying and reproducing national success stories. This approach underlines the need for sound MME, especially in the context of serious budget constraints, because it is essential to know whether a policy intervention was effective and whether the benefits outweighed the associated public costs (UNIDO 2013a).

International cooperation on technology and innovation policies

Technology and innovation policy-making is usually conducted at national level. As suggested by the European Union's subsidiarity principle,[5] interventions should be undertaken at the level where results are expected. International collaboration is needed with the trans-border and global problems driving collaboration in this area (Edler 2010). Globalized technology and innovation in general, the rise of emerging and developing economies as champions of globalization, and the growing roles of individuals, small firms and open modes are further reasons for international technology and innovation policy cooperation (Raunio, Kautonen and Saarinen 2013).

The OECD (2012c) emphasizes the need for effective international cooperation and the sharing of burdens and benefits to protect the global commons and the world's public goods, including technology and innovation. This implies not only pooling financial resources and sharing a large research infrastructure, but also improving the global knowledge base.

International policy collaboration is evident in three mutually reinforcing mega-trends:

- Increasing importance of international collaboration on technology and innovation substantiated by indicators like co-publications, co-inventions and joint research projects (Edler 2010 and Wagner 2006).
- Intensifying political ambitions to foster and use international technology and innovation to solve a range of global challenges (Wagner 2006).
- Enhanced international and transnational policy initiatives to shape collaboration. The policy and funding landscape has become more flexible, enabling more international collaboration in the scientific field (Edler 2010).

OECD (2012c) refers to three levels of interaction for technology and innovation: coordination for harmonious and efficient relations; collaboration to work jointly towards common goals; and integration, which implies a shift of competencies and autonomy. Further motivations for international collaboration among researchers, institutes, firms and public research institutes include (Edler 2010; OECD 2012b):

- Bundling financial and intellectual resources to exploit economies of scale and scope.
- Involving researchers from different countries and regions to shorten the innovation cycle.
- Gaining access to transnational knowledge networks.
- Accessing scientific talent and high-skilled workers, and additional research markets.
- Accessing or sharing the cost of major facilities.
- Achieving critical mass through cost sharing or combining datasets.
- Preparing the ground for innovative activities and markets abroad, linking to complementary skills and resources (such as in GVC integration by firms) or adjusting to local requirements.
- Benefiting from cost advantages elsewhere (mainly standardized activities of R&D, but increasingly also in specific scientific areas in which low-cost countries are highly competitive—such as nanotechnology in China and software development in India).
- Re-transferring knowledge to be used at the home location, and incorporating spillovers of international knowledge into national innovation systems.

This list shows that international cooperation and network building in technology and innovation are far more complicated than at the national level, largely reflecting higher transaction costs (for search, bargaining, enforcement, reporting and evaluating), greater risks of failure, and a broader range of actors. Moreover, international collaboration in technology and innovation occurs mostly among actors with

> **"Mainstreaming national and international planning and policy-making is central to identifying synergies, gaps and possible trade-offs**

equivalent capacities and seeks to avoid duplication, which implies a possible exclusion of actors with lower research capabilities from priority setting and collaboration, though such collaboration would be fundamental to achieve an inclusive innovation process globally.

Another challenge for policy-makers is that technology and innovation, as global public goods, are often characterized by market failures and long potential payback times for R&D expenditures. Similarly, the benefits of investing in technology and innovation as a global good are hard to trace back to the originators in the home country and do not provide clear metrics to donors and electorates (OECD 2012b). The flipside is that international cooperation on technology and innovation policy can enhance the visibility of domestic research efforts, and their greater recognition can lead to more government support.

International cooperation on technology and innovation policy involves a complex interrelation of elements with a broad range of actors, interests and resources to be aligned with good governance practice (Figure 6.2).

Figure 6.2
International cooperation on technology and innovation policy

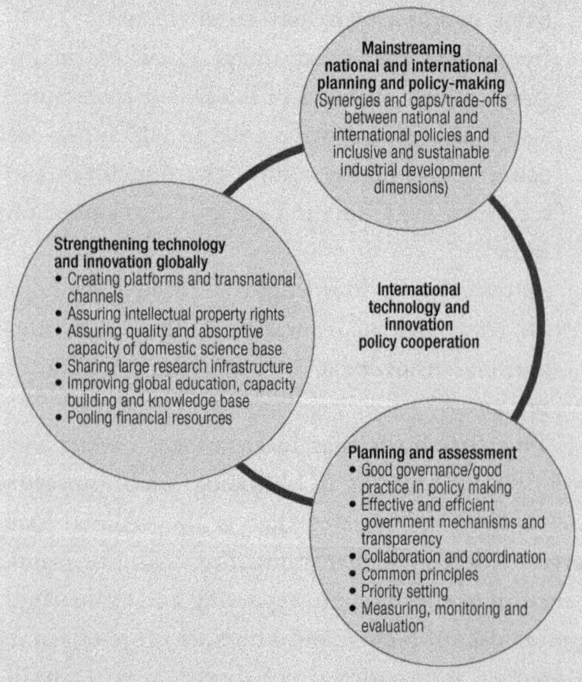

Source: Adapted from OECD (2012c).

The first key component in the figure—mainstreaming national and international planning and policy-making—is central to identifying synergies, gaps and possible trade-offs, not only between the national and international levels but also between the economic, social and environmental dimensions.

Creating platforms and transnational channels is another key factor, this time for strengthening technology and innovation globally, the second main component. The guarantee of IPRs of the cooperation partners on technology and innovation is also fundamental, as protected knowledge creates incentives to attract investment in R&D and helps to create further opportunities for technology and innovation. Even so, technology and innovation processes are becoming more open and sometimes IPRs are considered a barrier to international collaboration on technology and innovation, due to their exclusive character for other partners using IPRs-protected knowledge and exclusive rights. But once IPRs are licensed, that knowledge can be used simultaneously in multiple places.

OECD (2012c) suggests patent offices as a potential source for frameworks for international collaboration and to make information on IPRs accessible. But reducing transaction costs for IPRs licensing practices is fundamental to enhance technology transfer and the most effective transfer requires following three key factors: IPRs information, relevant know-how and guidance for implementing the technology. Policy-makers need to bring together a wide range of players in the public and private sectors to set up collaborative IPRs mechanisms for creating efficiencies in the IPRs exchange and for removing IPRs barriers (Box 6.7). Examination of governance of IPRs transactions and patent pools with respect to competition and anti-trust laws could be another supportive measure (OECD 2012b).

Auriol, Biancini and Paillacar (2012) highlight that global welfare and innovation are higher under the full protection regime if the developing country

“ Policy-makers need to bring together a wide range of players in the public and private sectors to set up collaborative intellectual property rights mechanisms

Box 6.7

Intellectual property rights for developing countries

The characteristics of the IPRs regimes are controversial for industrialized and industrializing countries. On the one hand, strong protection of IPRs offers firms the incentives to invest in innovation. On the other, IPRs protection represents a societal cost and can be considered a barrier due to their excluding character to other parties (Mazzoleni and Nelson 1998; OECD 2012b).

Intellectual property protection can incentivize investment in R&D and help to create opportunities for innovation in developing countries (OECD 2012b). The World Intellectual Property Organization (WIPO) stresses—as further reasons to protect intellectual property in developing countries—the importance of giving statutory expression to the moral and economic rights of creators and their creations, promoting creativity and the dissemination and application of results and encouraging fair trading (WIPO 2004). Nevertheless, the awareness of the usefulness of the patent system for technological development purposes and the existence of an adequate industrial property system providing patent information services are essential elements. There is an urgent need to coordinate this system and its patent information services with other branches of the government administration related to aspects of technology transfer and technological development.

Investigating incentives for developing countries to protect IPRs, Auriol, Biancini and Paillacar (2012) find that free-riding on rich countries' technology reduces the investment cost in R&D but yields a potential indirect cost, because a firm that violates IPRs cannot legally export to a country that enforces them. IPRs act as a barrier to entry to the advanced economy markets. Moreover, free riders cannot prevent others from copying their innovations. Their analysis, distinguishing between large and small developing countries, suggests that small ones should respect IPRs if they want to export and access advanced economies markets, while large emerging countries, such as China and India, will be more reluctant to do so as their huge domestic markets develop.

does not innovate. It is higher under a partial regime if both countries have access to similar R&D technology and the developing country market is large enough.

The quality and absorptive capacity of domestic science bases of countries has to be assured, as do sharing of large-scale research infrastructure, improving global education and building capacity and the knowledge base. Information sharing, technology transfer, and cooperation as well as collaboration are critical to enhance technology and innovation on a global level (OECD 2012b). The pooling of financial resources is another important but tricky prerequisite. Here, the characteristics of the collaboration and the cultural economic and political context matter. To tackle the financial challenge, the private sector needs to be engaged since the costs of technology and innovation are too high for the public sector alone. Because the private sector needs returns on investment in R&D, the demand, access to markets and the legal and institutional frameworks, including IPRs, have to be analysed and assured on a national level (OECD 2012b).

The third central component is planning and assessing international technology and innovation policy cooperation. This process should always consider and apply good governance and good practice. Effective and efficient government mechanisms as well as transparency are other preconditions. To collaborate on and coordinate technology and innovation internationally, internal domestic coordination has to be guaranteed. National and international coordination are interdependent and co-evolutionary but without a clear dominance. Therefore internal coordination, clarity and strategic actor capacity are needed (Edler 2010). The European Union develops multiannual roadmaps for easier international cooperation —mainstreaming research and innovation across other policies with strong international dimensions such as trade, the environment and education—and coordinates them with general external country strategies and internal policies (European Commission 2012a).

OECD (2012c) suggests keeping open channels of communication among experts, decision-makers and end-users to put knowledge into practice. This requires further involvement of participants in discussions, considering that mutual understanding is often

Technology and innovation now have a central role in the Sustainable Development Goals

hindered by jargon, language, training, expectations and experience. Common principles support international cooperation on technology and innovation policy through codes of conduct to engage the different stakeholders confidently. Several initiatives promote common principles in international technology and innovation policy coordination including: the UN Technology Facilitation Mechanism and Capacity Building, the Global Research Council (a voluntary forum set up to share best practice and establish common principles in international cooperation for the European Union), the Carnegie Group (working on establishing common principles for building large-scale research infrastructure) and contributions from UNESCO (United Nations Educational, Scientific and Cultural Organization) and OECD on access to research data from public funding.

The 2030 agenda for sustainable growth

The 2030 agenda for sustainable growth recognizes the crucial role of international cooperation in the technology and innovation field for promoting sustainable development in all countries. While technology and innovation were not defined as a priority and were clearly under-represented in the Millennium Development Goals, they now have a central role in the Sustainable Development Goals (UN 2015b):

- Technology development in order to enhance agricultural productive capacity in developing goals (Goal 2).
- Increasing support for scholarships available to developing countries for enrolment in higher education, including technology (Goal 4).
- Enhancing the use of enabling technology to promote the empowerment of women (Goal 5).
- Enhancing international cooperation and expanding infrastructure to facilitate clean energy research and technology (Goal 7).
- Increasing access to information and communications technology (Goal 9).
- Increasing scientific knowledge and research capacity and transfer of marine technology (Goal 14).
- Enhancing north-south, south-south, triangular regional and international cooperation on and access to science, technology and innovation (STI) as well as the promotion of environmentally sound technologies to developing countries and a fully operationalizing technology bank and STI capacity-building mechanism (Goal 17).

Priority setting is another fundamental component of planning and assessment; this refers to the negotiation process for diverse actors and stakeholders to agree on common objectives and actions. Different phases include forming priorities and implementing them, creating a framework, identifying and selecting priorities, and funding implementation of closer priorities. Priority setting is very challenging, because broad legitimacy and support among many actors are needed, so that priorities are often defined very broadly, leading to even more difficult translation into specific actions (OECD 2012b).

Technology, science and capacity building are major pillars for implementing the post-2015 agenda and the Rio+20 follow-up processes. The UN Secretary-General proposes to establish an online global platform, building on and complementing existing initiatives, to enhance international cooperation and coordination in this field and to promote networking, information sharing, knowledge transfer and technical assistance to advance the scaling up of clean technology initiatives (UN 2014a). The involvement of a wide range of actors is crucial. To implement the proposal, an informal inter-agency working group has been formed around several main work streams:

- Mapping existing technology facilitation initiatives including support for policy formulation and strengthening of technological capabilities and innovation systems.
- Identifying areas of synergy and areas of possible cooperation within the UN system on technology-related work.
- Developing options for a possible online knowledge hub and information-sharing platform, cooperating with relevant stakeholders on building or strengthening technology-focused partnerships

> **Technology, science and capacity building are major pillars for implementing the 2030 development agenda**

and collaborations, including on STI capacity building.

The working group was proposed to strengthen the coherence and synergies among science and technology–related capacity-building initiatives within the UN system, with a view to eliminating duplicated effort. A UN Interagency Task Team on Science, Technology and Innovation for the Sustainable Development Goals will be established, drawing on existing resources.

A coordinated approach to capacity building targets missing human capacities, restructuring current and establishing new institutions and supporting institutional capacity development. A Secretariat is likely to coordinate all existing efforts and organize new approaches. The Secretariat of the UN-wide Capacity Building in Technology for Development Initiative will play a catalytic role with national, subregional and international actors by delivering high-quality support to technology and innovation capacity development.

All these initiatives support the strengthening of a sustainable and coordinated development of technology and innovation for developed and developing countries—and are crucial in achieving the Sustainable Development Goals.

Notes

1. Carbon capture and storage is a process technology that allows capturing carbon dioxide emissions at their source of production, typically from the use of fossil fuels in electricity generation and industrial processes, transporting it via ships and pipelines, and ultimately storing it underground, thereby preventing such emissions from being released into the atmosphere (IEA 2015a).
2. "The choice of what is significant to measure, how and when to measure it, and how to interpret the results are dependent upon the underlying model of innovation that the evaluator is using, implicitly or explicitly. Much of the data collected by evaluators are themselves conditioned by the positioning of the evaluation and those who execute it. In consequence, it is usually necessary to understand the setting of the evaluation and the discourse in which the results are located before the choice of approach can be fully appreciated" (Georghiou and Roessner 2000, p. 658).
3. See for example Gereffi (1999), Gereffi, Humphrey and Sturgeon (2005), Sturgeon, Van Biesebroeck and Gereffi (2008).
4. Industrial symbiosis programmes are being piloted in Australia, Brazil, China, Mexico, the Republic of Korea and the United Kingdom.
5. In Europe the principle of subsidiarity aims to set the level of intervention most relevant in the areas of competences shared between the European Union and Member States. This may concern action at EU, national or local levels. In all cases, the European Union may only intervene if it is able to act more effectively than Member States (European Union 2015).

Chapter 7

Industrial trends: manufacturing valued added, exports, employment and energy and resource efficiency

As Part A of the *Industrial Development Report 2016* has made clear, manufacturing remains the main driving force of economic growth, largely attributable to its higher productivity and scope for innovation.

Over the past few decades, the majority of global manufacturing has steadily shifted from West to East and from North to South. Since the beginning of the century, rapid growth in manufacturing value added (MVA) has been a major source of poverty reduction in many developing and emerging industrial economies (DEIEs) through employment creation and income generation. Statistics suggest that they still have considerable capacity for manufacturing growth and technological progress in the coming decades.

Trends in manufacturing valued added

World MVA climbed strongly until the 2008–2009 global financial crisis (Figure 7.1). Industrialized countries contributed the highest share of world MVA, but along with DEIEs experienced a slowdown in MVA growth. Since 2010, MVA has recovered in both groups but has so far not reached the pre-crisis level within the industrialized country group.

Figure 7.1
World manufacturing value added, by country group and worldwide, 1990–2014

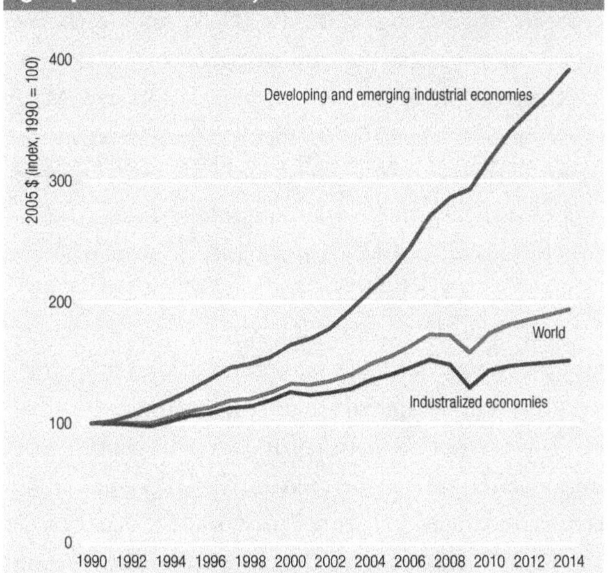

Note: Development level classification based on Annex B1, Table B1.2.
Source: UNIDO elaboration based on Manufacturing Value Added Database (UNIDO 2015e).

Global MVA reached an all-time high of $9,228 billion (at 2005 constant prices) in 2014. The MVA share of industrialized countries in gross domestic product (GDP) fell from 15.4 percent in 1990 to 14.5 percent in 2014; in DEIEs it increased from 16.2 percent in 1990 to 20.5 percent in 2014. The share of MVA in world GDP increased from 15.6 percent to 16.2 percent over the period.

Since 1990, MVA growth has remained consistently higher in DEIEs. By 2014, the MVA of DEIEs had expanded almost four times compared with 1990. Higher MVA growth has led to sustained economic growth in many developing countries.

Industrialized countries accounted for 64.1 percent of world MVA in 2014, but their share is falling due to lower manufacturing growth rates compared with major DEIEs.

China contributed 18.4 percent of world MVA in 2014 and is the second biggest manufacturing producer after the United States (Figure 7.2), which by itself contributes more than all non-China emerging countries combined. China's impact on the MVA growth rate in DEIEs is significant. The country's share in DEIEs' MVA increased from 15.8 percent in 1990 to 51.3 percent in 2014. China's manufacturing industry has become the largest sector in the economy and accounts for one-third of GDP.

Annual year-on-year MVA growth began to decline in 2008 and reached a low point in 2009 (Figure 7.3). The global crisis strongly affected industrialized economies, with MVA declining by about 13 percent. MVA growth of DEIEs slowed but stabilized at around a 5 percent a year.

Rising MVA growth rates in 2010 suggested the beginning of a significant recovery in manufacturing industry, only to be quashed in 2011 by a return to sluggish growth. Industrialized countries were affected by a

> **Global manufacturing value added reached an all-time high of $9,228 billion in 2014**

Figure 7.2
The 15 largest countries by manufacturing value added, 2014

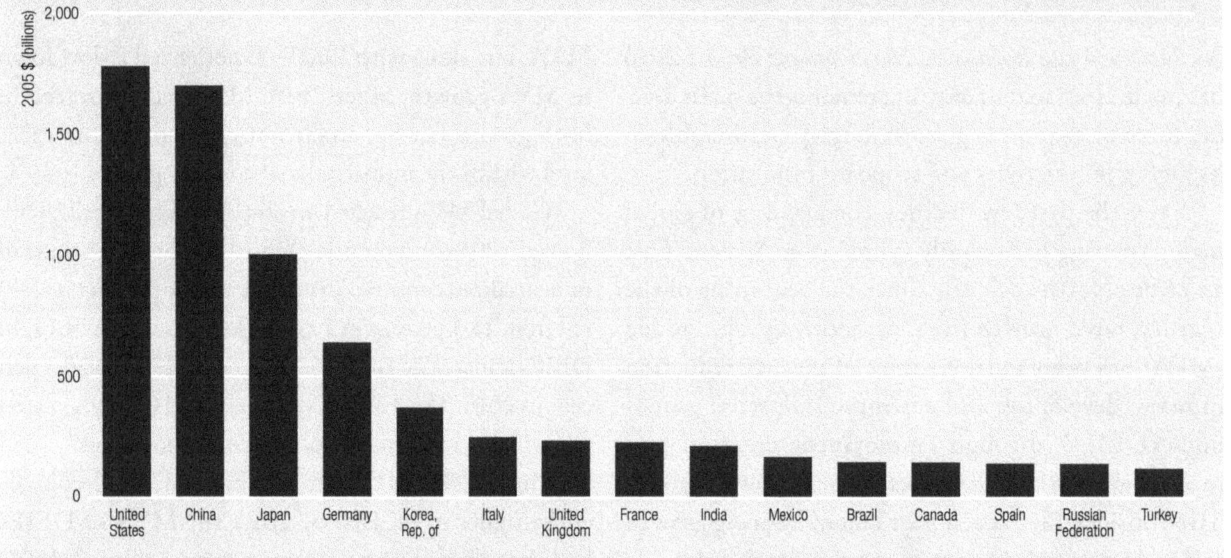

Source: UNIDO elaboration based on Manufacturing Value Added Database (UNIDO 2015e).

Figure 7.3
Annual growth of manufacturing value added, by country group, 2007–2014

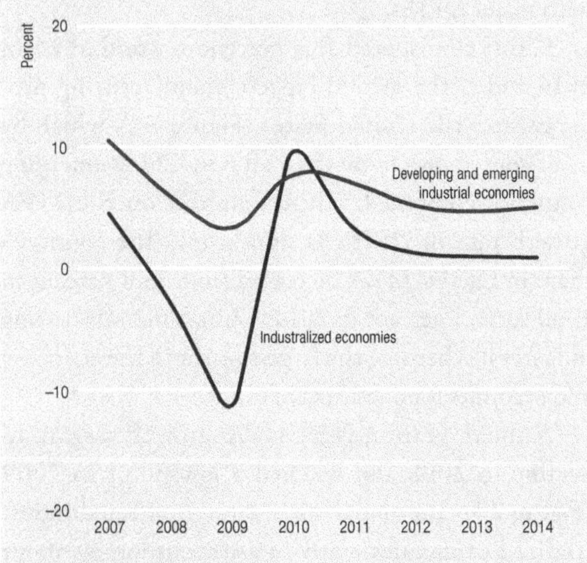

Note: Development level classification based on Annex B1, Table B1.2.
Source: UNIDO elaboration based on Manufacturing Value Added Database (UNIDO 2015e).

re-emerging recession, especially in Europe, and DEIEs felt the effects of continuing instability in world financial markets and of falling commodity prices. Global MVA grew by 2.3 percent in 2014, thanks mainly to the higher MVA growth in DEIEs.

Manufacturing value added per capita, 1990–2014

Instead of looking at total MVA, an alternative approach to comparing different groups is to normalize the relationships based on population size. MVA per capita—a measure of income generated by the manufacturing sector per person—provides a comprehensive insight into an economy's industrialization potential. The MVA per capita of industrialized economies is far higher than that of DEIEs (Figure 7.4). China's MVA per capita, which is higher than its group's average, still lags far behind that of industrialized countries despite its huge recent gains.

The further potential of industrialization in economic catch-up can be seen in the MVA per capita trends illustrated in Figure 7.4. World MVA per capita was $1,277 in 2014, with the group of industrial economies enjoying an MVA per capita of $4,773, against the $553 of DEIEs. Despite the trebling of MVA per capita in DEIEs since 1990, these countries still

> **The time lags for developing countries to reduce disparities and structurally transform their economies have been getting shorter**

Figure 7.4
Trends in manufacturing per capita, by country group and worldwide, 1990–2014

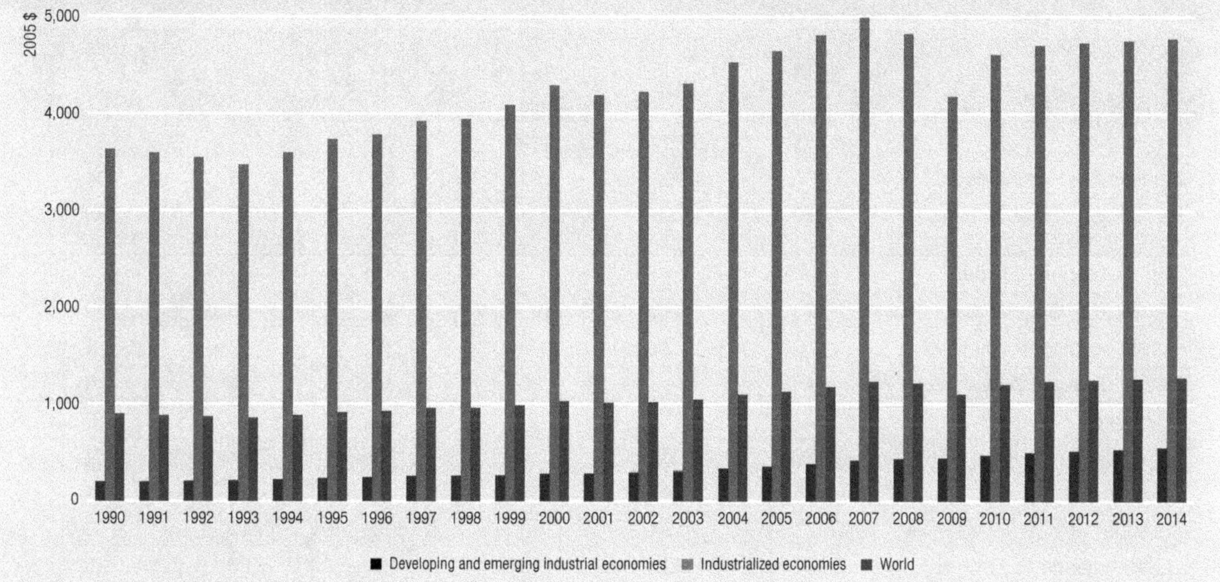

Note: Development level classification based on Annex B1, Table B1.2.
Source: UNIDO elaboration based on Manufacturing Value Added Database (UNIDO 2015e).

remain at about half of the world MVA per capita average and less than one-eighth of the MVA per capita of industrialized economies. While closing this gap will be no easy feat, the experience of recently industrializing countries suggests that the time lags for developing countries to reduce disparities and structurally transform their economies have been getting shorter, especially for late-industrializers that have pursued active industrial strategies and policies.

On industrial productivity (measured by MVA per capita), China remains far behind the leading industrialized countries, even if it is the fastest-growing and best industrial performer among the DEIEs. Still, China has improved in all pillars of industrial competitiveness (see next chapter).

Manufacturing in developing and emerging industrial economies, 1990–2014

Manufacturing not only produces essential commodities for domestic consumption and export, but also provides new technologies for other sectors of the economy. Higher MVA growth has led to sustained economic growth in many developing countries. Long-term stable MVA growth allows countries to employ a much larger workforce in manufacturing activities, contributing to a rise in income. Growth in manufacturing also helps progressively diffuse new technologies to other sectors of the economy, such as agriculture, transport and services, driving economic growth.

Manufacturing remains a key driving force of overall economic growth of DEIEs. From 1990 to 2014, global MVA doubled from $4,753 billion to $9,228 billion at 2005 constant prices (Table 7.1). MVA growth has stayed consistently higher than GDP growth (that is, aggregate economic output) in DEIEs since 1992. By 2014, the MVA of DEIEs had increased 2.4 times from 2000 at 2005 constant prices, while their GDP doubled; industrialized countries saw their MVA increase overall by only 51.3 percent (Figure 7.5).

The average annual growth of global MVA reveals a stark change of pattern around the turn of the century. Over both subperiods 1990–2000 and 2000–2014, it averaged 2.8 percent. In DEIEs it surged from 5.1 percent in 1990–2000 to 6.4 percent in

> **"Manufacturing remains a key driving force of overall economic growth of developing and emerging industrial economies**

Table 7.1
Manufacturing value added in developing and emerging industrial economies by development group and region, 1990, 2000 and 2014

	Manufacturing value added (billions, constant $ 2005)			Percentage of manufacturing value added		
	1990	2000	2014	1990	2000	2014
World	4,753	6,295	9,228	100	100	100
Industrialized countries	3,907	4,902	5,914	82	78	64
Developing and emerging industrial economies	846	1,393	3,314	18	22	36
By development group						
Emerging industrial countries	708	1,222	2,994	84	88	90
Least developed countries	20	22	54	2	2	2
Other developing countries	118	148	266	14	11	8
By region						
Africa	79	92	144	9	7	4
Asia and Pacific	315	746	2,362	37	54	71
Europe	151	164	300	18	12	9
Latin America	301	391	508	36	28	15

Note: Regional and development level classification based on Annex B1, Tables B1.1 and B1.2.
Source: UNIDO elaboration based on Manufacturing Value Added Database (UNIDO 2015e).

Figure 7.5
Economic and industrial growth trends by country group, 1990–2014

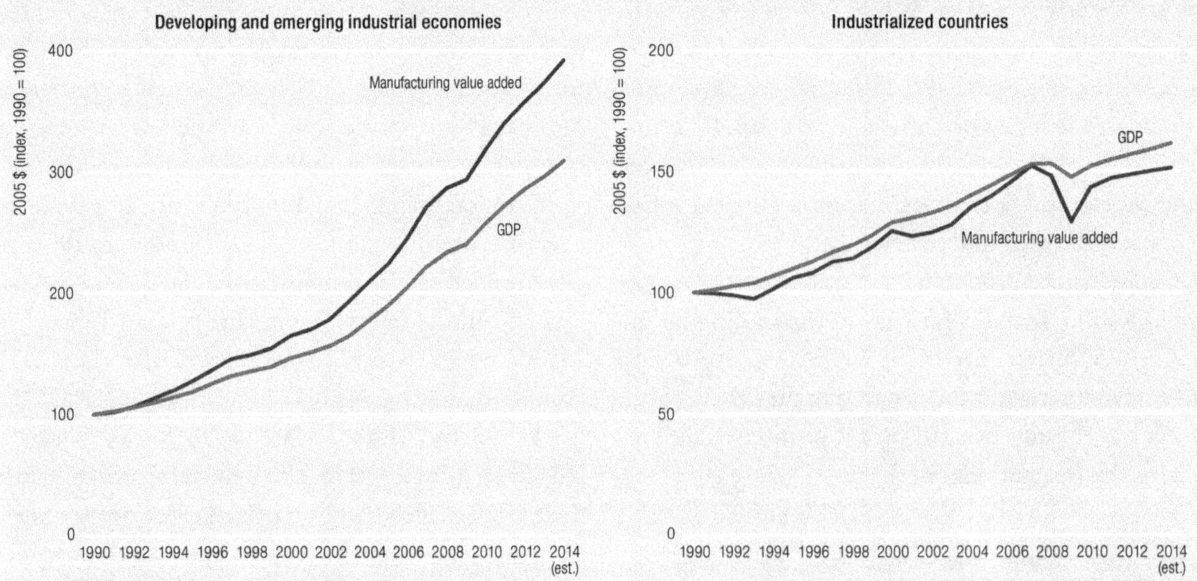

Note: GDP is gross domestic product. Development level classification based on Annex B1, Table B1.2.
Source: UNIDO elaboration based on Manufacturing Value Added Database (UNIDO 2015e).

“The dominant global manufacturing industries were food and beverages, chemicals and chemical products and machinery and equipment

2000–2014, while in industrialized countries it fell from 2.3 percent to 1.3 percent.

China sustained exceptionally high annual MVA growth throughout both subperiods, averaging 12.8 and 10.0 percent over 1990–2000 and 2000–2014. Some other DEIEs—Belarus (–0.2 percent over 1990–2000; 8.1 percent over 2000–2014), Bulgaria (–8.9 percent; 4.1 percent), Chile (1.3 percent; 4.9 percent), Peru (3.5 percent; 5.2 percent) and Romania (–2.8 percent; 3.6 percent)—saw a sharp acceleration in MVA growth after 2000.

DEIEs as a whole improved their share in total MVA but performance varied widely. Among the top five, China's share in world MVA increased by 6.5 times over 1990–2014. China's manufacturing industry has become the largest sector in the country and accounted in 2012 for one-third of GDP and more than 18 percent of global MVA, second only to the United States. Although China—and India—improved their DEIE group share, the other three of the five faltered, particularly Brazil (Figure 7.6).

Sectoral composition of world manufacturing value added

In 2013, the dominant global manufacturing industries were food and beverages (12.0 percent), chemicals and chemical products (11.7 percent) and machinery and equipment (8.5 percent; Table 7.2).

Between 2000 and 2013, eight manufacturing sectors registered an increase in their MVA share worldwide, from a combined 39.7 percent to 46.6 percent. Significant increases were recorded in manufacture of radio, television and communication equipment; basic metals, chemicals and chemical products; and motor vehicles, trailers and semi-trailers. The increase in basic metals was driven mainly by the rapid growth in MVA of DEIEs, as well as country investments in infrastructure.

Decreases were observed in 14 manufacturing sectors from a combined 60.3 percent in 2000 to 53.4 in 2013, including traditional industries such as textiles, wearing apparel, fur and wood products and paper and printing. Manufacture of fabricated metal products and machinery and equipment each also witnessed share decreases in manufacturing structure.

The manufacturing structure of industrialized countries tends to be made up less of primary and secondary goods, with more emphasis on medium- and high-tech sectors (medium and high tech classification is based on UNIDO (2010b), while traditional industries contribute a higher share in DEIEs. Yet the share of medium- and high-tech industries in

Figure 7.6
Manufacturing value added share of the five largest countries in developing and emerging industrial economies' group total, 1990, 2000 and 2014

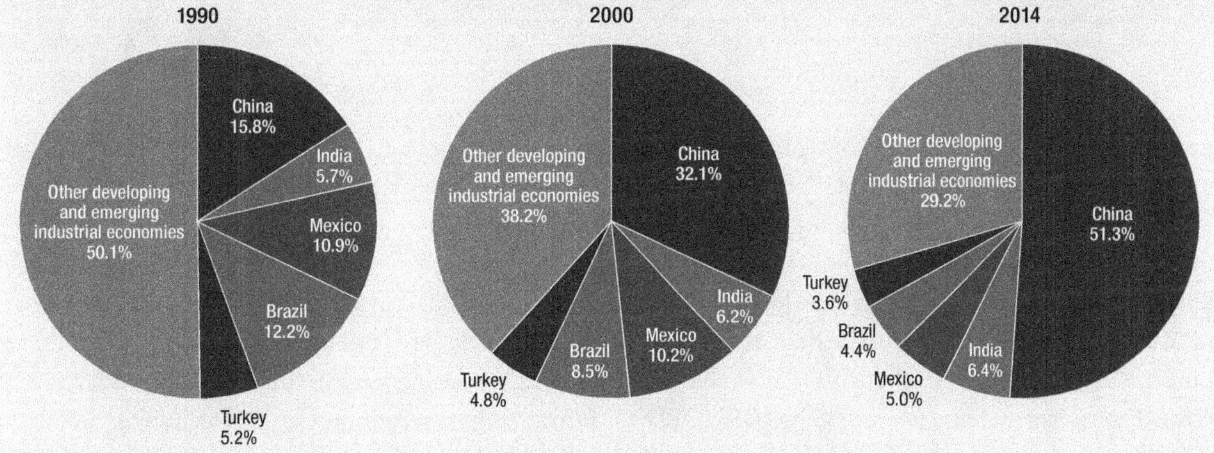

Source: UNIDO elaboration based on Manufacturing Value Added Database (UNIDO 2015e).

> **“The share of medium- and high-tech industries in total manufacturing increased sharply in developing and emerging industrial economies**

Table 7.2

Share of manufacturing value added, by industry group within country groups and worldwide, 2000, 2005 and 2013 (percent)

	Industrialized countries			Developing and emerging industrial economies			World		
ISIC description	2000	2005	2013	2000	2005	2013	2000	2005	2013
Food and beverages	11.0	11.4	11.2	16.6	14.9	13.3	12.1	12.2	12.0
Tobacco products	1.1	0.9	0.7	3.5	2.9	2.5	1.6	1.4	1.4
Textiles	2.3	1.8	1.1	5.9	5.3	4.5	3.0	2.7	2.5
Wearing apparel, fur	1.7	1.0	0.7	3.8	3.4	2.9	2.1	1.6	1.6
Leather, leather products and footwear	0.7	0.4	0.3	2.0	1.4	1.3	0.9	0.7	0.7
Wood products (excluding furniture)	2.0	2.0	1.5	1.7	1.4	1.4	1.9	1.8	1.5
Paper and paper products	3.0	2.8	2.4	3.0	2.9	2.7	3.0	2.9	2.5
Printing and publishing	4.6	4.2	3.4	2.1	1.8	1.4	4.1	3.6	2.6
Coke, refined petroleum products, nuclear fuel	3.1	3.5	3.2	6.2	5.2	3.7	3.7	3.9	3.4
Chemicals and chemical products	11.0	12.0	12.0	11.2	11.2	11.1	11.0	11.8	11.7
Rubber and plastic products	4.7	4.6	4.4	3.4	3.5	3.3	4.5	4.3	3.9
Non-metallic mineral products	4.0	3.8	3.1	5.5	5.4	5.8	4.3	4.2	4.2
Basic metals	5.0	5.0	4.5	7.8	10.0	11.2	5.5	6.2	7.1
Fabricated metal products	8.0	7.5	7.1	3.9	4.0	4.6	7.2	6.6	6.1
Machinery and equipment n.e.c.	9.7	9.7	9.2	4.8	5.9	7.4	8.8	8.7	8.5
Office, accounting and computing machinery	1.5	1.4	2.0	1.2	1.5	1.5	1.5	1.5	1.8
Electrical machinery and apparatus	4.0	3.8	3.9	2.8	3.3	4.5	3.8	3.6	4.1
Radio, television and communication equipment	5.2	6.2	9.7	3.7	4.6	5.1	4.9	5.8	7.9
Medical, precision and optical instruments	3.5	3.9	4.7	0.7	0.9	1.1	3.0	3.1	3.3
Motor vehicles, trailers, semi-trailers	7.7	8.3	8.3	6.2	6.4	6.7	7.4	7.8	7.7
Other transport equipment	2.9	3.0	3.8	1.5	1.7	1.9	2.6	2.7	3.1
Furniture; manufacturing n.e.c.	3.3	3.0	2.7	2.3	2.3	2.1	3.1	2.8	2.5
Total	100.0	100.0	100.0	100.0	100.0	100.0	100.0	100.0	100.0

Note: ISIC is International Standard Industrial Classification; n.e.c. is not elsewhere classified. Development level classification based on Annex B1, Table B1.2.
Source: UNIDO elaboration based on INDSTAT2 (UNIDO 2015g).

total manufacturing increased sharply in DEIEs over 2000–2013 (see Table 7.3 and Table 7.4).

In 2000, industrialized economies were the main producers in all manufacturing industries by a wide margin (Table 7.3). Growth in developing economies, coupled with the effects of the financial crisis on industrialized countries, resulted in a remarkable shift in the distribution of manufacturing. By 2013, DEIEs were the main producers of most basic consumer goods and manufacturers of basic metals.

Over the period, the MVA share of DEIEs by industry group has doubled in several cases, and in a few trebled, indicating a relocation of some industries from the industrialized to the developing world and the strengthening of manufacturing within DEIEs. Yet industrialized countries are leading in technologically complex and high value-added activities, such as production of information and communications technology goods.

In 2013, China ranked either first or second worldwide in 19 out of 22 manufacturing industries.

> **Growth in developing economies resulted in a remarkable shift in the distribution of manufacturing; developing and emerging industrial economies were the main producers of most basic consumer goods and manufacturers of basic metals**

Table 7.3
Share of manufacturing value added, by industry group within country groups, 2000, 2005 and 2013 (percent)

	Industrialized countries			Developing and emerging industrial economies		
ISIC description	2000	2005	2013	2000	2005	2013
Food and beverages	74	69.5	56.6	26	30.5	43.4
Tobacco products	57.5	48.3	30.7	42.5	51.7	69.3
Textiles	62.8	49.7	28.2	37.2	50.3	71.8
Wearing apparel, fur	65	46.7	28.1	35	53.3	71.9
Leather, leather products and footwear	58.4	46.6	26.4	41.6	53.4	73.6
Wood products (excluding furniture)	83.6	80.6	62.2	16.4	19.4	37.8
Paper and paper products	81.2	74.4	58.3	18.8	25.6	41.7
Printing and publishing	90.3	87.5	79.2	9.7	12.5	20.8
Coke, refined petroleum products, nuclear fuel	68.1	66.4	57.2	31.9	33.6	42.8
Chemicals and chemical products	80.7	76	62.7	19.3	24	37.3
Rubber and plastic products	85.3	79.8	67.4	14.7	20.2	32.6
Non-metallic mineral products	75.4	67.9	45.4	24.6	32.1	54.6
Basic metals	73.2	59.5	38.4	26.8	40.5	61.6
Fabricated metal products	89.8	84.9	70.3	10.2	15.1	29.7
Machinery and equipment n.e.c.	89.6	83.1	65.7	10.4	16.9	34.3
Office, accounting and computing machinery	84.6	73.5	66.7	15.4	26.5	33.3
Electrical machinery and apparatus	86	77	57.2	14	23	42.8
Radio, television and communication equipment	85.6	80.1	74.7	14.4	19.9	25.3
Medical, precision and optical instruments	95.3	92.5	87.2	4.7	7.5	12.8
Motor vehicles, trailers, semi-trailers	84.1	79.4	65.8	15.9	20.6	34.2
Other transport equipment	89.3	83.9	75.1	10.7	16.1	24.9
Furniture; manufacturing n.e.c.	86	79.8	66.8	14	20.2	33.2

Note: ISIC is International Standard Industrial Classification; n.e.c. is not elsewhere classified. Development level classification based on Annex B1, Table B1.2.
Source: UNIDO elaboration based on INDSTAT2 (UNIDO 2015g).

Table 7.4
Medium- and high-tech industry group

ISIC description	ISIC code Rev. 3
Chemicals and chemical products	24
Machinery and equipment	29
Office, accounting and computing machinery	30
Electrical machinery and apparatus	31
Radio, television and communication equipment and apparatus	32
Medical, precision and optical instruments, matches and clocks	33
Motor vehicles, trailers and semi-trailers	34
Other transport equipment*	35

* The subsector 351 is subtracted.
Note: ISIC is International Standard Industrial Classification. For medium and high tech classification details please see UNIDO (2010b).
Source: UNIDO (2010b).

> **China ranked either first or second worldwide in 19 out of 22 manufacturing industries**

Among DEIEs, China is the undisputed leader in all 22. When China is excluded, India, Mexico and Brazil are the leaders in most industrial sectors.

Share of medium- and high-tech industries in manufacturing value added

The share of medium- and high-tech industries in a country's MVA captures the technological complexity of manufacturing. Development generally entails a structural transition from resource-based and low-tech activities to medium- and high-tech activities. The more complex the production structures of a country becomes, the higher the opportunities for learning and technological innovation at sectoral and intersectoral levels. In addition, medium- and high-tech manufacturers add greater value than low-tech manufacturers and contribute considerably to MVA.

The change in manufacturing structure is best evident in shift of the industry towards more technologically complex products. The share of resource-based industry in global manufacturing fell from 33 percent in 1990 to 28 percent in 2013, while the share of medium- and high-tech products increased from 44.6 percent to 46.7 percent (Table 7.5). In 2013, medium- and high-tech manufacturing accounted for more than half the MVA in industrialized countries, increasing from 46.3 percent in 1990 to 50.1 percent in 2013. Over 1990–2013, medium- and high-tech manufacturing was the dominant manufacturing sector in industrialized economies. In DEIEs, the share of medium- and high-tech manufacturing picked up marginally, from 33.6 percent to 34.0 percent.

Low-tech manufacturing maintained its share in MVA manufacturing to 25.3 percent over 1990–2013 worldwide. China's manufacturing structure remained stable, with an almost 42 percent share of medium- and high-tech manufacturing in total MVA. On technological intensity, China positioned itself more prominently in the medium- and high-tech segment between 2000 and 2013 than the other BRICS countries (Brazil, Russian Federation, India and South Africa). However, China has yet to close the gap with industrialized leaders. The Russian Federation and South Africa are facing difficulties in expanding their medium- and high-tech sectors.

The rate at which industrialized economies changed their structure and, in particular, shifted from resource-based manufacturing over 1990–2013 is impressive: MVA from medium- and high-tech manufacturing grew at 4.1 percent over 1990–2000 (see Table 7.5). The average was slower (3.6 percent) in 2001–2013 due to the financial crisis. Comparing these results with the average growth rates of overall MVA, and the other two technology sectors, medium- and high-tech manufacturing stands out as the main engine generating MVA.

The results for DEIEs are different. Over 2001–2013, the average growth rate of MVA generated by medium- and high-tech manufacturing was higher than in 1990–2000. In contrast, average MVA growth of resource-based and low-tech activities over 2001–2013 were significantly lower than in 1990–2000.

On a regional perspective, Asia and Pacific shows the highest share of medium- and high-tech manufacturing in MVA. The region already had a fairly high share of medium- and high-tech products starting in 1990, increasing to 49.3 percent in 2013. China is a heavy contributor to this high medium- and high-tech share, thanks to the scale of the country, a rapid proliferation of policy initiatives—high-tech parks, government funding and so on—its rapidly growing international connections and the return of skilled personnel from abroad. The relocation of low-tech manufacturing from industrialized economies, mainly to China, as a more cost-efficient economy, explains the expansion of the MVA share of those products in the region.

Europe exhibits a similar pattern of reducing its resource-based industry and maintaining its medium- and high-tech shares. Its medium- and high-tech manufacturing share rose from 40.3 percent in 1990 to 46.6 percent in 2013. The region has the second highest share of medium- and high-tech manufacturing in its MVA.

Structural change in Latin America and Africa has not been evident, while the region's figures do not illustrate a clear pattern among industrial segments over the period studied. Africa exhibits the lowest regional medium- and high-tech share among

“The more complex the production structures of a country becomes, the higher the opportunities for learning and technological innovation

Table 7.5
Technology composition of manufacturing value added, by development group, region and income, 1990, 2000, 2010 and 2013

	1990			2000			2010			2013		
	Resource based	Low tech	Medium and high tech	Resource based	Low tech	Medium and high tech	Resource based	Low tech	Medium and high tech	Resource based	Low tech	Medium and high tech
World	33.0	22.5	44.6	32.0	24.0	44.1	28.1	26.0	46.0	28.0	25.3	46.7
Industrialized countries	32.0	21.8	46.3	29.0	22.1	49.0	25.7	23.3	51.1	25.7	24.2	50.1
Developing and emerging industrial economies	39.5	26.9	33.6	45.5	30.3	24.2	31.6	30.0	38.6	36.6	29.4	34.0
By development group												
Emerging industrial countries	38.3	26.5	35.3	36.2	27.4	36.5	31.0	29.6	39.5	36.0	28.4	35.5
Least developed countries	71.5	12.1	16.4	71.1	14.2	14.7	67.5	24.1	8.4	66.8	24.3	8.9
Other developing countries	47.8	31.9	20.4	55.6	33.4	11.0	39.8	31.4	29.0	35.7	34.4	29.8
By region												
Africa	42.2	36.1	21.7	44.0	33.0	23.1	45.0	31.6	23.5	44.7	32.9	22.4
South Africa	36.6	35.5	27.8	38.0	35.7	26.3	38.6	34.9	26.6	41.3	34.3	24.4
Asia and Pacific	29.8	24.2	46.1	34.5	26.2	39.5	26.0	27.0	47.0	25.1	25.6	49.3
China	36.1	26.1	37.8	31.4	25.7	42.9	28.6	30.0	41.4	28.6	30.0	41.4
India	31.4	28.6	40.0	31.4	27.5	41.2	22.7	38.1	39.2	21.2	38.0	40.8
Europe	34.8	25.0	40.3	31.0	25.1	44.0	28.7	25.0	46.3	27.9	25.5	46.6
Poland	35.9	30.4	33.7	43.2	27.5	29.3	32.7	28.2	39.1	34.8	28.1	37.1
Turkey	35.5	38.1	26.3	41.1	30.9	28.0	40.2	27.1	32.7	40.2	27.1	32.7
Latin America	34.3	24.8	40.9	37.8	24.3	38.0	36.2	29.3	34.6	37.9	29.1	33.0
Mexico	31.1	26.8	42.1	36.6	20.8	42.6	33.4	29.9	36.9	33.2	29.9	37.0
By income												
High income	42.8	27.9	29.3	34.3	39.7	26.0	33.7	31.6	34.7	33.0	29.3	37.6
Upper middle income	37.8	27.2	35.0	35.7	27.5	36.8	30.9	29.8	39.3	37.0	29.4	33.6
Lower middle income	46.7	25.7	27.6	56.2	32.0	11.8	35.4	31.6	33.0	34.9	29.7	35.4
Low income	70.6	12.1	17.4	69.5	14.8	15.8	63.2	26.1	10.7	64.6	26.0	9.4

Note: Tech classification based on Table 7.4. Regional, development level and income classification based on Annex B1 and Annex B5, Table B5.1.
Source: UNIDO elaboration based on INDSTAT2 (UNIDO 2015g).

the other regions, and manufacturing there is highly dependent on resource-based products.

Regional manufacturing value added trends

Starting from 2004, the Asia and Pacific region has become the biggest manufacturing region in the world, driven mainly by China (Figures 7.7 and 7.8).

Europe’s share in global MVA tumbled considerably from 1990 to 2014. From 1990 to 2000, Europe was the biggest manufacturing region in the world, but lost that position to the Asia and Pacific region. MVA in Europe is highly concentrated in the euro zone. European Union countries accounted for 23.0 and 83.6 percent of world and European MVA, respectively, in 2014. In Europe, the pace of decline has at least slowed, but still leaves the region’s MVA below it its peak in 2007.

Africa’s MVA remains very low and accounts for only 1.6 percent of world MVA. Industrial and manufacturing development has not improved over time. The share of African MVA in GDP fell from 12.8 percent in 1990 to a low of 10.1 percent in 2014.

> The Asia and Pacific region has become the biggest manufacturing region in the world

Figure 7.7
Regional shares in total world manufacturing value added, 1990, 2000, 2010 and 2014

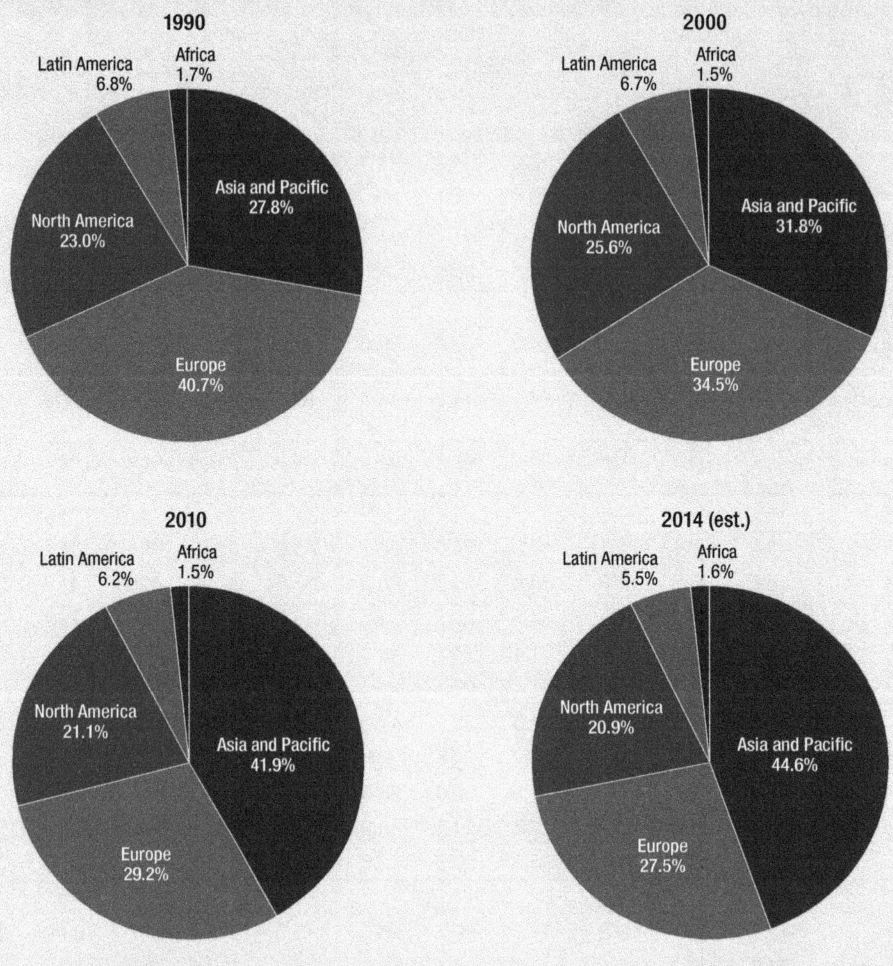

Note: Regional classification based on Annex B1, Table B1.1.
Source: UNIDO elaboration based on Manufacturing Value Added Database (UNIDO 2015e).

The Latin American region's MVA increased by a moderate annual average of 1.4 percent throughout 2000–2014. However, the region experienced a slight decline in world MVA share, losing to the Asia and Pacific region. The strongest manufacturers in the region according to MVA growth are Peru, Chile and Argentina.

Trends in manufactured exports

An increasingly export-oriented manufacturing sector, accompanied by a rising share of manufacturing in total exports, is part of a normal pattern of structural change in the growth process of DEIEs. Following this pattern, developing countries today have increased their presence in the export of manufactured goods. A growing number of developing economies are now benefiting from integration into the global economy through manufactured export growth and diversification. In most of these instances, export promotion has played a critical role in long-run growth by supporting a virtuous circle of investment, innovation and poverty reduction.

It is widely recognized that the benefits of the export of manufactured goods are greater than those from that

> **World export growth rates averaged 7.5 percent over 2005–2013**

Figure 7.8

Regional trends in manufacturing value added, 1990–2014

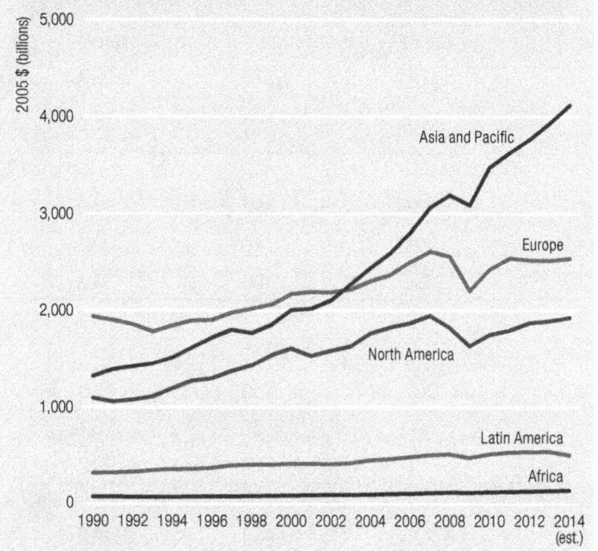

Note: Regional classification based on Annex B1, Table B1.1.
Source: UNIDO elaboration based on Manufacturing Value Added Database (UNIDO 2015e).

of primary commodities, largely due to the higher value added. Successful DEIEs have pursued export-led economic growth policies, diversifying from primary commodities to manufactured goods. Like their industrialized peers, the success of these economies stems from concentrating on manufactured exports.

World export growth rates averaged 7.5 percent over 2005–2013; in 2013 world trade reached a peak of $17 trillion, with 83.0 percent comprising manufactured products (Table 7.6). Over the period, world output expanded at an average 2.3 percent a year, though many countries saw a decline during the crisis.

Overall trend

In 2013, world manufacturing trade reached a peak at $13,866 billion, on the whole growing faster than MVA and GDP over 2005–2013 (see Table 7.6). Global manufacturing trade recovered fully after a sharp decline during 2007–2009, largely due to the fast expanding DEIEs. Indeed, their relative weight has grown enormously, mainly due to China's meteoric rise as an exporter. Exports of primary products surged, but still only account for 16 percent of world trade.

Manufactured exports by industrialized economies expanded by an annual average 5.2 percent over 2005–2013, reaching $8,929 billion in 2013 (Table 7.7). Over the same period, DEIEs expanded their manufactured exports by an average 11.5 percent to peak at $4,937 billion, 2.3 times more than in 2005.

Their fast-growing share in world manufactured exports reflects the dynamism of DEIEs. In 1990, the group accounted for 11.5 percent of world manufacturing trade, 20.9 percent in 2000 and 35.6 percent in 2013 (Figure 7.9). The emerging industrial economies contributed most to the DEIE growth path by increasing their share in global manufactured exports to 18.5 percent and 32.6 percent in 2000 and 2013, respectively, up from 9.5 percent in 1990. It is expected that the role as exporters for DEIEs will increase significantly over the next years, reflecting their high growth rate and the development of the middle class.

Table 7.6

World exports by product category, 2005–2013

Category	Exports (billions, current $)									Average growth rate 2005–2013 (percent)
	2005	2006	2007	2008	2009	2010	2011	2012	2013	
Manufacturing	8,130	9,367	10,772	12,050	9,421	11,409	13,422	13,363	13,866	6.9
Primary	1,146	1,411	1,543	2,197	1,422	1,939	2,511	2,442	2,620	10.9
Other	102	137	163	193	141	185	224	214	196	8.5
Total trade	9,378	10,915	12,478	14,440	10,984	13,533	16,157	16,018	16,682	7.5

Note: Product category classification based on ISIC Rev. 3, ITC (2015).
Source: UNIDO elaboration based on United Nations Comtrade database (UNSD 2015a).

> **Manufactured exports by industrialized economies expanded by an annual average 5.2 percent over 2005–2013, reaching $8,929 billion in 2013**

Table 7.7
World manufacturing exports by development group, region and income, selected years, 1995–2013 (billions, current $)

	1995	2000	2005	2010	2013
World	3,901	5,079	8,130	11,409	13,866
Industrialized countries	3,218	4,015	5,967	7,579	8,929
Developing and emerging industrial economies	683	1,064	2,163	3,831	4,937
By development group					
Emerging industrial countries	653	938	1,944	3,451	4,526
Least developed countries	7	14	24	49	39
Other developing countries	24	113	195	330	372
By region					
Asia and Pacific	346	566	1,291	2,509	3,371
Europe	83	127	302	483	620
Latin America	213	309	460	632	733
Africa	41	62	110	207	212
By income (world)					
High income	3,407	4,221	6,225	7,914	9,269
Upper middle income	417	669	1,570	2,872	3,771
Lower middle income	72	178	313	578	794
Low income	6	12	22	45	33

Note: Regional, development level and income classification based on Annex B1.
Source: UNIDO elaboration based on United Nations Comtrade database (UNSD 2015a).

Figure 7.9
Share in world manufactured export by country group, 1990–2013

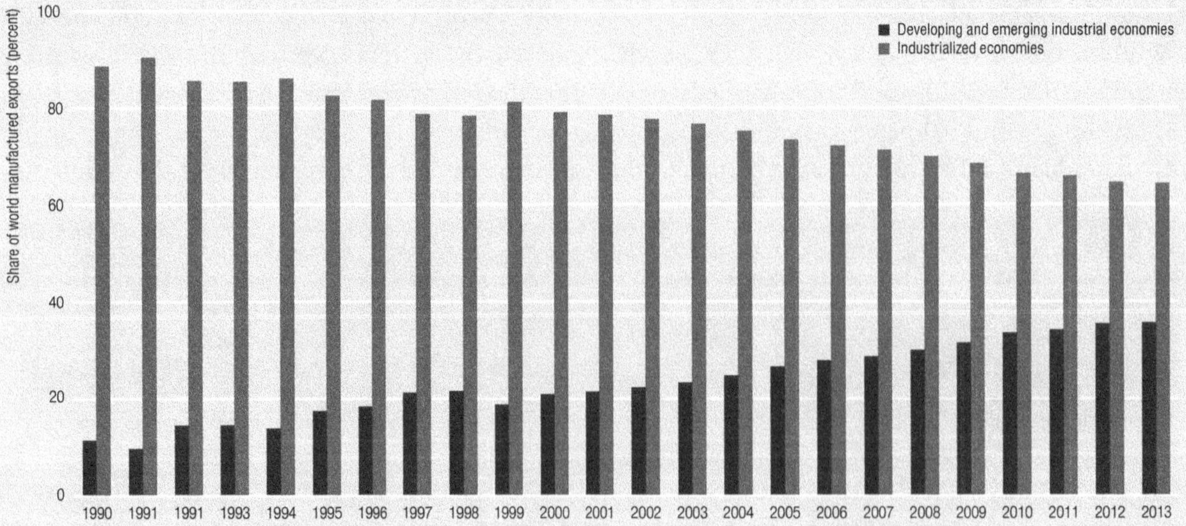

Note: Development level classification based on Annex B1, Table B1.2.
Source: UNIDO elaboration based on United Nations Comtrade database (UNSD 2015a).

> **The high-tech sector reached its peak, 25 percent, in 2000, and fell to 20 percent in 2013**

In addition, their dependence on developed-country markets is expected to weaken as these economies move towards more advanced manufactured exports.

The three largest manufacturing exporters in the DEIE group—China, Mexico and India—accounted for 62.1 percent of the total of the country group in 2013, up from 55.3 percent in 2000, indicating the rapid growth of larger economies and the increasing gap with smaller economies.

Around 58 percent of the world manufactured exports consists of medium- and high-tech products such as chemical machinery and equipment, communications equipment and motor vehicles. The high-tech sector reached its peak, 25 percent, in 2000, and fell to 20 percent in 2013 (Figure 7.10). This could be due to the high investment risk in the sector, which can hold markets back. While the export share of low- and medium-tech products fell during 2000–2013, the share of resource-based goods increased from 17.8 percent to 23.7 percent. The increasing size of the middle classes in industrialized and developing countries has generated higher demand for processed food.

Figure 7.10
Technology composition of manufactured exports, worldwide, 1990–2013

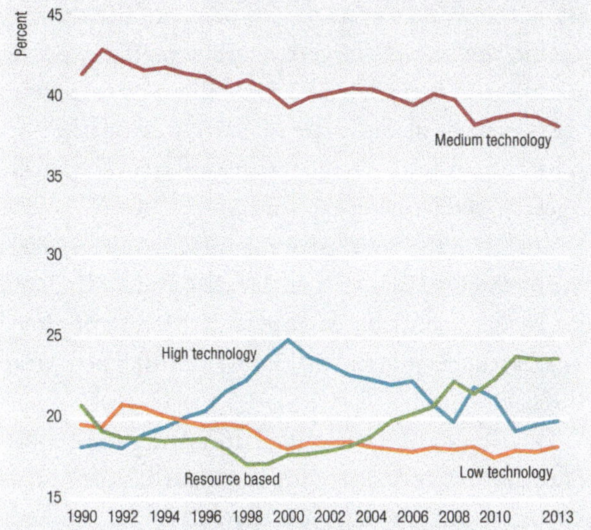

Note: Tech classification based on Annex B5, Table B5.1.
Source: UNIDO elaboration based on United Nations Comtrade database (UNSD 2015a).

In most but not all industrialized economies, export promotion was critical in long-term competitiveness by supporting investment and maintaining technology. Most of those industrialized economies originally transitioned from dependence on primary products to becoming important manufacturing exporters. DEIEs are moving along the same path by increasing their export quality and accelerating the speed of production of medium- and high-tech manufactured goods. Over 2005–2013, DEIEs expanded their share in world exports of medium- and high-tech products by an average annual 6.4 percent, which was faster than the 2.3 percent and 3.7 percent in resource-based and low-tech manufactured products. Over the same period the corresponding shares of industrialized countries in world manufactured exports fell by 0.9 percent and 2.8 percent each year (Figure 7.11).

Manufactured exports per capita

Manufactured exports per capita capture the ability of a country to produce goods competitively and to keep pace with technological changes. Data on manufactured exports indicate international efficiency, other things being equal, and reveal structural trends. But data for large economies are biased by large internal demand and incentives towards domestic markets.

Trends in manufactured exports per capita are depicted in Figure 7.12. Growth for DEIEs is impressive, yet their levels are far below those of industrialized countries. Industrialized and developing manufactured-goods exporters suffered during the global economic crisis, but the developing world was hit harder because of the large decline in demand from industrialized partners.

In line with its rank of the lowest MVA, the least developed countries (LDCs) group accounts for only some 2.7 percent of the world's manufactured exports, and lags behind all other country groups per capita.

Developing and emerging industrial economies manufactured exports

The share in world manufactured exports of DEIEs has soared since 2000, but not all countries have

> **Over 2005–2013, developing and emerging industrial economies expanded their share in world exports of medium- and high-tech products by an average annual 6.4 percent**

Figure 7.11
Average change in world market share of manufactured exports, by technology level and country group, 2005–2013

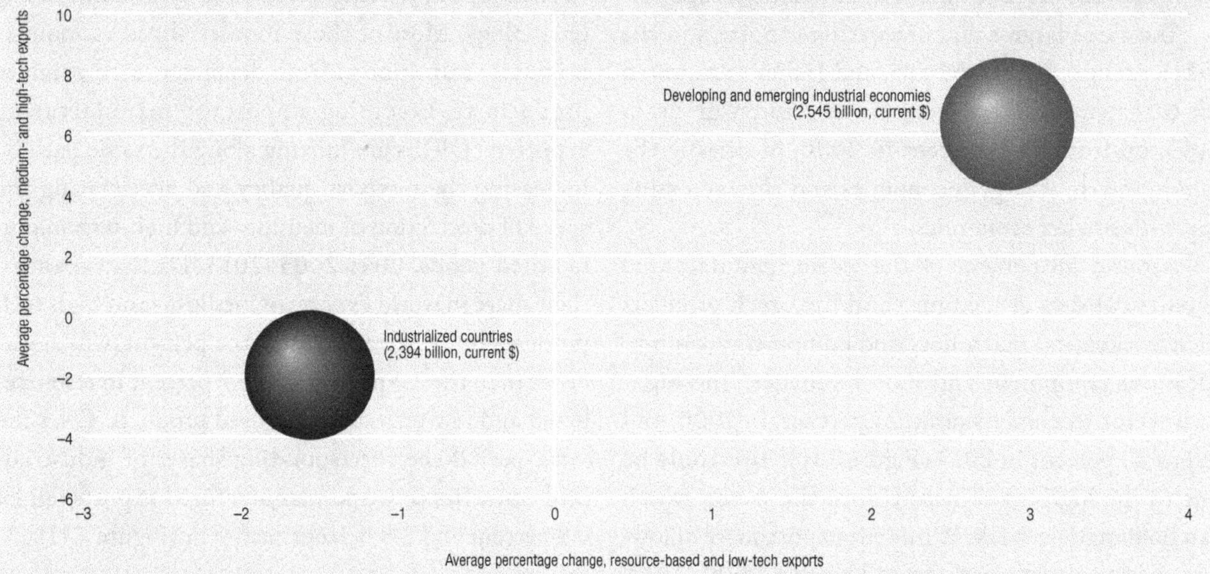

Note: Bubble size indicates the change in the value of manufactured exports (in parentheses). Tech classification based on Annex B5, Table B5.1. Development level classification based on Annex B1, Table B1.2.
Source: UNIDO elaboration based on United Nations Comtrade database (UNSD 2015a).

Figure 7.12
Growth trends in manufactured exports per capita, by country group and worldwide, 1990–2013

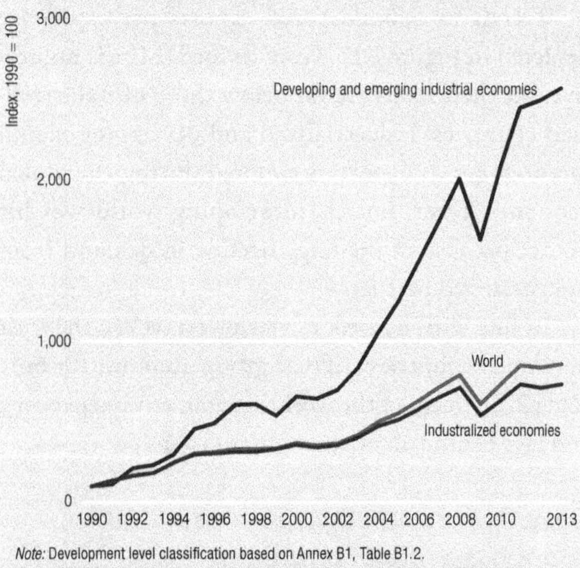

Note: Development level classification based on Annex B1, Table B1.2.
Source: UNIDO elaboration based on United Nations Comtrade database (UNSD 2015a).

contributed equally (Table 7.7). China's performance is key, as a large producer and exporter. The country's exports of manufactured goods grew at an annual average 18.8 percent over 2000–2013, more than twice as fast as the world's average of 8 percent. The country became the world's largest manufacturing exporter in 2008, and it exported $2,329 billion by value with a 16.8 percent global share in 2013. Regionally, Asia and Pacific is both the largest manufacturer and the largest manufacturing exporter.

Together, the five biggest manufacturing exporters within the DEIEs—China, Mexico, India, Thailand and Brazil—contributed almost 25 percent of global manufacture exports in 2013, up from 10.5 percent in 2000.

Developing and emerging Latin American exporters could not compete with Asia and Pacific's rapid expansion and gradually lost share in DEIE manufactured exports, which fell from 31.2 percent in 1995 to 29.0 percent in 2000 and to 14.8 percent in 2013. The export dynamism of Mexico—the largest exporter in

> **Developing and emerging industrial economies saw a progressive move into the production and export of more complex, medium- and high-tech manufactures**

the region—depends heavily on the North American market, largely due to the North American Free Trade Agreement. The market was hit strongly during the global economic crisis due to the sudden drop in demand from North American importers.

Thanks to Poland, European exporters succeeded in maintaining their share in the DEIEs' exports. Poland's share in developing countries' exports increased from 3.2 percent in 2000 to 4.1 percent in 2013. Even though Poland's export market is mainly within the European Union, the country benefits from a diversified export portfolio, largely concentrated in medium-tech products (45.8 percent in 2013).

Despite a high growth rate in manufactured exports in Africa, the region's share in the global market remained marginal (1.5 percent in 2013). The region concentrated mainly on resource-based and medium-tech manufactured products.

More than half (52 percent) of manufactured exports from the developing world consist of medium- and high-tech products, up from 29 percent and 46 percent in 1990 and 2000. The structural change in exports of manufactured goods is in progress in DEIEs. The DEIE country group saw a progressive move into the production and export of more complex, medium- and high-tech manufactures (Figure 7.13).

The recovery of global manufactured exports

World manufactured exports grew annually by 11.2 percent over 2000–2007, reaching around \$12,000 billion in 2008, with the growth rate in DEIEs (17.7 percent) far higher than that in industrialized countries (9.5 percent) (Table 7.8). After 2008, economic recession in the United States, the European Union and Japan hit the DEIEs through sharp falls in demand, investment, tourism, and in changed policy development goals. The global fall in manufactured exports in 2009 (down 21.8 percent) hit all DEIEs (which saw a 17.9 percent drop).

By 2013, world manufactured exports appeared to have fully recovered by setting a new record of \$13,866 billion, 15.1 percent larger than the 2008 peak. Though DEIEs have a smaller share in overall manufacturing output and manufactured exports than industrialized countries, they have played a major role in the recent revival in trade. Of the \$1,816 billion additional exports in 2013 relative to 2008, DEIEs accounted for \$1,347 billion (74.2 percent) coupled with an average growth rate of 14.0 percent, while the industrialized economies registered \$469 billion in extra manufactured exports (25.8 percent) with a growth rate of 8.5 percent a year over 2007–2013.

Figure 7.13
Technology structure of manufactured exports in developing and emerging industrial economies, 1990, 2000 and 2013

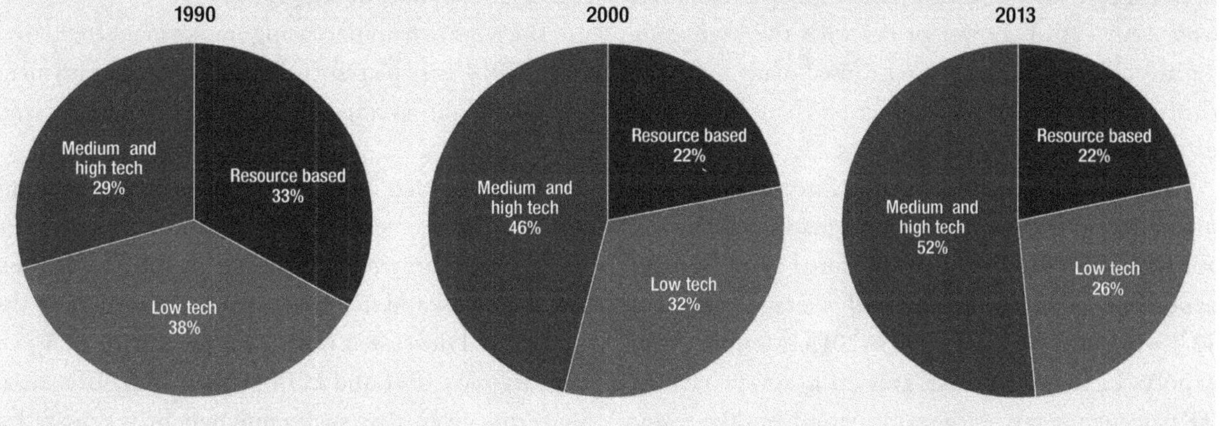

Note: Tech classification based on Annex B5, Table B5.1.
Source: UNIDO elaboration based on United Nations Comtrade database (UNSD 2015a).

> The Asia and Pacific region, led by China, recorded a new peak of $3,371 billion in manufactured exports in 2013

Table 7.8
Average annual growth rate by development group, region and income, 1990-2013 (percent)

	1990–2000	2000–2007	2007–2013
By industrialization level			
World	9.2	11.2	3.7
Industrialized countries	7.8	9.5	1.6
Developing and emerging industrial economies	21.4	17.7	8.7
By development group			
Emerging industrial countries	20.8	18.4	8.9
Least developed countries	19.1	14.1	-0.9
Other developing countries	28.1	11.9	7.0
By region (world)			
Africa	23.8	19.7	6.0
Asia and Pacific	22.7	18.6	8.7
Europe	17.8	13.1	10.0
Latin America	13.8	13.7	-2.2
By income			
High income	24.5	13.0	6.5
Upper middle income	10.6	12.7	6.3
Lower middle income	8.2	12.3	1.8
Low income	19.1	9.3	5.6

Note: Geometric means are used to calculate average growth rates. Regional, development level and income classification based on Annex B1.
Source: UNIDO elaboration based on United Nations Comtrade database (UNSD 2015a).

The Asia and Pacific region, led by China, recorded a new peak of $3,371 billion in manufactured exports in 2013, with an average growth of 15.7 percent a year over 2009–2013. Lower prices with the high competitiveness of China's market caused many manufacturing firms to relocate production there from more expensive, industrialized countries.

The developing and emerging European region is maintaining its share in the global manufactures market with a 4.5 percent share in global manufactured exports in 2013, recovering with average growth of 11.3 percent a year over 2009–2013. Manufactured exports in Latin America grew at a slower pace, by 9.8 percent a year over the same period, but the region failed to maintain its share of world manufactured exports, contributing a low of 5.3 percent in 2013.

Africa followed a similar pattern to Latin America, but with a growth rate of 10.2 percent, taking its share to a low of 1.5 percent in 2013. The region concentrates on resource-based manufactured exports, which are the key factor in overall growth as product prices and the demand from industrializing countries have increased. High-tech products account for only 7.2 percent of manufactured exports.

Despite some signs of progress, LDCs remain highly vulnerable to geopolitical tensions and political instabilities. Lack of proper infrastructure to support manufacturing adds to the problem. In 2013, LDCs accounted for 0.3 percent of world manufactured exports. The group traditionally concentrated on low-tech manufactured products, but in the past few years that share has dropped dramatically due to lack of support in industry and some countries' struggles with war. LDCs' manufactured exports grew by a low average of 1.8 percent a year over 2009–2013.

Manufacturing employment trends

Manufacturing remains a crucial economic sector in advanced and developing economies. In economic development, manufacturing has often played a key role in job creation, attracting workers from agricultural activities towards production labour with higher wages. This structural change has lifted many countries from a low- to a middle-income (and sometimes higher) group, indicating the importance of manufacturing in economic development.

The world's manufacturing employment trend over 1991–2014 is depicted in Figure 7.14. Employment declined by an average of 0.4 percent annually over 2008–2013.

Although manufacturing has recovered in some advanced and emerging economies in the past few years, manufacturing industries continue to lose jobs. The share of world manufacturing employment in the global total decreased from 14.4 percent to 11.5 percent between 1991 and 2014 (Figure 7.15). This raises concerns, given that such employment is crucial for poverty reduction in many countries. Furthermore, this trend may also point towards premature

"The share of world manufacturing employment in the global total decreased from 14.4 percent to 11.5 percent between 1991 and 2014

Figure 7.14
World manufacturing employment, 1991–2014

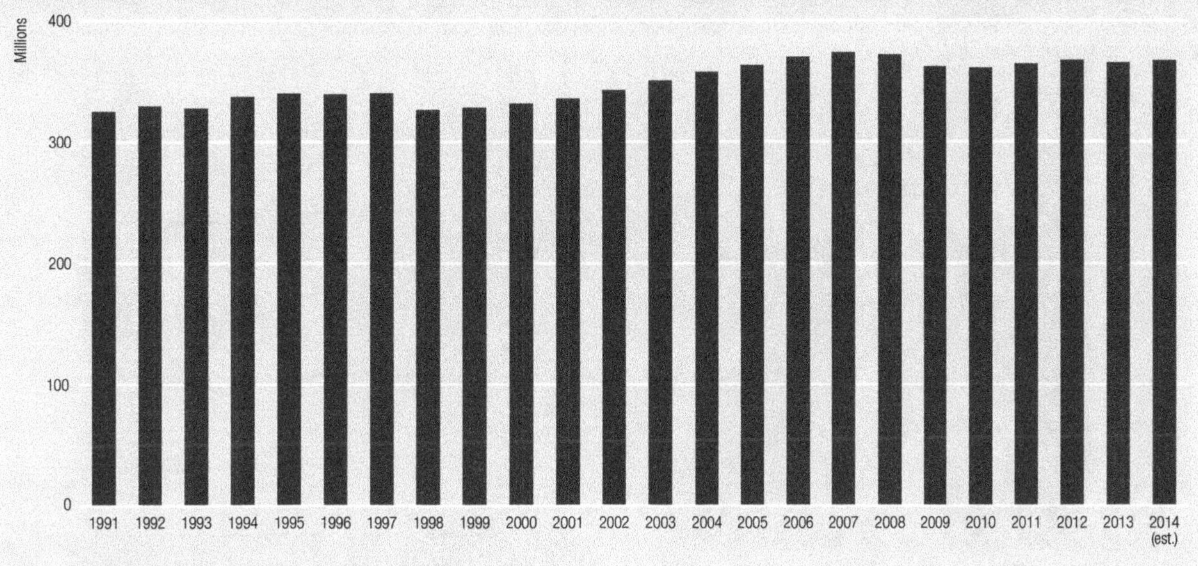

Source: UNIDO elaboration based on Trends Econometric Models database (ILO 2014).

Figure 7.15
Share of manufacturing employment in total employment, worldwide, 1991–2014

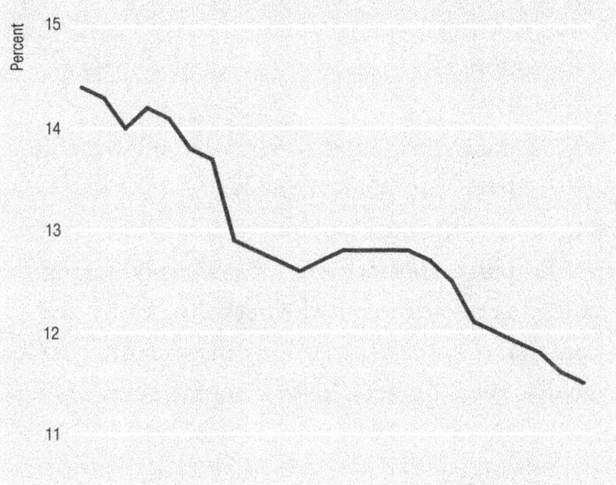

Source: UNIDO elaboration based on Trends Econometric Models database (ILO 2014).

deindustrialization of developing countries. If this is so, it would be difficult for those countries to establish the same levels of economic development recorded earlier by industrialized economies. One argument is that the increase in income may have changed the pattern of demand, from manufactured goods towards services. However, the extent to which the service sector is capable of replacing manufacturing is open to doubt.

Manufacturers are already struggling as a result of not having enough skilled workers—both men and women. With women representing more than half the population in 2014, but a far lower share of the manufacturing workforce, females are underrepresented in manufacturing. Women are critical to filling manufacturing's skills gap. The female share of employment in agriculture grew from 40 percent in 1991 to 44 percent in 2014 (Figure 7.16). However, the share of female employees in manufacturing dropped sharply, from 50 percent in 1991 to 38 percent in 2014. Due to higher wages in manufacturing jobs, this sector would be important for reducing the wage gap between men and women.

Industrialized countries

Manufacturing's role in the global economy is changing, and it shifts as nations mature. In today's advanced economies manufacturing promotes

> **Manufacturing employment declined in major industrialized countries**

Figure 7.16
Share of female employment in total employment by sector, 1991 and 2014

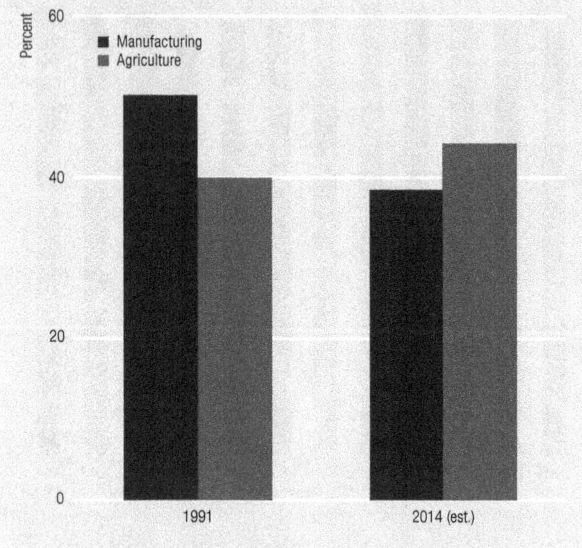

Source: UNIDO elaboration based on Trends Econometric Models database (ILO 2014).

Figure 7.18
Manufacturing employment in the top five industrialized economies, 1991–2014

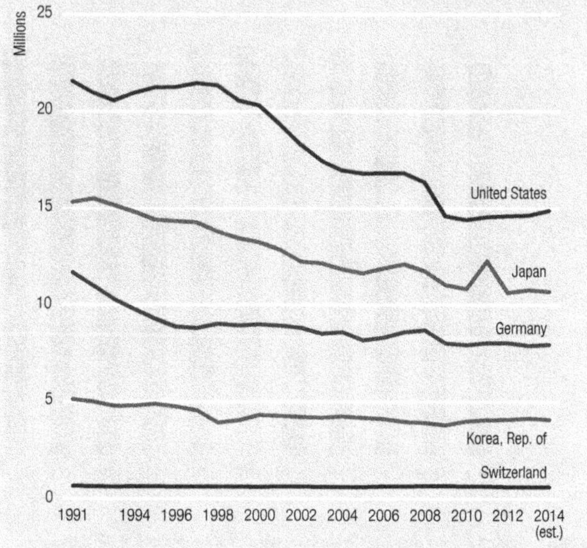

Source: UNIDO elaboration based on Trends Econometric Models database (ILO 2014).

Figure 7.17
Manufacturing employment and its share of total employment in industrialized countries, 1991–2014

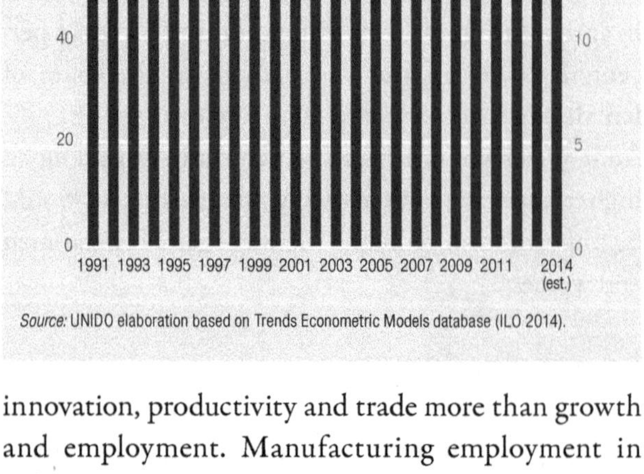

Source: UNIDO elaboration based on Trends Econometric Models database (ILO 2014).

innovation, productivity and trade more than growth and employment. Manufacturing employment in industrialized countries fell from 91 million jobs in 1991 to 63 million in 2014, and from 21.8 percent of total employment to 13.2 percent (Figure 7.17). Manufacturing employment in industrialized countries accounted for around 2 percent of global employment in 2014.

Manufacturing employment declined in major industrialized countries, namely the United States, Japan, Germany, the Republic of Korea and Switzerland. Employment declined most sharply during the financial crisis, although the declining trend started before that (Figure 7.18). Among these countries, Germany has the highest share of manufacturing employment in the total (19.4 percent in 2014), the United States the lowest (9.8 percent), down respectively from 30.6 percent and 17.7 percent in 1991. Switzerland shows a more stable trend, with almost 14.0 percent of jobs still in manufacturing (Table 7.9).

This phase of deindustrialization in industrialized countries is related to the rapidly growing share of services, exacerbated by the transfer of industrial technologies to the developing world and the establishment of factories there to reduce labour costs. Advanced economies now focus on research and development

“ Developing and emerging industrial economies have seen their manufacturing employment climb

Table 7.9
Share of manufacturing employment in total employment, selected countries, 1991–2014 (percent)

	Germany	Japan	Korea, Rep. of	Switzer-land	United States
1991	30.6	24.3	26.7	17.1	17.7
1992	29.2	24.3	25.6	16.8	17.1
1993	27.8	23.7	24.2	16.6	16.5
1994	26.5	23.1	23.7	16.9	16.5
1995	25.0	22.4	23.5	16.7	16.4
1996	24.0	22.2	22.5	16.4	16.1
1997	23.9	21.9	21.2	16.4	15.9
1998	24.1	21.3	19.5	16.1	15.5
1999	23.8	21.0	19.8	15.9	14.7
2000	23.8	20.7	20.3	15.4	14.3
2001	23.6	20.2	19.8	15.8	13.6
2002	23.5	19.5	19.1	15.6	12.9
2003	23.0	19.3	19.0	15.4	12.3
2004	23.1	18.8	19.0	15.3	11.8
2005	22.0	18.4	18.5	15.3	11.5
2006	21.9	18.7	18.0	15.4	11.3
2007	22.1	18.9	17.1	15.2	11.2
2008	22.1	18.4	16.7	15.4	10.9
2009	20.4	17.5	16.3	15.0	10.1
2010	20.0	17.2	16.9	14.9	10.0
2011	19.9	19.7	16.9	14.5	10.0
2012	19.8	17.1	16.6	13.6	9.9
2013	19.4	17.1	16.7	13.5	9.8
2014 (est.)	19.4	17.0	16.4	13.5	9.8

Source: UNIDO elaboration based on Trends Econometric Models database (ILO 2014).

Figure 7.19
Manufacturing employment and share of manufacturing jobs in total employment in developing and emerging industrial economies, 1991–2014

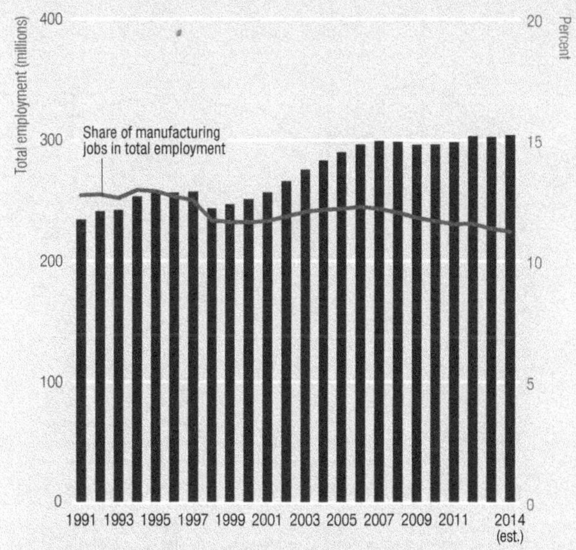

Note: Development level classification based on Annex B1, Table B1.2.
Source: UNIDO elaboration based on Trends Econometric Models database (ILO 2014).

(R&D), as they can no longer compete with the low-cost manufacturing in emerging markets.

Developing and emerging industrial countries

Balancing the loss of manufacturing employment in developed countries, DEIEs have seen their manufacturing employment climb. In the last 10 years, many manufacturing jobs have been shifting from Western Europe and North America to the emerging countries, particularly in the Asia and Pacific region. By 2014, 304 million jobs had been created within DEIE manufacturing—in 1991, manufacturing employment in DEIEs accounted for only 234 million jobs (Figure 7.19). The share of manufacturing employment in total employment in developing countries decreased slightly from 12.7 percent in 1991 to 11.2 percent in 2014. Today, developing countries' manufacturing employment accounts for around 9.5 percent of global employment.

The decline in manufacturing jobs is not confined to the industrialized world. Figure 7.20 depicts manufacturing employment in five major DEIEs. China has the highest number of employees among them, though its share of manufacturing jobs as a share of total employment dropped from a high of 14.9 percent in 1995 to 11.7 percent in 2014 (Table 7.10). The highest rate of manufacturing employment was registered in Poland, with 25.2 percent in 1991 but down to 19.1 percent in 2014. The share increased in Brazil and Mexico at the beginning of the century but they, too, have experienced a decline in recent years. Among the five, only India managed to raise its level and share of

> **Manufacturing remains critically important to both developing and emerging industrial economies and industrialized countries**

7

Figure 7.20
Manufacturing employment in the top five developing and emerging industrial economies, 1991–2014

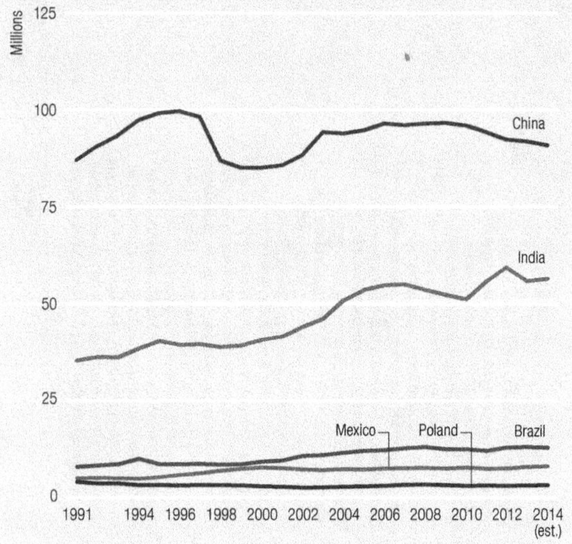

Source: UNIDO elaboration based on Trends Econometric Models database (ILO 2014).

Table 7.10
Share of manufacturing employment in total employment, selected countries, 1991–2014 (percent)

	Poland	China	India	Brazil	Mexico
1991	25.2	13.9	10.9	12.9	16.1
1992	23.7	14.2	10.9	12.7	15.9
1993	23.5	14.4	10.7	12.8	14.9
1994	21.2	14.8	11	14.5	14.4
1995	21.1	14.9	11.4	12.2	15.6
1996	20.9	14.9	10.9	12.3	16.7
1997	20.9	14.5	10.8	12.2	16.9
1998	20.9	12.7	10.3	11.8	18.4
1999	20.7	12.3	10.3	11.6	19.2
2000	20	12.1	10.5	12.2	19.7
2001	19.9	12.1	10.4	12.5	19.2
2002	18.7	12.4	10.7	13.5	18
2003	19	13.1	10.9	13.7	17.5
2004	19.9	12.9	11.7	13.9	17.6
2005	20.1	12.9	12.1	14.1	17
2006	20.5	13	12.3	14	16.9
2007	20.7	12.9	12.3	14.4	16.8
2008	20.4	12.9	12	14.4	16.6
2009	19.3	12.9	11.7	13.7	16.4
2010	18.5	12.7	11.4	13.4	16.3
2011	18.8	12.4	12.3	12.6	15.6
2012	18.6	12	13	13.2	15.2
2013	19.1	11.9	12.1	13.2	15.6
2014 (est.)	19.1	11.7	12	12.9	15.7

Source: UNIDO elaboration based on Trends Econometric Models database (ILO 2014).

manufacturing employment, from 35 million in 1991 to more than 56 million in 2014, largely explained by increased openness to trade and rising labour productivity.

In sum, the global manufacturing sector experienced uncertain years, with major DEIEs leaping into the top tier of manufacturing economies, the 2008 economic recession's negative effect on demand, and the dramatic fall in manufacturing employment in industrialized economies. However, manufacturing remains critically important to both the DEIEs and industrialized countries. In DEIEs, it helps countries move from a low-income, agriculture-based economy towards rising incomes and, with them, rising living standards. In the industrialized world, it remains a key source of innovation and competitiveness, making outsized contributions to R&D, exports and productivity growth. Labour absorption in the manufacturing sector may accelerate and manufacturers will continue to hire workers, but in the long run manufacturing's share of employment declines, and is compensated by job growth in the service sector. Changes in demand structure due to rising incomes and the impact of global industrial competitiveness push economies to specialize in medium- and high-tech activities and to increase the demand for a highly skilled workforce.

Resource efficiency and energy intensity in manufacturing

Resource efficiency is an important strategy for sustainable growth. Manufacturing production processes today have to be highly productive, less energy intensive, and more resource efficient. Materials are by far the most critical cost factor for the sector, at over 70 percent on average.

"The current trend for input and manufacturing value added growth shows manufacturing value added growing faster than input

Figure 7.21
Resource efficiency in manufacturing, 1995–2011

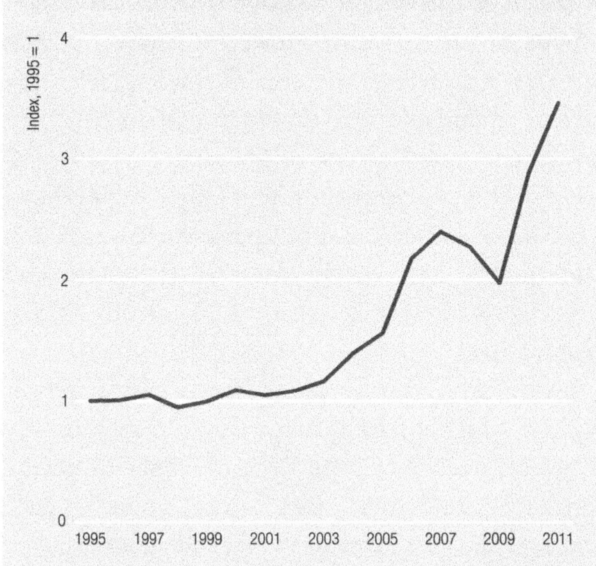

Source: UNIDO elaboration based on World Input-Output Database (Timmer and others 2015).

Figure 7.22
Changes in resource efficiency, selected countries, 1995 and 2011

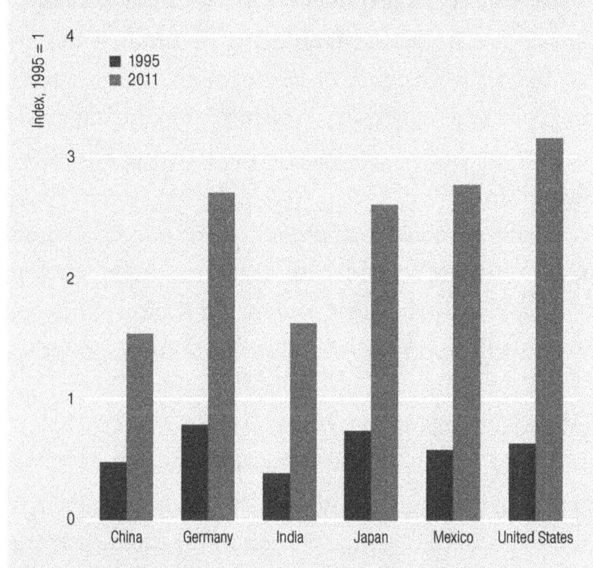

Source: UNIDO elaboration based on World Input-Output Database (Timmer and others 2015).

An input-output approach to resource efficiency

The total input in manufacturing in current prices (based on input-output tables) is used to estimate resource efficiency. Figure 7.21 illustrates the ratio of value added globally at basic prices generated, divided by the input over 1995–2011—in other words, how much value added has been generated per unit of input. Performance was not very impressive between 1995 and 2000, but then started to pick up until the global crisis, when it turned down due to a large decline in consumption and a massive decline in global MVA, before sharply recovering.

The increase in manufacturing resource efficiency varies among countries. Among the major industrialized and developing economies, it climbed sharply from 1995 to 2011 in the United States, Mexico, Germany and Japan, reflecting shifts in the composition of manufacturing activity (Figure 7.22). Impressive growth in the indicator was also observed in India and China.

A lower resource efficiency index does not necessarily mean that a country is less resource efficient; in some cases more sophisticated products still remain highly resource intensive in manufacturing. For instance, production of a modern wind turbine—a high-tech product—requires more than 8,000 different components, many of which are made from steel, cast iron and concrete. While most economies worldwide are becoming more resource efficient, there is still a large gap in the indicator between developing and industrialized economies. The current manufacturing production and consumption pattern in industrialized economies cannot be replicated in fast-growing developing nations. In developing countries, manufacturing policy-makers are faced with two contradictory challenges: increasing economic development to the level of industrialized countries, which requires a rise in consumption, but simultaneously reducing manufacturing's environmental impact.

The current trend for input and MVA growth shows MVA growing faster than input. To sustain manufacturing production and environmentally sound resource consumption, it is necessary not only to control consumption (that is, grow it less quickly

Global final total energy consumption rose from 5,495 million tons of oil equivalent in 1990 to 7,950 million tons of oil equivalent in 2012, a 45 percent increase

than production) but also to increase production while lowering resource consumption.

Technological upgrades are one solution to balancing these challenges. Manufacturing product design also has a direct effect on resource efficiency, which can result in a reduction in consumption during product use, as well as efficient manufacturing and recycling options.

A comprehensive approach to resource-efficient manufacturing would help countries to maintain economic growth while benefiting society. Indeed, resource efficiency is essential for sustainable growth.

Global trends in manufacturing energy intensity

Energy intensity is defined as the amount of energy used to produce one unit of economic output. It is the inverse of energy efficiency: less energy intensity means more energy efficiency. Energy intensity is measured by dividing the amount of energy used (in physical terms, millions of tonnes of oil equivalent, or MTOE) by the MVA in monetary terms (in constant 2005 $). The energy intensity of manufacturing is the amount of energy used to produce one unit of value added.

Global final total energy consumption rose from 5,495 MTOE in 1990 to 7,950 MTOE in 2012, a 45 percent increase (Figure 7.23). Worldwide energy consumption grew most quickly in the transport and service sectors, both showing an increase of over 56 percent, mainly due to the rapid increase in number of travellers and worldwide cargo, and the fast expansion of the service sector. However, despite a smaller increase, manufacturing remains the largest energy user, at around one-third (2,404 MTOE in 2012), for an average annual increase of 2 percent.

World final energy consumption per capita shows a much less sharp increase of 9 percent over the period —driven by demand for energy-using goods and services—and only 4.5 percent in manufacturing (Figure 7.24).

Although a growing economy generally correlates with growing absolute energy use, energy intensity may well decline (Figure 7.25). Declines in energy intensity can be a proxy for efficiency improvements, and over the past 22 years global energy intensity in manufacturing has fallen by more than 25 percent, in part reflecting efficiency gains from modern technologies and processes. Other factors are also at play. For example, structural change in manufacturing—shifting from energy-intensive sectors such as basic metals, chemicals and wood products to less energy-intensive industries such as transport equipment or food products—would cause a decline in the energy intensity index that does not necessarily reflect an increase in energy efficiency. The global energy intensity trend faced an upward surge over 2008–2010.

Figure 7.23
World final energy consumption by sector, 1990–2012

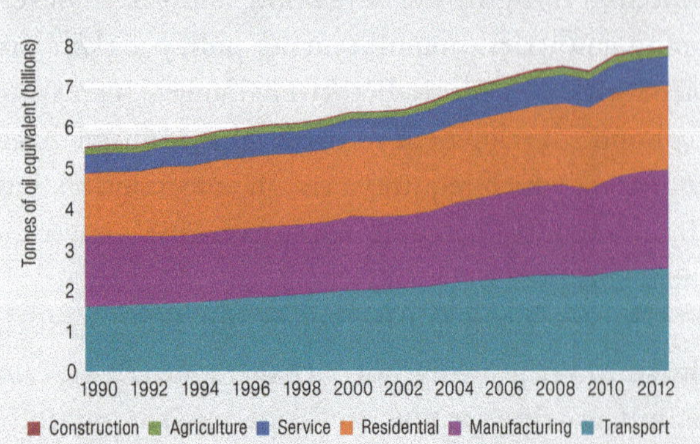

Source: UNIDO elaboration based on IEA Energy Flow Charts (IEA 2013).

Figure 7.24
World final energy consumption per capita by industry, 1990–2012

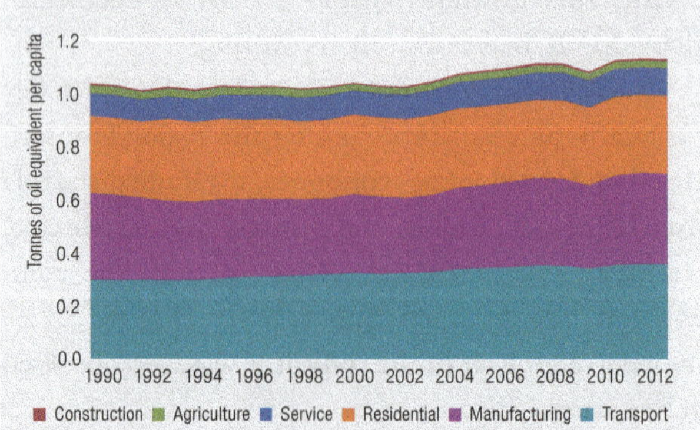

Source: UNIDO elaboration based on IEA Energy Flow Charts (IEA 2013).

“ **Global energy intensity and manufacturing value added trends are diverging**

Figure 7.25
World manufacturing energy intensity, 1990–2012

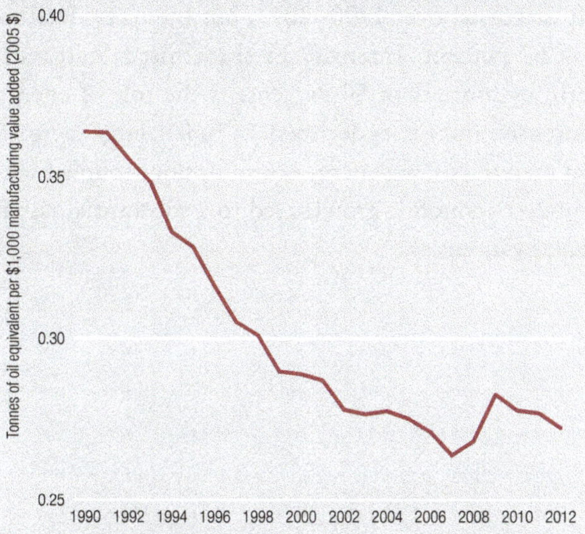

Source: UNIDO elaboration based on IEA Energy Flow Charts (IEA 2013) and Manufacturing Value Added Database (UNIDO 2015e).

Figure 7.26
Global trends in manufacturing value added, manufacturing energy consumption, and manufacturing energy intensity, 1990–2012

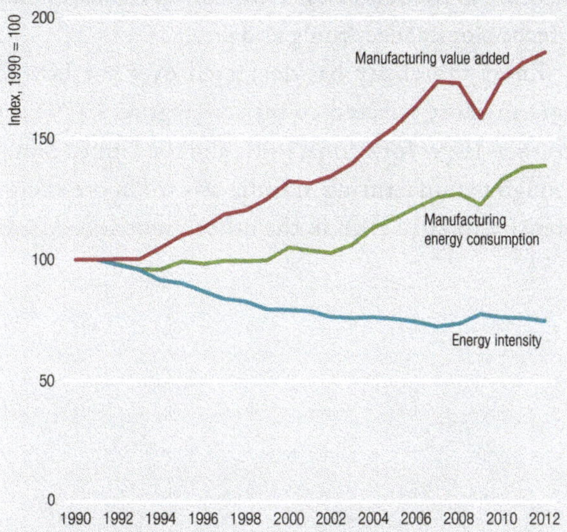

Source: UNIDO elaboration based on IEA Energy Flow Charts (IEA 2013) and Manufacturing Value Added Database (UNIDO 2015e).

After the start of the global financial crisis in 2008, many countries adopted stimulus packages focused on energy-intensive sectors such as manufacturing, which led to a jump in manufacturing energy intensity. However, as the world economy began to recover after 2010, the previous pattern of easing global energy intensity in manufacturing resumed.

Global energy intensity and MVA trends are diverging (Figure 7.26): manufacturing is becoming more productive while keeping the sector's energy intensity at a lower level. Furthermore, while MVA and energy consumption levels are growing, the gap between the two is increasing, and growth in consumption is lower than growth of MVA. Over 1990–2012, MVA increased 86 percent, more than twice the 39 percent growth in energy consumption. Detailed data are required to conduct sectoral analysis to identify which drivers—technological efficiency or structural change—were the main drivers in energy intensity's decline.

Comparing manufacturing efficiency by country based on the energy intensity indicator is not

Figure 7.27
Manufacturing energy intensity growth, selected industrialized and developing countries, 1990–2012

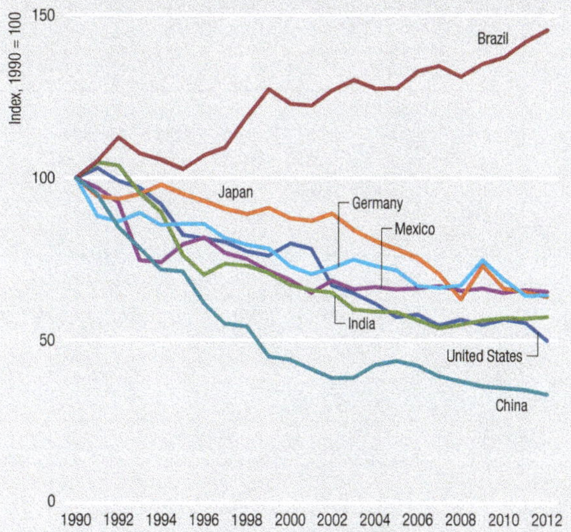

Source: UNIDO elaboration based on IEA Energy Flow Charts (IEA 2013) and Manufacturing Value Added Database (UNIDO 2015e).

> **Over 1990–2012, manufacturing value added increased 86 percent, more than twice the 39 percent growth in energy consumption**

straightforward, as many characteristics define the efficient level of energy consumed. Furthermore, such comparisons require an analysis of the manufacturing structure of the countries, their natural resources, rate of technological deepening and prices.

Energy intensity has decreased over the last 22 years in most selected countries (Figure 7.27). The drop was larger for China, India and the United States (though manufacturing in India was still more energy intensive in 2012 than in the other countries). Again, this is not direct evidence of inefficiency in India's manufacturing. In level terms, China has the highest consumption of energy among the other countries. It increased over 1990–2012, but intensity declined by 67 percent. Intensity in the United States also fell, by more than 50 percent, as the role of energy-intensive industries declined. In Brazil, large increases in manufacturing energy consumption, coupled with modest economic growth, led to a substantial rise in energy intensity.

Chapter 8
The Competitive Industrial Performance index

The competitiveness of manufacturing industry is one of the basic determinants of long-run sustainable growth, so it is important to understand countries' relative positions on this metric and on the determinants of competitive ability. Shifts in the relative position of industrialized and emerging industrial economies in manufacturing value added (MVA) and industrial exports are, to a large extent, attributable to changes in individual countries' industrial competitiveness.

UNIDO assesses and benchmarks industrial competitiveness through its Competitive Industrial Performance (CIP) index, building on a meso-concept of competitiveness that emphasizes countries' manufacturing development and implies that industrial competitiveness is multidimensional. Industrial competitiveness is defined as the capacity of countries to increase their presence in international and domestic markets while developing industrial sectors and activities with higher value added and higher technological level.

Countries can learn in international markets and become more industrially competitive if they develop technological capability, expand production capacity and invest in infrastructure. Hence, increasing industrial competitiveness requires selective policy interventions through which comparative advantages are exploited while new competitive advantages are created.

The index

The CIP index is a performance (or "outcome") indicator rather than a potential (or "process") indicator. It consists of output sub-indicators only. Focusing on industrial competitiveness and structural economic variables, it provides country rankings that tend to remain relatively stable over short periods of time, because processes of technological learning are cumulative and take time. The effects of learning are reflected in industrial statistics and structural economic variables only in the medium to long term and can be captured through detailed longitudinal studies, in particular by tracking changes in key dimensions over time. The CIP index allows us to observe not only the absolute levels of key indicators at any particular time, but also their rates of change.

The CIP index consists of eight sub-indicators along three dimensions of industrial competitiveness (Figure 8.1). The first dimension describes a country's capacity to produce and export manufactures and is captured by its MVA per capita (MVApc) and its manufactured exports per capita (MXpc).

The second dimension covers a country's levels of technological deepening and upgrading. To proxy for this complex dimension, two composite sub-indicators —industrialization intensity and export quality—are constructed. The degree of industrialization intensity is computed as a linear aggregation of medium- and high-tech MVA share in total MVA (MHVAsh) and MVA share in total gross domestic product (GDP)

Figure 8.1
Components of the CIP index

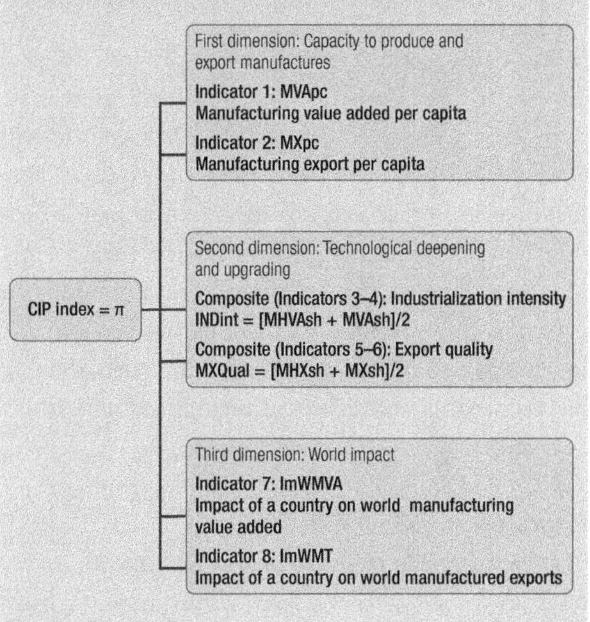

Source: UNIDO (2013c).

"The CIP index consists of eight sub-indicators along three dimensions of industrial competitiveness

(MVAsh). Country export quality is obtained as a linear aggregation of medium- and high-tech manufactured export share in total manufactured exports (MHXsh) and manufactured export share in total exports (MXsh).

The third dimension of competitiveness is the country impact on world manufacturing, both in terms of value-added share in world MVA (ImWMVA) and in world manufacturing trade (ImWMT).

Definition of sub-indicators

The revised version of the CIP index (CIP.8) encompasses three dimensions captured by four individual and two composite sub-indicators (eight sub-indicators in total).

Indicator 1: MVApc

MVApc captures the level of a country's industrialization and is expressed per capita to adjust for country size. The MVApc is the relative value of the total net manufacturing output to population size. Unlike gross output, MVA is free of double counting, as the cost of intermediate consumption is excluded. Also, it is measured at basic prices to avoid tax distortions.

Indicator 2: MXpc

MXpc captures the ability of a country to produce goods competitively and implicitly to keep up with technological changes. Like MVApc, MXpc is expressed per capita to adjust for country size. Data on manufactured exports indicate prima facie international efficiency and reveal structural trends. However, data on the MXpc of large economies are biased by the existence of large internal demand and incentives towards domestic markets. Also, data on re-exports are not available at regular intervals for all countries.

Indicator 3: MHVAsh

This captures the technological complexity of manufacturing. The higher the share of medium- and high-tech MVA in total MVA, the more technologically complex the industrial structure of a given country and its overall industrial competitiveness. Empirical analyses have shown that development generally entails a structural transition from resource-based and low-tech activities to medium- and high-tech activities. The more complex the production structures of a given country become, the higher the opportunities for learning and technological innovation at the sectoral and intersectoral levels.

Indicator 4: MVAsh

This indicator captures manufacturing weight within an economy. In other words, MVAsh indicates the contribution of the manufacturing sector to total production.

Indicator 5: MHXsh

MHXsh captures the technological content and complexity of exports. The share of medium- and high-tech products in manufactured exports is considered jointly with the previous indicator, because MHXsh might differ substantially from MHVAsh in certain circumstances. For example, large import-substituting developing countries are characterized by a relatively more complex structure of MVA than of manufactured exports.

Indicator 6: MXsh

This indicator captures manufacturing weight in export activity.

Indicator 7: ImWMVA

This indicator is measured by a country's share in world MVA, which indicates a country's relative performance and impact in overall manufacturing.

Indicator 8: ImWMT

ImWMT is measured by a country's share in world manufactured exports. It shows a country's competitive status relative to other countries in international markets: gains in world market share reflect greater competitiveness, while losses signal deterioration of competitiveness.

> **Among the five most competitive are four high-income countries (Germany, Japan, the Republic of Korea and the United States), along with China ranking fifth**

Final index composition

In the construction of the final composite index, three fundamental dimensions are considered:

First dimension: Capacity to produce and export manufactured goods

- Indicator 1: MVA per capita (MVApc)
- Indicator 2: Manufactured exports per capita (MXpc)

Second dimension: Technological deepening and upgrading

- Composite indicator combining indicators 3 and 4: Industrialization intensity, INDint = [MHVAsh + MVAsh]/2
- Composite indicator combining indicators 5 and 6: Manufactured exports quality, MXQual = [MHXsh + MXsh]/2

Third dimension: World impact

- Indicator 7: Impact of a country on world MVA (ImWMVA)
- Indicator 8: Impact of a country on world manufactures trade (ImWMT)

The composite index is then computed as the equal-weighted geometric average of MVApc, MXpc, INDint, MXQual, ImWMVA and ImWMT.

The 2013 CIP ranking

The 2013 CIP rankings are shown in Table 8.1. (The underlying indicators are presented in Annex B2, Table B2.1.) Countries are divided based on CIP values into five colour-highlighted quintiles: top, upper middle, middle, lower middle and bottom.

Countries in the top quintile account for nearly 83 percent of world MVA and more than 85 percent

Table 8.1
CIP index, 2013

Top quintile · Upper middle quintile · Middle quintile · Lower middle quintile · Bottom quintile

CIP ranking 2013	CIP index 2013	Country	CIP ranking 2013	CIP index 2013	Country
1	0.576	Germany	20	0.190	Mexico
2	0.466	Japan	21	0.188	Poland
3	0.442	Korea, Rep. of	22	0.186	Denmark
4	0.442	United States	23	0.183	Finland
5	0.366	China	24	0.176	Malaysia
6	0.345	Switzerland	25	0.170	Slovakia
7	0.341	Singapore	26	0.167	Thailand
8	0.321	Netherlands	27	0.164	Hungary
9	0.313	Belgium	28	0.147	Australia
10	0.309	Italy	29	0.145	Israel
11	0.300	France	30	0.143	Turkey
12	0.297	Taiwan Province of China	31	0.130	Norway
13	0.268	Austria	32	0.124	Russian Federation
14	0.259	Sweden	33	0.119	Slovenia
15	0.237	Canada	34	0.118	Portugal
16	0.231	Ireland	35	0.112	Brazil
17	0.231	Czech Republic	36	0.108	Saudi Arabia
18	0.217	Spain	37	0.095	Belarus
19	0.210	United Kingdom	38	0.092	Lithuania

> "Germany's manufacturing sector is a key factor in its macroeconomic performance, with a strong industrial core and an ability to control complex industrial value creation chains

Table 8.1 (continued)
CIP index, 2013

CIP ranking 2013	CIP index 2013	Country
39	0.090	Romania
40	0.089	Argentina
41	0.088	South Africa
42	0.087	Indonesia
43	0.083	India
44	0.080	Bahrain
45	0.079	New Zealand
46	0.077	Estonia
47	0.073	Kuwait
48	0.073	Qatar
49	0.072	Greece
50	0.071	Viet Nam
51	0.069	Chile
52	0.069	Luxembourg
53	0.067	Philippines
54	0.066	United Arab Emirates
55	0.062	Ukraine
56	0.058	Croatia
57	0.058	Bulgaria
58	0.051	Tunisia
59	0.049	Latvia
60	0.049	Costa Rica
61	0.048	Malta
62	0.047	Oman
63	0.046	Peru
64	0.042	Trinidad and Tobago
65	0.042	Iran, Islamic Rep. of
66	0.042	Kazakhstan
67	0.041	Morocco
68	0.040	Colombia
69	0.039	Venezuela, Bolivarian Rep. of
70	0.037	Serbia
71	0.037	Egypt
72	0.036	Iceland
73	0.034	El Salvador
74	0.031	Hong Kong SAR China
75	0.031	Pakistan
76	0.031	Guatemala
77	0.030	Bangladesh
78	0.030	Jordan
79	0.030	Botswana
80	0.029	Uruguay
81	0.029	Sri Lanka
82	0.026	Mauritius
83	0.026	Macedonia, Former Yugoslav Rep. of
84	0.025	Namibia
85	0.024	Lebanon
86	0.023	Bosnia and Herzegovina
87	0.022	Algeria
88	0.021	Swaziland
89	0.021	Ecuador
90	0.020	Cambodia
91	0.019	Brunei Darussalam
92	0.018	Honduras
93	0.018	Côte d'Ivoire
94	0.017	Georgia
95	0.017	Cyprus
96	0.014	Jamaica
97	0.013	Bahamas
98	0.013	Albania
99	0.013	Syrian Arab Republic
100	0.013	Armenia
101	0.012	Bolivia, Plurinational State of
102	0.012	Congo, Rep. of the
103	0.011	Paraguay
104	0.011	Barbados
105	0.011	Azerbaijan
106	0.011	Senegal
107	0.011	Cameroon
108	0.011	Fiji
109	0.011	Zambia
110	0.011	Nigeria
111	0.010	Moldova, Rep. of
112	0.010	Suriname
113	0.010	Kenya
114	0.010	Gabon
115	0.010	Papua New Guinea
116	0.009	Mongolia
117	0.008	Panama
118	0.008	State of Palestine

“ Japan’s industrial competitiveness is supported by its large manufacturing base, high-tech exports and high manufacturing per capita

Table 8.1 (continued)
CIP index, 2013

CIP ranking 2013	CIP index 2013	Country	CIP ranking 2013	CIP index 2013	Country
119	0.007	Ghana	132	0.004	China, Macao SAR
120	0.007	Mozambique	133	0.003	Cabo Verde
121	0.007	Tanzania, United Rep. of	134	0.003	Haiti
122	0.007	Belize	135	0.003	Malawi
123	0.006	Madagascar	136	0.003	Rwanda
124	0.005	Bermuda	137	0.003	Iraq
125	0.005	Kyrgyzstan	138	0.002	Ethiopia
126	0.005	Niger	139	0.001	Central African Republic
127	0.005	Yemen	140	0.001	Burundi
128	0.005	Nepal	141	0.000	Eritrea
129	0.004	Uganda	141	0.000	Gambia
130	0.004	Tajikistan	141	0.000	Tonga
131	0.004	Saint Lucia			

Source: UNIDO elaboration based on Competitive Industrial Performance index database (UNIDO 2015a).

of global manufactures trade. Among the five most competitive are four high-income countries (Germany, Japan, the Republic of Korea and the United States), along with China ranking fifth. Four of these five countries are among the world’s most industrialized countries, and jointly with China account for 59 percent of world MVA.

Germany’s manufacturing sector is a key factor in its macroeconomic performance, with a strong industrial core and an ability to control complex industrial value creation chains. Its medium- and high-tech exports accounts for 73 percent of its total manufactured exports, and it has maintained its technological lead against newcomers in the global economy. Germany thus has strong technological upgrading and deepening on both the production and trade sides.

Japan’s industrial competitiveness is supported by its large manufacturing base, high-tech exports and high manufacturing per capita. The United States’ industrial competitiveness arises from its large manufacturing base, although it is more aimed at the domestic market than Japan or any other developed country. The United States alone accounts for nearly 20 percent of world MVA. The Republic of Korea has a competitive manufacturing sector based on a high share of medium- and high-tech industries.

In the top quintile, given the population size and stage of development, China has the lowest per capita values on both trade and production sides. China’s position in the ranking is attributable to its high share in global trade (though low per capita values indicate that manufacturing has the potential to grow further). China has increased its share of manufacturing exports to 17 percent of global manufacturing trade in 2013 and is the largest exporter in the world today. It has also started positioning itself as a high-tech manufacturing exporter: the export share of medium- and high-tech products almost doubled over 1995–2013 (Figure 8.2). China’s manufacturing industry has become the largest sector in the economy and accounted for more than one-third of GDP and 18 percent of global MVA in 2013, second only to the United States.

Others in the top quintile include Switzerland, Singapore and the Netherlands on account of their very high exports per capita in general and high-tech exports in particular. Other top-quintile members include major European Union transition economies

Grouping by stage of development is tightly related to a country's per capita manufacturing value added and its share in world manufacturing value added

Figure 8.2
Trade and production structure in China and the Republic of Korea, selected years, 1995–2013

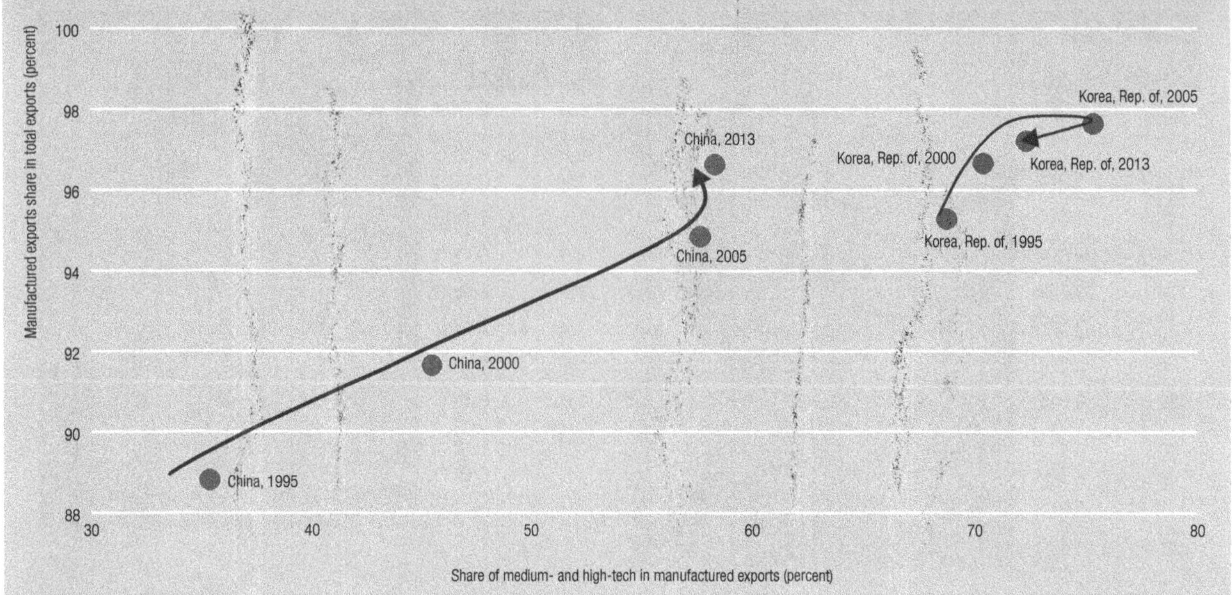

Note: Tech classification based on Annex B5, Table B5.1.
Source: UNIDO estimate based on Competitive Industrial Performance index database (UNIDO 2015a).

such as the Czech Republic, Poland, Slovakia and Hungary—due to their export orientation—which are more focused on the European market. The list is completed by Mexico, Malaysia and Thailand, whose competitiveness arises from their participation in global value chains.

The upper middle quintile includes some of the most populous countries in the world, such as Turkey, the Russian Federation, Brazil, Indonesia, South Africa, India and the Philippines. The production and export performance of high-tech products in the Philippines and Indonesia is strong, while the Russian Federation and South Africa have higher MVAs per capita but low manufacturing exports due to their dependence on foreign sales of natural resources. India and Brazil accounted for 2.2 percent and 1.7 percent of global MVA in 2013 respectively.

The middle quintile has populous countries like Iran, Egypt and Bangladesh and some less populous nations like Costa Rica, Iceland, Oman and Uruguay. Countries in the lower middle and bottom quintiles include less developed countries by income, accounting for roughly 0.8 percent of world MVA in 2013. Their level of industrialization is on average less than one-third that of countries in the middle quintile.

The industrial competitiveness of nations by industrial comparator

We can compare and track a country's stage of industrial performance relative to other countries at the same level of industrial development.

Grouping by stage of development is tightly related to a country's per capita MVA and its share of MVA in world MVA. Thus (except for China) the CIP ranking also reflects a country's stage of development, and the CIP ranking of countries within the group does not differ dramatically from the overall CIP ranking (Table 8.2). Industrialized countries account for 64.6 percent of global MVA and 67.3 percent of world manufacturing trade.

The CIP rankings for 2013 show that most industrialized countries have lost ground from the 2010 rankings. Denmark and Finland have been replaced by Mexico and Poland (see Table 8.2) during the past three years in the 20th and 21st rank of the world CIP index. Germany, Japan, the United States and

"Most industrialized countries have lost ground from the 2010 rankings

Table 8.2

Industrial competitiveness ranking and selected indicators for industrialized countries and world ranking comparison, 2013

■ Top quintile ■ Upper middle quintile ■ Middle quintile ■ Lower middle quintile ■ Bottom quintile

Group ranking 2013	World ranking 2010	World ranking 2013	Country	MVA per capita (2005 $) 2013	Manufactured exports per capita (current $) 2013	Impact of a country on world MVA (percent) 2013	Impact of a country on world manufactures trade (percent) 2013
1	1	1	Germany	7,655.8	15,504.2	7.0	10.1
2	2	2	Japan	7,820.7	5,163.5	11.0	5.2
3	3	3	Korea, Rep. of	7,180.7	11,043.4	3.9	4.3
4	4	4	United States	5,464.5	3,229.0	19.4	8.1
5	6	6	Switzerland	10,147.3	25,700.8	0.9	1.6
6	5	7	Singapore	9,700.0	32,285.9	0.6	1.5
7	11	8	Netherlands	4,813.2	27,818.9	0.9	3.7
8	12	9	Belgium	4,531.6	40,287.4	0.6	3.5
9	7	10	Italy	4,151.6	7,540.9	2.8	3.6
10	10	11	France	3,568.3	7,724.3	2.5	3.9
11	9	12	Taiwan Province of China	4,517.0	11,765.5	1.2	2.2
12	15	13	Austria	7,680.6	17,251.7	0.7	1.2
13	14	14	Sweden	6,896.7	15,530.8	0.7	1.2
14	17	15	Canada	4,092.4	7,791.7	1.6	2.2
15	16	16	Ireland	6,736.3	23,133.4	0.3	0.8
16	18	17	Czech Republic	4,039.8	14,074.5	0.5	1.2
17	19	18	Spain	2,960.7	5,425.1	1.5	2.0
18	13	19	United Kingdom	3,671.4	2,844.5	2.6	1.4
19	21	22	Denmark	5,508.0	14,248.6	0.3	0.6
20	20	23	Finland	6,168.4	12,407.2	0.4	0.5
21	23	24	Malaysia	1,717.0	6,201.9	0.6	1.5
22	27	25	Slovakia	3,125.9	14,745.8	0.2	0.6
23	26	27	Hungary	2,365.7	9,634.3	0.3	0.8
24	29	28	Australia	3,050.9	5,399.5	0.8	1.0
25	28	29	Israel	3,232.5	8,265.7	0.3	0.5
26	30	31	Norway	5,211.5	8,101.9	0.3	0.3
27	35	32	Russian Federation	968.1	1,532.1	1.5	1.7
28	32	33	Slovenia	3,659.3	12,485.4	0.1	0.2
29	34	34	Portugal	2,280.2	5,489.6	0.3	0.5
30	42	38	Lithuania	2,102.7	9,209.2	0.1	0.2
31	47	44	Bahrain	2,502.3	15,242.2	0.0	0.2
32	43	45	New Zealand	3,574.8	3,844.8	0.2	0.1
33	50	46	Estonia	2,099.7	12,056.6	0.0	0.1
34	48	47	Kuwait	1,785.7	10,400.5	0.1	0.3
35	55	48	Qatar	4,595.1	5,693.4	0.1	0.1

> **China, Brazil, Mexico and South Africa are in the group of emerging industrial countries that are changing their industrial structures in the right direction**

Table 8.2 (continued)

Industrial competitiveness ranking and selected indicators for industrialized countries and world ranking comparison, 2013

Group ranking 2013	World ranking 2010	World ranking 2013	Country	MVA per capita (2005 $) 2013	Manufactured exports per capita (current $) 2013	Impact of a country on world MVA (percent) 2013	Impact of a country on world manufactures trade (percent) 2013
36	44	52	Luxembourg	3,719.1	22,502.8	0.0	0.1
37	53	54	United Arab Emirates	2,612.9	2,314.1	0.3	0.2
38	62	61	Malta	1,842.9	11,318.0	0.0	0.0
39	71	72	Iceland	5,393.9	3,934.2	0.0	0.0
40	68	74	China, Hong Kong SAR	631.1	877.3	0.1	0.1
41	126	124	Bermuda	851.0	329.3	0.0	0.0
42	129	132	China, Macao SAR	297.2	140.2	0.0	0.0

Note: MVA is manufacturing value added. Red indicates a fall in the rankings from 2010, green is a rise.
Source: UNIDO elaboration based on Competitive Industrial Performance index database (UNIDO 2015a).

the Republic of Korea, although not among the winners, show very stable and enduring industrial competitiveness that relies on long-term advantages such as high technology, good education and advanced infrastructure.

Industry needs trade to show how internationally competitive the productive sector is, while trade needs industry to show that export performance has (or does not have) deep industrial roots. Let us consider Germany. It increased its industrial deepening significantly over 1990–2013, and changed its trade structure towards manufactured exports over 2000–2013 (Figure 8.3). The bubble size shows that it increased its export capacity strongly, with $15,504 in manufactured exports per capita in 2013. On the production side, Germany is on an ideal path. The manufacturing share in total GDP increased over 1990–2013 while the country's production deepening and production capacity increased simultaneously. This reflects the fact that technology-intensive sectors are less affected by declining prices of manufactured goods. More widely, the economies that shifted their manufacturing towards medium- and high-tech activities seem less vulnerable and more competitive.

Among other industrialized winners, Lithuania improved its ranking by four places over 2010–2013, increasing the share of MVA in overall GDP while boosting manufacturing production and exports. Singapore, the United Kingdom and Italy moved down in the ranking by two, six and three places.

Several transition economies as well as large countries such as China, Brazil, Mexico and South Africa are in the group of emerging industrial countries that are changing their industrial structures in the right direction. Emerging industrial countries in 2013 accounted for 31.6 percent and 31.7 percent of global MVA and global manufacturing trade respectively, marking gains largely due to China's fast expanding market.

Table 8.3 ranks the emerging industrial countries and shows indicators 1, 2, 7 and 8 for each country. Relative to 2010, more countries from emerging economies are now among the top global players. China, Mexico and Poland are the top five industrially competitive performers among emerging industrial countries; they also belong to the top quintile in the world CIP ranking. China and Poland have been the biggest winners since the beginning of the century.

Mexico and Poland saw a slight revival in competitiveness, of two and three places respectively. But Chile, Argentina and Venezuela are losing ground,

“ China, Mexico and Poland are the top five industrially competitive performers among emerging industrial countries

Figure 8.3
Trade and production structure in Germany, 1995, 2000 and 2013

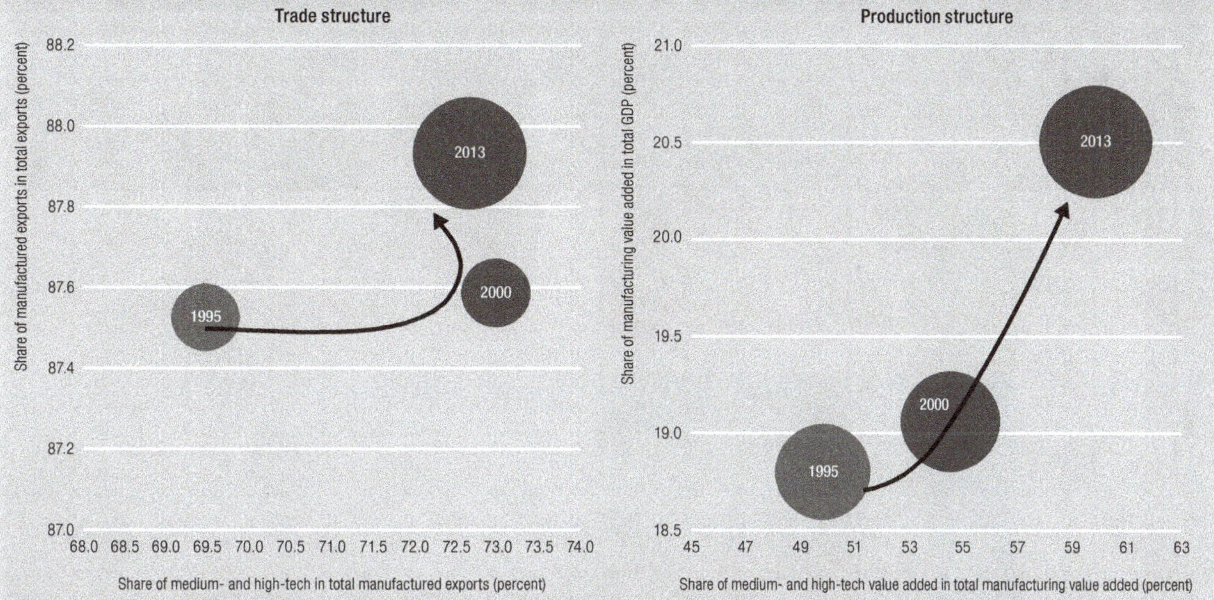

Note: GDP is gross domestic product. Bubble size represents manufactured exports and manufacturing value added per capita. Tech classification based on Annex B5, Table B5.1.
Source: UNIDO elaboration based on Competitive Industrial Performance index database (UNIDO 2015a).

Table 8.3
Industrial competitiveness ranking and selected indicators for emerging industrial countries and world ranking comparison, 2013

■ Top quintile ■ Upper middle quintile ■ Middle quintile ■ Lower middle quintile ■ Bottom quintile

Group ranking 2013	World ranking 2010	World ranking 2013	Country	MVA per capita (2005 $) 2013	Manufactured exports per capita (current $) 2013	Impact of a country on world MVA (percent) 2013	Impact of a country on world manufactures trade (percent) 2013
1	8	5	China	1,142.6	1,540.5	17.5	16.8
2	22	20	Mexico	1,340.9	2,514.4	1.8	2.4
3	24	21	Poland	2,323.6	4,656.8	1.0	1.4
4	25	26	Thailand	1,168.4	2,998.6	0.9	1.6
5	31	30	Turkey	1,548.3	1,778.4	1.3	1.1
6	33	35	Brazil	756.7	766.8	1.7	1.2
7	36	36	Saudi Arabia	2,046.1	2,429.9	0.7	0.6
8	40	37	Belarus	1,551.2	3,325.9	0.2	0.2
9	39	39	Romania	854.6	2,625.9	0.2	0.4
10	37	40	Argentina	1,524.6	908.9	0.7	0.3
11	38	41	South Africa	894.0	1,208.9	0.5	0.5
12	41	42	Indonesia	451.3	438.8	1.3	0.9
13	45	43	India	161.7	223.3	2.2	2.2

“China and India climbed in the rankings, while Brazil and South Africa both fell

Table 8.3 (continued)

Industrial competitiveness ranking and selected indicators for emerging industrial countries and world ranking comparison, 2013

Group ranking 2013	World ranking 2010	World ranking 2013	Country	MVA per capita (2005 $) 2013	Manufactured exports per capita (current $) 2013	Impact of a country on world MVA (percent) 2013	Impact of a country on world manufactures trade (percent) 2013
14	46	49	Greece	1,395.6	2,623.3	0.2	0.2
15	49	51	Chile	1,129.3	2,243.6	0.2	0.3
16	51	55	Ukraine	358.7	1,088.5	0.2	0.4
17	52	56	Croatia	1,350.0	2,550.8	0.1	0.1
18	57	57	Bulgaria	753.8	2,895.1	0.1	0.2
19	59	58	Tunisia	652.8	1,317.7	0.1	0.1
20	65	59	Latvia	1,057.1	5,028.2	0.0	0.1
21	61	60	Costa Rica	1,048.2	1,763.8	0.1	0.1
22	66	62	Oman	1,297.8	2,308.2	0.1	0.1
23	70	66	Kazakhstan	605.9	1,042.8	0.1	0.1
24	67	68	Colombia	493.2	333.4	0.3	0.1
25	56	69	Venezuela, Bolivarian Rep. of	806.9	425.4	0.3	0.1
26	74	70	Serbia	361.3	1,289.2	0.0	0.1
27	78	80	Uruguay	979.7	994.7	0.0	0.0
28	83	82	Mauritius	1,065.9	1,468.5	0.0	0.0
29	84	83	Macedonia, Former Yugoslav Rep. of	415.5	1,828.1	0.0	0.0
30	87	91	Brunei Darussalam	2,740.2	887.4	0.0	0.0
31	91	95	Cyprus	871.8	583.4	0.0	0.0
32	109	112	Suriname	599.2	625.0	0.0	0.0

Note: MVA is manufacturing value added. Red represents a fall in the rankings from year 2010, while green is a rise.
Source: UNIDO elaboration based on Competitive Industrial Performance index database (UNIDO 2015a).

challenged by the emerging competitiveness of Asian nations. The BRICS (Brazil, Russian Federation, India, China and South Africa) economies have mixed patterns: China and India climbed in the rankings, while Brazil and South Africa both fell.

Among other developing countries (excluding least developed countries), most countries maintained their rankings (Table 8.4). Viet Nam, Philippines, Peru, Trinidad and Tobago, and Iran are the five most competitive nations in this group. Viet Nam has improved its ranking by an impressive eight places. Its growth is supported by manufactured exports from mainly foreign companies alongside technological upgrading of its industries, a result of government policy to overhaul the financial system and encourage foreign investment. The Philippines advanced in all pillars except for a minor decline in medium- and high-tech share in industry. It edged up one place to 53 among 142 countries in 2013.

Iran lost five places, mainly because it lost share in world manufactured exports due to the effects of European Union sanctions from January 2013.

The contribution of the least developed countries to world MVA and world manufactured exports is very weak, and group totals—quite marginal—are dominated by a few countries such as Bangladesh and Cambodia. In many competitive industrial markets, export promotion plays a critical role in long-term

“Among other developing countries, Viet Nam, Philippines, Peru, Trinidad and Tobago, and Iran are the five most competitive

Table 8.4

Industrial competitiveness ranking and selected indicators for other industrial countries (excluding least developed countries) and world ranking comparison, 2013

■ Top quintile ■ Upper middle quintile ■ Middle quintile ■ Lower middle quintile ■ Bottom quintile

Group ranking 2013	World ranking 2010	World ranking 2013	Country	MVA per capita (2005 $) 2013	Manufactured exports per capita (current $) 2013	Impact of a country on world MVA (percent) 2013	Impact of a country on world manufactures trade (percent) 2013
1	58	50	Viet Nam	235.6	1,128.9	0.2	0.8
2	54	53	Philippines	353.4	495.6	0.4	0.4
3	63	63	Peru	604.7	714.6	0.2	0.2
4	64	64	Trinidad and Tobago	941.4	5,564.4	0.0	0.1
5	60	65	Iran, Islamic Rep. of	324.9	340.7	0.3	0.2
6	69	67	Morocco	323.6	534.8	0.1	0.1
7	72	71	Egypt	242.2	228.2	0.2	0.1
8	73	73	El Salvador	647.8	789.7	0.0	0.0
9	75	75	Pakistan	139.1	112.1	0.3	0.2
10	77	76	Guatemala	404.6	448.0	0.1	0.1
11	76	78	Jordan	398.7	730.9	0.0	0.0
12	85	79	Botswana	465.3	3,573.4	0.0	0.1
13	79	81	Sri Lanka	357.2	345.5	0.1	0.1
14	86	84	Namibia	491.1	1,713.0	0.0	0.0
15	81	85	Lebanon	480.6	620.7	0.0	0.0
16	89	86	Bosnia and Herzegovina	323.6	1,149.7	0.0	0.0
17	82	87	Algeria	183.1	377.6	0.1	0.1
18	88	88	Swaziland	641.5	888.8	0.0	0.0
19	90	89	Ecuador	408.8	284.4	0.1	0.0
20	92	92	Honduras	270.1	301.3	0.0	0.0
21	94	93	Côte d'Ivoire	115.5	268.4	0.0	0.0
22	95	94	Georgia	342.8	317.9	0.0	0.0
23	96	96	Jamaica	274.2	487.0	0.0	0.0
24	102	97	Bahamas	909.6	657.4	0.0	0.0
25	101	98	Albania	224.9	465.0	0.0	0.0
26	97	99	Syrian Arab Republic	65.2	231.3	0.0	0.0
27	105	100	Armenia	284.4	318.0	0.0	0.0
28	99	101	Bolivia, Plurinational State of	152.6	281.9	0.0	0.0
29	104	102	Congo, Rep. of the	96.9	625.3	0.0	0.0
30	110	103	Paraguay	179.0	233.9	0.0	0.0
31	100	104	Barbados	645.0	764.3	0.0	0.0
32	111	105	Azerbaijan	173.8	251.5	0.0	0.0
33	106	107	Cameroon	153.8	65.1	0.0	0.0
34	108	108	Fiji	445.3	457.6	0.0	0.0
35	98	110	Nigeria	44.6	35.1	0.1	0.0

> **The Philippines advanced in all pillars except for a minor decline in medium- and high-tech share in industry**

Table 8.4 (continued)

Industrial competitiveness ranking and selected indicators for other industrial countries (excluding least developed countries) and world ranking comparison, 2013

Group ranking 2013	World ranking 2010	World ranking 2013	Country	MVA per capita (2005 $) 2013	Manufactured exports per capita (current $) 2013	Impact of a country on world MVA (percent) 2013	Impact of a country on world manufactures trade (percent) 2013
36	117	111	Moldova, Rep. of	118.0	312.1	0.0	0.0
37	107	113	Kenya	61.1	58.1	0.0	0.0
38	112	114	Gabon	274.5	642.8	0.0	0.0
39	113	115	Papua New Guinea	71.8	324.0	0.0	0.0
40	115	116	Mongolia	91.9	680.1	0.0	0.0
41	118	117	Panama	338.1	90.2	0.0	0.0
42	119	118	State of Palestine	148.9	114.1	0.0	0.0
43	123	119	Ghana	52.9	79.8	0.0	0.0
44	120	122	Belize	440.0	376.8	0.0	0.0
45	122	125	Kyrgyzstan	54.0	94.9	0.0	0.0
46	127	130	Tajikistan	59.1	15.5	0.0	0.0
47	128	131	Saint Lucia	276.3	244.2	0.0	0.0
48	132	133	Cabo Verde	166.3	75.6	0.0	0.0
49	137	137	Iraq	36.9	14.9	0.0	0.0
50	142	141	Tonga	168.7	14.9	0.0	0.0

Note: MVA is manufacturing value added. Red represents a fall in the rankings from 2010, green is a rise. Selected countries belong to the category "Other developing countries" based on Annex B1, Table B1.2.
Source: UNIDO elaboration based on Competitive Industrial Performance index database (UNIDO 2015a).

competitiveness by supporting investment and maintaining technology. This seems to need more attention among least developed countries.

Most of the countries belong to the bottom quintile of the CIP ranking (Table 8.5). In 2013 the average MVA per capita in this group was less than $48, and average manufacturing exports per capita were only $67. These countries lack capacity to produce and export manufactured goods, and their economic structures (share of MVA in total GDP) rely very little on manufacturing.

Changes in industrial competitiveness, 1990–2013 and 2000–2013

Data for long-term changes in industrial competitiveness for top-quintile countries in the CIP ranking (Figure 8.4) suggest that Poland, China and the Republic of Korea were experiencing the rapid and cumulative process of rising industrial competitiveness before the turn of the century.

Changes between 1990 and 2013

Poland saw the biggest change, jumping 29 positions after 1990 to rank 21 in 2013. Second was China, which leaped 28 positions, leading the BRICS countries in global competitiveness. (Indeed, the gap in competitiveness between China and the other BRICS countries widened sharply, with China overtaking the Russian Federation and setting a 27-place lead.) The Czech Republic and the Republic of Korea both registered notable jumps of 13 places. Among the South-East and East Asian economies, Thailand and Malaysia climbed 12 and 5 places. Changes were also seen in European Union manufacturing-led exporters such as Slovakia, Hungary and Ireland, as well as Mexico in Latin America.

"The contribution of the least developed countries to world manufacturing value added and world manufactured exports is very weak

Table 8.5

Industrial competitiveness ranking and selected indicators for least developed countries and world ranking comparison, 2013

Group ranking 2013	World ranking 2010	World ranking 2013	Country	MVA per capita (2005 $) 2013	Manufactured exports per capita (current $) 2013	Impact of a country on world MVA (percent) 2013	Impact of a country on world manufactures trade (percent) 2013
1	80	77	Bangladesh	118.28	152.13	0.21	0.19
2	93	90	Cambodia	146.84	428.64	0.03	0.05
3	103	106	Senegal	98.86	117.20	0.02	0.01
4	114	109	Zambia	76.93	182.34	0.01	0.02
5	133	120	Mozambique	50.95	44.62	0.02	0.01
6	116	121	Tanzania, United Rep. of	43.04	32.95	0.02	0.01
7	121	123	Madagascar	37.22	42.51	0.01	0.01
8	134	126	Niger	18.40	66.96	0.00	0.01
9	125	127	Yemen	59.45	36.04	0.02	0.01
10	124	128	Nepal	26.32	23.97	0.01	0.01
11	130	129	Uganda	27.28	17.28	0.01	0.01
12	135	134	Haiti	50.58	6.19	0.01	0.00
13	131	135	Malawi	22.59	22.21	0.00	0.00
14	136	136	Rwanda	22.03	26.96	0.00	0.00
15	139	138	Ethiopia	13.33	6.81	0.01	0.01
16	138	139	Central African Republic	15.92	4.96	0.00	0.00
17	142	141	Eritrea	10.91	0.44	0.00	0.00
18	141	141	Gambia	22.53	0.64	0.00	0.00

Note: MVA is manufacturing value added. Red indicates a fall in the rankings from year 2010, while green is a rise. Group ranking is based on CIP ranking for 2013.
Source: UNIDO elaboration based on Competitive Industrial Performance index database (UNIDO 2015a).

Other countries went the other way—some sharply, including the United Kingdom, Canada, Italy, Israel and Australia. The United Kingdom slid the most, by 14 positions, mainly due to declining export market share. The economy has changed its structure from manufacturing since 1990 and industrial output and employment fell relative to services. The share of manufacturing in GDP tumbled from 16 percent in 1990 to 10 percent in 2013, and MVA per capita also decreased.

Yet despite the significant gains and losses over time in country rankings generally, the three top positions have not changed much since the early 1990s, being shared among Germany, Japan and the United States.

For countries outside the top quintile, Viet Nam was a big winner, gaining 42 CIP index places to 50 in 2013. Its industry has benefited from global trends such as increased foreign direct investment, globalized value chains and greater regional dynamism.

In the other direction went China, Macao SAR and China, Hong Kong SAR, due to severe deindustrialization and a shift to services. Portugal also lost industrial competitiveness as manufacturing exports fell. Among the BRICS (excluding China), the Russian Federation slid seven positions to 32, reflecting reductions in its capacity to innovate and its declining share of manufacturing in output. Brazil, India and South Africa also lost ground on the CIP index over 1990–2013. By contrast, India improved 19 places to 43, becoming a key global player. Still, the country needs to overcome technology barriers and invest more in medium- and high-tech R&D activities. That may well be problematic over

Top-quintile economies that witnessed remarkable gains in industrial competitiveness include China, Poland, Slovakia and the Republic of Korea

Figure 8.4
Changes in the CIP index for countries in the top quintile, from 1990 to 2013 and from 2000 to 2013

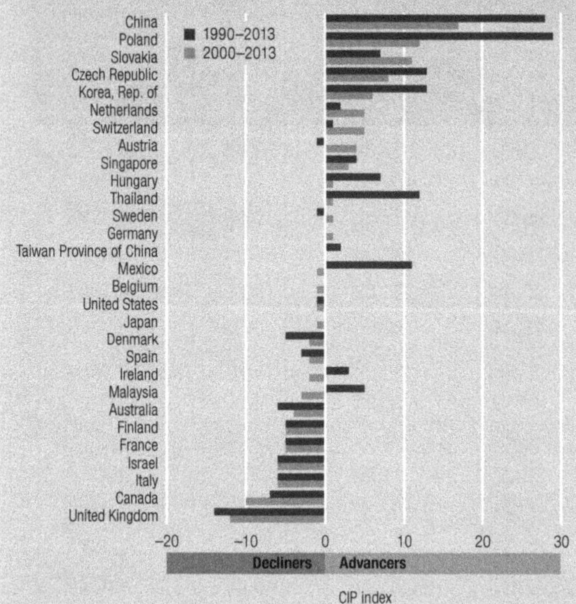

Source: UNIDO elaboration based on Competitive Industrial Performance index database (UNIDO 2015a).

the long term, as its growth model has not followed the manufacturing path but has gone straight to a services economy, without strengthening its industrial base.

Changes between 2000 and 2013

Since the turn of the century, top-quintile economies that witnessed remarkable gains in industrial competitiveness include China, Poland, Slovakia and the Republic of Korea. Slovakia gained by 11 places since 2000 due to an increase in per capita exports following its entry into the European Union.

Those moving down in the top quintile were the United Kingdom, Canada, Italy and Israel, which lost between 12 and 6 positions. Among the largest emerging industrial countries, Mexico was penalized due to the decline in its MVA world market share as it fell back slightly in industrial competitiveness to 20. Malaysia dropped from 21 to 24, reflecting a lack of progress in entering knowledge-based industries, which is beginning to undermine its industrial competitiveness.

Other countries that improved competitiveness strongly but were not in the top quintile were Viet Nam, Nigeria and Iran, which rose by 29, 26 and 21 places, respectively. Nigeria's improvement was based on expanding manufacturing production and manufactured exports; however, the medium- and high-tech sector's share in MVA and manufacturing exports declined. Iran and Viet Nam improved all CIP subcomponents substantially.

Lithuania, Albania, Kazakhstan and Peru (up 18, 18, 16 and 13 places, respectively) also saw large gains. Lithuania's improved competitiveness came from its advances in high-tech manufacturing, despite a decline in the share of manufactured goods in total exports. Gains for Kazakhstan, Albania and Peru involved expanded industrial activity.

From short-term and zero-sum to long-term and win-win

Strong competitiveness is vital in the new global market. But some commentators regard competitiveness as a zero-sum game among nations in a short-term struggle to sustain their positions. Such an outlook is partly responsible for economic recessions, which need to be countered by industrial policies that ensure sustained industrial competitiveness. The 2008 financial crisis proved that short-term actions may produce quick and lucrative results, but the long-term consequences can be disastrous.

Germany, Japan and the United States are examples of countries that kept their positions in industrial competitiveness over the long term. They did this by maintaining knowledge and high-tech industries that are resource efficient and energy efficient. These countries enjoy higher, and more sustainable, industrial performance.

The success of these countries' industrial policies may seem readily apparent, but it is not always seen this way by politicians. To do this, a long-term vision is required. Sustainable industrial competitiveness may involve sacrificing current benefits and spending more on the foundations of the industry, providing support and infrastructure, for the goal of long-term gains in inclusive well-being and prosperity via industrialization.

Annexes

Annex A1

World Bank country and economy classification

Table A1.1
World Bank countries and economies by income classification (gross national income per capita)

High income ($12,746 or more)				
Andorra	Curaçao	Ireland	New Zealand	St. Kitts and Nevis
Antigua and Barbuda	Cyprus	Isle of Man	Northern Mariana Islands	St. Martin (French)
Aruba	Czech Rep.	Israel	Norway	Sweden
Australia	Denmark	Italy	Oman	Switzerland
Austria	Equatorial Guinea	Japan	Poland	Taiwan Province of China
Bahamas	Estonia	Korea, Rep. of	Portugal	Trinidad and Tobago
Bahrain	Faeroe Islands	Kuwait	Puerto Rico	Turks and Caicos Islands
Barbados	Finland	Latvia	Qatar	United Arab Emirates
Belgium	France	Liechtenstein	Russian Federation	United Kingdom
Bermuda	French Polynesia	Lithuania	San Marino	United States
Brunei Darussalam	Germany	Luxembourg	Saudi Arabia	Uruguay
Canada	Greece	Macao SAR, China	Singapore	Virgin Islands (United States)
Cayman Islands	Greenland	Malta	Sint Maarten (Dutch)	
Channel Islands	Guam	Monaco	Slovakia	
Chile	Hong Kong SAR, China	Netherlands	Slovenia	
Croatia	Iceland	New Caledonia	Spain	
Upper middle income ($12,475–$4,126)				
Albania	Bulgaria	Hungary	Marshall Islands	South Africa
Algeria	China	Iran, Islamic Rep. of	Mauritius	St. Lucia
American Samoa	Colombia	Iraq	Mexico	St. Vincent and the Grenadines
Angola	Costa Rica	Jamaica	Montenegro	Suriname
Argentina	Cuba	Jordan	Namibia	Thailand
Azerbaijan	Dominica	Kazakhstan, Rep. of	Palau	Tonga
Belarus	Dominican Rep.	Lebanon	Panama	Tunisia
Belize	Ecuador	Libya	Peru	Turkey
Bosnia and Herzegovina	Fiji	Macedonia, Former Yugoslav Rep. of	Romania	Turkmenistan
Botswana	Gabon	Malaysia	Serbia	Tuvalu
Brazil	Grenada	Maldives	Seychelles	Venezuela, Bolivarian Rep. of

Lower middle income ($4,125–$1,046)				
Armenia	Ghana	Mauritania	Samoa	Uzbekistan
Bhutan	Guatemala	Micronesia, Federated States of	São Tomé and Principe	Vanuatu
Bolivia, Plurinational State of	Guyana	Moldova, Rep. of	Senegal	Viet Nam
Cabo Verde	Honduras	Mongolia	Solomon Islands	West Bank and Gaza
Cameroon	India	Morocco	South Sudan	Yemen
Congo, Rep. of the	Indonesia	Nicaragua	Sri Lanka	Zambia
Côte d'Ivoire	Kiribati	Nigeria	Sudan	
Djibouti	Kosovo	Pakistan	Swaziland	
Egypt	Kyrzygzstan	Papua New Guinea	Syrian Arab Rep.	
El Salvador	Lao People's Dem. Rep.	Paraguay	Timor-Leste	
Georgia	Lesotho	Philippines	Ukraine	
Low income ($1,045 or less)				
Afghanistan	Comoros	Kenya	Nepal	Uganda
Bangladesh	Congo, Dem. Rep. of the	Korea, Dem. People's Rep. of	Niger	Zimbabwe
Benin	Eritrea	Liberia	Rwanda	
Burkina Faso	Ethiopia	Madagascar	Sierra Leone	
Burundi	Gambia	Malawi	Somalia	
Cambodia	Guinea	Mali	Tajikistan	
Central African Rep.	Guinea-Bissau	Mozambique	Tanzania, United Rep. of	
Chad	Haiti	Myanmar	Togo	

Table A1.2
World Bank countries and economies by region classification

Industrialized countries and economies				
Americas				
Aruba	Bermuda	Canada	United States	
Bahamas	British Virgin Islands	Greenland		
Asia				
Brunei Darussalam	Hong Kong SAR, China	New Zealand	Taiwan Province of China	
French Polynesia	Japan	Singapore		
Europe				
Andorra	Finland	Ireland	Monaco	Spain
Austria	France	Italy	Netherlands	Sweden
Belgium	Germany	Liechtenstein	Norway	Switzerland
Denmark	Iceland	Luxembourg	San Marino	United Kingdom
North Africa and Middle East				
Cyprus	Israel	Kuwait	Qatar	United Arab Emirates
Oceania				
Australia	New Zealand			
Industrializing countries and economies				
Central America and Caribbean				
Antigua and Barbuda	Cuba	Grenada	Jamaica	Saint Kitts and Nevis
Barbados	Dominica	Guatemala	Nicaragua	Saint Lucia
Belize	Dominican Rep.	Haiti	Panama	St. Vincent and the Grenadines
Costa Rica	El Salvador	Honduras	Puerto Rico	Trinidad and Tobago
East Asia				
China	Macao SAR, China	Korea, Dem. People's Rep. of	Korea, Rep. of	Mongolia
Eastern Europe (excluding USSR)				
Albania	Czech Rep.	Kosovo	Poland	Slovakia
Bosnia and Herzegovina	Czechoslovakia, Former	Macedonia, Former Yugoslav Rep. of	Romania	Slovenia
Bulgaria	Hungary	Montenegro	Serbia	Yugoslavia, Former
Croatia				
Former Union of Soviet Socialist Republics (USSR)				
Armenia	Georgia	Latvia	Russian Federation	Ukraine
Azerbaijan	Kazakhstan	Lithuania	Tajikistan	USSR, Former
Belarus	Kyrgyzstan	Moldova, Rep. of	Turkmenistan	Uzbekistan
Estonia				
North America				
Mexico				
North Africa and Middle East				
Algeria	Iraq	Libya	Saudi Arabia	Tunisia
Bahrain	Jordan	Morocco	Sudan	Turkey
Egypt	Lebanon	Oman	Syrian Arab Rep.	Yemen

Oceania				
Fiji	New Caledonia	Samoa	Tonga	
Kiribati	Papua New Guinea	Solomon Islands	Vanuatu	
South Asia				
Afghanistan	Bhutan	Iran, Islamic Rep. of	Nepal	Sri Lanka
Bangladesh	India	Maldives	Pakistan	
South America				
Argentina	Chile	Guyana	Suriname	
Bolivia, Plurinational State of	Colombia	Paraguay	Uruguay	
Brazil	Ecuador	Peru	Venezuela, Bolivarian Rep. of	
South-East Asia				
Cambodia	Lao People's Dem. Rep.	Myanmar	Thailand	
Indonesia	Malaysia	Philippines	Viet Nam	
Sub-Saharan Africa				
Angola	Congo, Rep. of the	Guinea-Bissau	Namibia	Swaziland
Benin	Congo, Dem. Rep. of the	Kenya	Niger	Tanzania, United Rep. of
Botswana	Côte d'Ivoire	Lesotho	Nigeria	Togo
Burkina Faso	Djibouti	Liberia	Rwanda	Uganda
Burundi	Equatorial Guinea	Madagascar	São Tomé and Principe	Zambia
Cabo Verde	Ethiopia	Malawi	Senegal	Zimbabwe
Cameroon	Gabon	Mali	Seychelles	
Central African Rep.	Gambia	Mauritania	Sierra Leone	
Chad	Ghana	Mauritius	Somalia	
Comoros	Guinea	Mozambique	South Africa	
Western Europe				
Greece	Malta	Portugal		

Source: UNIDO's elaboration based on World Bank (2015b).
Note: World Bank GNI per capita operational guidelines and analytical classifications as per 1990.

Annex A2

Classification of manufacturing industries by technology group

ISIC full description	Abbreviation used in this report	ISIC code rev. 3	Technology group
Food and beverages	Food and beverages	15	Low tech
Tobacco products	Tobacco	16	Low tech
Textiles	Textiles	17	Low tech
Wearing apparel, fur and leather products and footwear	Wearing apparel	18 and 19	Low tech
Wood products (excluding furniture)	Wood products	20	Low tech
Paper and paper products	Paper	21	Low tech
Printing and publishing	Printing and publishing	22	Low tech
Furniture, manufacturing n.e.c.	Furniture, n.e.c.	36	Low tech
Coke, refined petroleum products and nuclear fuel	Coke and refined petroleum	23	Medium tech
Rubber and plastic products	Rubber and plastic	25	Medium tech
Non-metallic mineral products	Non-metallic minerals	26	Medium tech
Basic metals	Basic metals	27	Medium tech
Fabricated metal products	Fabricated metals	28	Medium tech
Chemicals and chemical products	Chemicals	24	High tech
Machinery and equipment n.e.c. and office, accounting, computing machinery	Machinery and equipment	29 and 30	High tech
Electrical machinery and apparatus and radio, television and communication equipment	Electrical machinery and apparatus	31 and 32	High tech
Medical, precision and optical instruments	Precision instruments	33	High tech
Motor vehicles, trailers, semi-trailers and other transport equipment	Motor vehicles	34 and 35	High tech

Note: ISIC is International Standard Industrial Classification; n.e.c. is not elsewhere classified. The three technology groups follow OECD (2005) technology classification based on R&D intensity relative to value added and gross production statistics.
Source: UNIDO's elaboration based on INDSTAT2 (UNIDO 2012).

Annex A3

Sectoral disaggregation and definition of modern market activities

International Standard Industrial Classification rev. 3 code	Sector name
C	Mining and quarrying
D	Manufacturing
E	Electricity, gas and water
F	Construction
I	Transport, storage and communication
JtK-70	Finance, insurance, and business services
CtF+I+JtK-70	Modern market activities
AtB	Agriculture, hunting, forestry and fishing
GtH	Wholesale and retail trade and restaurants and hotels
70	Real estate
LtQ	Community, social and personal services
AtB, GtH, 70, LtQ	Non-modern activities

Source: Adapted from Lavopa and Szirmai (2014).

Annex A4

A stochastic frontier approach for Figure 5.9

This annex reviews the nature of the stochastic frontier problem. Suppose that a country has a production function $f(z_{it}; \beta)$. In a world without error or inefficiency, in time t, the ith country would produce:

1) $q_{it} = f(z_{it}, \beta)\varepsilon_{it}$

where z is inputs and ε_{it} is the level of efficiency of country i at time t. ε_{it} must be in the interval (0,1]. If $\varepsilon_{it} = 1$, the country is making the most from inputs z and thereby achieving the maximal output with the technology embodied in the production function. In this case a country achieves maximum efficiency. When $\varepsilon_{it} < 1$, the country is not making the most of its inputs and hence production is inefficient.

Output is also assumed to be subject to random shocks. Taking natural logarithms the equation becomes:

2) $\ln(q_{it}) = \log [f(z_{it}, \beta)] + \log \varepsilon_{it} + v_{it}$.

When considering emissions, the interpretation of q can be interpreted as undesirable outputs. If $\varepsilon_{it} = 1$, the country is achieving the maximal undesirable output with the technology embodied in the production function. In this case we say that the country achieves the maximum inefficiency.

In Chapter 5, z is understood as a vector of drivers of carbon dioxide emissions. Borrowing from Stern (2002), we select the simple model specification:

3) $\ln (E_{it}) = \log [f(GDP_{it}, Coal_{it}, Natgas_{it}, Oil_{it}, Nuclear_{it}, Hydroelectric_{it}, \beta_{kit})] + \log \varepsilon_{it} + v_{it}$

where Emissions depends on the level of GDP and on consumption of inputs in the production process.

Annex B1

Country and economy groups

Table B1.1
Countries and economies by region

Industrialized countries and economies				
Asia and the Pacific				
Bahrain	Korea, Rep. of	Macao SAR, China	Qatar	Taiwan Province of China
Hong Kong SAR, China	Kuwait	Malaysia	Singapore	United Arab Emirates
Japan				
Europe				
Austria	France	Iceland	Portugal	Switzerland
Belgium	Germany	Lithuania	Russian Federation	United Kingdom
Czech Rep.	Hungary	Luxembourg	Slovakia	Liechtenstein
Denmark	Andorra	Malta	Slovenia	Monaco
Estonia	Ireland	Netherlands	Spain	San Marino
Finland	Italy	Norway	Sweden	
North America				
Bermuda	Canada	Greenland	United States	
Others				
Aruba	Cayman Islands	Guam	New Zealand	
Australia	French Guiana	Israel	Puerto Rico	
British Virgin Islands	French Polynesia	New Caledonia	Virgin Islands (United States)	
Industrializing countries and economies				
Africa				
Algeria	Congo, Rep. of the	Kenya	Niger	Swaziland
Angola	Côte d'Ivoire	Lesotho	Nigeria	Tanzania, United Rep. of
Benin	Djibouti	Liberia	Réunion	Togo
Botswana	Egypt	Libya	Rwanda	Tunisia
Burkina Faso	Equatorial Guinea	Madagascar	Sao Tome and Principe	Uganda
Burundi	Eritrea	Malawi	Senegal	Zambia
Cabo Verde	Ethiopia	Mali	Seychelles	Zimbabwe
Cameroon	Gabon	Mauritania	Sierra Leone	
Central African Rep.	Gambia	Mauritius	Somalia	
Chad	Ghana	Morocco	South Africa	
Comoros	Guinea	Mozambique	South Sudan	
Congo, Dem. Rep. of the	Guinea-Bissau	Namibia	Sudan	

B1

Asia and the Pacific				
Afghanistan	India	Lebanon	Papua New Guinea	Timor-Leste
Armenia	Indonesia	Maldives	Philippines	Tonga
Azerbaijan	Iran, Islamic Rep. of	Marshall Islands	Samoa	Turkmenistan
Bangladesh	Iraq	Micronesia, Federated States of	Saudi Arabia	Tuvalu
Bhutan	Jordan	Mongolia	Solomon Islands	Uzbekistan
Brunei Darussalam	Kazakhstan	Myanmar	Sri Lanka	Vanuatu
Cambodia	Kiribati	Nepal	State of Palestine	Viet Nam
China	Korea, Dem. People's Rep. of	Oman	Syrian Arab Rep.	Yemen
Cook Islands	Kyrgyzstan	Pakistan	Tajikistan	
Fiji	Lao People's Dem. Rep.	Palau	Thailand	
Europe				
Albania	Croatia	Latvia	Romania	Ukraine
Belarus	Cyprus	Moldova, Rep. of	Serbia	
Bosnia and Herzegovina	Georgia	Montenegro	Macedonia, Former Yugoslav Rep. of	
Bulgaria	Greece	Poland	Turkey	
Latin America and the Caribbean				
Anguilla	Chile	Grenada	Mexico	Saint Vincent and the Grenadines
Antigua and Barbuda	Colombia	Guadeloupe	Montserrat	Suriname
Argentina	Costa Rica	Guatemala	Nicaragua	Trinidad and Tobago
Bahamas	Cuba	Guyana	Panama	Uruguay
Barbados	Dominica	Haiti	Paraguay	Venezuela, Bolivarian Rep. of
Belize	Dominican Rep.	Honduras	Peru	
Bolivia, Plurinational State of	Ecuador	Jamaica	Saint Kitts and Nevis	
Brazil	El Salvador	Martinique	Saint Lucia	

Table B1.2
Countries and economies by industrialization level

Industrialized countries and economies				
Andorra	Taiwan Province of China	Iceland	Monaco	Slovenia
Aruba	Czech Rep.	Ireland	Netherlands	Spain
Australia	Denmark	Israel	New Caledonia	Sweden
Austria	Estonia	Italy	New Zealand	Switzerland
Bahrain	Finland	Japan	Norway	United Arab Emirates
Belgium	France	Korea, Rep. of	Portugal	United Kingdom
Bermuda	French Guiana	Kuwait	Puerto Rico	United States
British Virgin Islands	French Polynesia	Liechtenstein	Qatar	Virgin Islands (United States)
Canada	Germany	Lithuania	Russian Federation	
Cayman Islands	Greenland	Luxembourg	San Marino	
Hong Kong SAR, China	Guam	Malaysia	Singapore	
Macao SAR, China	Hungary	Malta	Slovakia	
Industrializing countries and economies (or developing and emerging industrial economies)				
Emerging industrial countries and economies				
Argentina	Colombia	Kazakhstan	Saudi Arabia	Turkey
Belarus	Costa Rica	Latvia	Serbia	Ukraine
Brazil	Croatia	Mauritius	South Africa	Uruguay
Brunei Darussalam	Cyprus	Mexico	Suriname	Venezuela, Bolivarian Rep. of
Bulgaria	Greece	Oman	Thailand	
Chile	India	Poland	Macedonia, Former Yugoslav Rep. of	
China	Indonesia	Romania	Tunisia	

Other developing countries and economies				
Albania	Cook Islands	Guyana	Mongolia	Saint Lucia
Algeria	Cuba	Honduras	Montenegro	Saint Vincent and the Grenadines
Angola	Côte d'Ivoire	Iran, Islamic Rep. of	Montserrat	Seychelles
Anguilla	Dominica	Iraq	Morocco	Sri Lanka
Antigua and Barbuda	Dominican Rep.	Jamaica	Namibia	State of Palestine
Armenia	Ecuador	Jordan	Nicaragua	Swaziland
Azerbaijan	Egypt	Kenya	Nigeria	Syrian Arab Rep.
Bahamas	El Salvador	Korea, Dem. People's Rep. of	Pakistan	Tajikistan
Barbados	Equatorial Guinea	Kyrgyzstan	Palau	Tonga
Belize	Fiji	Lebanon	Panama	Trinidad and Tobago
Bolivia, Plurinational State of	Gabon	Libya	Papua New Guinea	Turkmenistan
Bosnia and Herzegovina	Georgia	Maldives	Paraguay	Uzbekistan
Botswana	Ghana	Marshall Islands	Peru	Viet Nam
Cameroon	Grenada	Martinique	Philippines	Zimbabwe
Cape Verde	Guadeloupe	Micronesia, Federated States of	Réunion	
Congo, Rep. of the	Guatemala	Moldova, Rep. of	Saint Kitts and Nevis	
Least developed countries and economies				
Afghanistan	Congo, Dem. Rep. of the	Lesotho	Rwanda	Timor-Leste
Bangladesh	Djibouti	Liberia	Samoa	Togo
Benin	Eritrea	Madagascar	Sao Tome and Principe	Tuvalu
Bhutan	Ethiopia	Malawi	Senegal	Uganda
Burkina Faso	Gambia	Mali	Sierra Leone	Vanuatu
Burundi	Guinea	Mauritania	Solomon Islands	Yemen
Cambodia	Guinea-Bissau	Mozambique	Somalia	Zambia
Central African Rep.	Haiti	Myanmar	South Sudan	
Chad	Kiribati	Nepal	Sudan	
Comoros	Lao People's Dem. Rep.	Niger	Tanzania, United Rep. of	

Table B1.3
Countries and economies by income

High income				
Andorra	Curacao	Hong Kong SAR, China	Netherlands	Slovenia
Anguilla	Cyprus	Hungary	New Caledonia	Spain
Aruba	Czech Rep.	Iceland	New Zealand	Sweden
Australia	Denmark	Ireland	Norway	Switzerland
Austria	Equatorial Guinea	Israel	Oman	Taiwan Province of China
Bahamas	Estonia	Italy	Poland	Trinidad and Tobago
Bahrain	Finland	Japan	Portugal	United Arab Emirates
Barbados	France	Korea, Rep. of	Puerto Rico	United Kingdom
Belgium	French Polynesia	Kuwait	Qatar	United States

Bermuda	Germany	Liechtenstein	Saint Kitts and Nevis	Virgin Islands (United States)
Brunei Darussalam	Greece	Luxembourg	Saudi Arabia	
Canada	Greenland	Macao SAR, China	Singapore	
Croatia	Guam	Malta	Slovakia	
Upper middle income				
Algeria	Chile	Jamaica	Mexico	Seychelles
American Samoa	China	Jordan	Montenegro	South Africa
Angola	Colombia	Kazakhstan	Namibia	Suriname
Antigua and Barbuda	Costa Rica	Latvia	Palau	Thailand
Argentina	Cuba	Lebanon	Panama	Tunisia
Azerbaijan	Dominica	Libya	Peru	Turkey
Belarus	Dominican Rep.	Lithuania	Romania	Turkmenistan
Bosnia and Herzegovina	Ecuador	Macedonia, Former Yugoslav Rep. of	Russian Federation	Uruguay
Botswana	Gabon	Malaysia	Saint Lucia	Venezuela, Bolivarian Rep. of
Brazil	Grenada	Maldives	Saint Vincent and the Grenadines	
Bulgaria	Iran, Islamic Rep. of	Mauritius	Serbia	
Lower middle income				
Albania	El Salvador	Lao People's Dem. Rep.	Paraguay	Syrian Arab Rep.
Armenia	Fiji	Lesotho	Philippines	Timor-Leste
Belize	Georgia	Marshall Islands	Samoa	Tonga
Bhutan	Ghana	Micronesia, Federated States of	São Tomé and Principe	Tuvalu
Bolivia, Plurinational State of	Guatemala	Moldova, Rep. of	Senegal	Ukraine
Cabo Verde	Guyana	Mongolia	Solomon Islands	Uzbekistan
Cameroon	Honduras	Morocco	South Sudan	Vanuatu
Congo, Rep. of the	India	Nicaragua	Sri Lanka	Viet Nam
Côte d'Ivoire	Indonesia	Nigeria	State of Palestine	Yemen
Djibouti	Iraq	Pakistan	Sudan	Zambia
Egypt	Kiribati	Papua New Guinea	Swaziland	
Low income				
Afghanistan	Comoros	Haiti	Mali	Sierra Leone
Bangladesh	Congo, Dem. Rep. of the	Kenya	Mauritania	Somalia
Benin	Eritrea	Korea, Dem. People's Rep. of	Mozambique	Tajikistan
Burkina Faso	Ethiopia	Kyrgyzstan	Myanmar	Tanzania, United Rep. of
Burundi	Gambia	Liberia	Nepal	Togo
Cambodia	Guinea	Madagascar	Niger	Uganda
Central African Rep.	Guinea-Bissau	Malawi	Rwanda	Zimbabwe
Chad				

Source: UNIDO (2015a)

Annex B2

Indicators of competitive industrial performance by economy

Table B2.1

Competitive industrial performance, 2008 and 2013

Country	MVA per capita (2005 $)		Manufactured exports per capita (current $)		Medium- and high-tech MVA share in total manufacturing (percent)		Share of MVA in GDP (percent)		Medium- and high-tech manufactured exports share in total manufactured exports (percent)		Manufactured exports share in total exports (percent)		Impact of a country on world MVA (percent)		Impact of a country on world manufactures trade (percent)	
	2008	2013	2008	2013	2008	2013	2008	2013	2008	2013	2008	2013	2008	2013	2008	2013
Albania	213.2	224.9	121.4	465.0	14.1	17.3	7	6	14.3	10.8	70.4	63.3	0.00	0.01	0.00	0.01
Algeria	173.1	183.1	407.2	377.6	15.2	27.2	6	6	0.7	0.8	18.3	22.4	0.00	0.08	0.00	0.12
Argentina	1,328.4	1,524.6	975.4	908.9	26.0	26.0	20	19	38.6	50.6	55.3	49.2	0.01	0.70	0.00	0.30
Armenia	227.5	284.4	263.6	318.0	7.4	4.3	10	12	30.9	15.5	86.6	69.7	0.00	0.01	0.00	0.01
Australia	3,503.1	3,050.9	3,708.7	5,399.6	21.5	29.7	10	8	26.3	17.1	43.0	50.0	0.01	0.79	0.01	0.99
Austria	7,442.6	7,680.6	18,240.3	17,251.7	43.8	45.1	19	19	59.3	62.0	88.3	88.1	0.01	0.72	0.01	1.16
Azerbaijan	124.5	173.8	346.2	251.6	13.2	10.5	4	5	11.2	15.5	6.4	9.9	0.00	0.02	0.00	0.02
Bahamas	836.8	909.6	807.4	657.4	27.3	27.3	4	4	56.2	74.0	68.6	68.0	0.00	0.00	0.00	0.00
Bahrain	2,395.5	2,502.3	9,886.7	15,242.2	28.2	22.4	14	14	6.5	1.7	87.7	91.0	0.00	0.04	0.00	0.16
Bangladesh	84.2	118.3	99.1	152.1	17.5	9.5	17	19	2.8	2.0	94.5	95.7	0.00	0.21	0.00	0.19
Barbados	925.7	645.0	785.0	764.3	38.1	38.1	6	5	28.7	29.6	83.0	81.6	0.00	0.00	0.00	0.00
Belarus	1,233.9	1,551.2	3,121.7	3,325.9	42.0	47.1	30	31	39.2	35.6	91.6	83.7	0.00	0.16	0.00	0.25
Belgium	5,403.0	4,531.6	38,347.7	40,287.5	43.4	47.3	14	12	54.7	51.9	87.6	87.5	0.01	0.56	0.04	3.53
Belize	411.9	440.0	347.2	376.8	18.5	18.5	10	11	0.1	0.1	34.6	37.9	0.00	0.00	0.00	0.00
Bermuda	1,148.1	851.0	144.5	329.3	18.9	25.6	1	1	43.7	34.2	97.6	98.6	0.00	0.00	0.00	0.00
Bolivia, Plurinational State of	134.2	152.6	245.5	281.9	5.1	5.1	12	12	3.4	4.1	35.0	24.6	0.00	0.02	0.00	0.02
Bosnia and Herzegovina	341.4	323.6	999.0	1,149.7	14.9	15.1	10	10	25.4	23.6	76.8	77.4	0.00	0.01	0.00	0.04
Botswana	373.6	465.3	2,407.1	3,573.4	8.0	16.8	6	7	4.8	3.4	94.0	95.4	0.00	0.01	0.00	0.06
Brazil	783.9	756.7	691.7	766.8	37.2	35.1	15	13	46.3	40.1	67.0	63.4	0.02	1.68	0.01	1.21
Brunei Darussalam	3,037.4	2,740.2	701.4	887.4	3.3	3.3	12	11	38.6	69.6	3.4	3.2	0.00	0.01	0.00	0.00
Bulgaria	675.0	753.8	2,175.1	2,895.1	29.3	28.9	15	16	32.4	36.8	72.6	70.9	0.00	0.06	0.00	0.17
Burundi	15.7	12.8	3.3	3.8	2.2	2.8	10	8	44.8	28.0	19.9	18.9	0.00	0.00	0.00	0.00
Cabo Verde	150.9	166.3	25.0	75.6	27.1	27.1	6	6	0.0	1.1	63.8	54.5	0.00	0.00	0.00	0.00
Cambodia	106.0	146.8	235.0	428.6	0.3	0.3	18	21	3.4	11.3	75.2	70.2	0.00	0.03	0.00	0.05
Cameroon	151.0	153.8	59.3	65.1	5.1	5.0	16	16	24.7	14.3	54.6	33.0	0.00	0.04	0.00	0.01
Canada	4,504.5	4,092.4	7,668.7	7,791.7	37.9	30.6	12	11	55.3	56.7	59.6	60.1	0.02	1.60	0.02	2.16
Central African Rep.	25.7	15.9	17.0	5.0	9.2	9.3	7	6	4.7	14.0	62.2	47.2	0.00	0.00	0.00	0.00
Chile	1,079.2	1,129.3	1,899.1	2,243.6	16.2	16.3	13	12	15.0	11.1	49.5	51.6	0.00	0.22	0.00	0.31
China	788.4	1,142.6	1,020.4	1,540.5	44.0	44.0	33	33	58.1	58.3	95.8	96.6	0.13	17.55	0.12	16.83
China, Hong Kong SAR	688.7	631.1	1,531.5	877.3	32.7	34.1	2	2	38.4	40.9	62.9	31.9	0.00	0.05	0.00	0.05
China, Macao SAR	450.0	297.2	2,262.8	140.2	2.0	6.5	1	1	3.3	2.6	96.1	27.3	0.00	0.00	0.00	0.00
Taiwan Province of China	4,231.8	4,517.0	10,158.1	11,765.5	62.5	66.2	25	24	68.3	70.3	96.1	95.9	0.01	1.17	0.02	2.16
Colombia	528.4	493.2	358.5	333.4	20.0	21.4	14	11	34.6	38.0	43.0	27.4	0.00	0.26	0.00	0.13
Congo, Rep. of the	79.1	96.9	433.8	625.3	1.1	5.6	5	5	67.8	85.9	18.3	26.6	0.00	0.01	0.00	0.02

Country	MVA per capita (2005 $)		Manufactured exports per capita (current $)		Medium- and high-tech MVA share in total manufacturing (percent)		Share of MVA in GDP (percent)		Medium- and high-tech manufactured exports share in total manufactured exports (percent)		Manufactured exports share in total exports (percent)		Impact of a country on world MVA (percent)		Impact of a country on world manufactures trade (percent)	
	2008	2013	2008	2013	2008	2013	2008	2013	2008	2013	2008	2013	2008	2013	2008	2013
Costa Rica	986.7	1,048.2	1,609.8	1,763.8	17.9	14.4	19	18	60.7	62.5	74.9	74.9	0.00	0.06	0.00	0.07
Côte d'Ivoire	110.4	115.5	235.5	268.4	15.0	15.0	12	11	16.7	45.6	44.0	45.1	0.00	0.03	0.00	0.04
Croatia	1,661.7	1,350.0	2,937.5	2,550.8	31.8	31.8	14	13	49.6	44.3	90.7	85.9	0.00	0.06	0.00	0.09
Cyprus	1,251.5	871.9	508.5	583.4	12.9	17.0	7	6	63.2	53.1	68.2	70.1	0.00	0.01	0.00	0.01
Czech Rep.	4,045.1	4,039.8	12,844.8	14,074.5	40.9	48.9	28	29	67.0	67.3	91.5	93.3	0.01	0.48	0.01	1.19
Denmark	5,576.2	5,508.0	16,039.2	14,248.6	44.6	52.4	11	12	52.8	52.1	76.1	72.5	0.00	0.34	0.01	0.63
Ecuador	414.8	408.9	297.2	284.4	8.1	8.1	13	11	19.9	15.8	22.9	17.9	0.00	0.07	0.00	0.04
Egypt	233.3	242.2	212.7	228.2	22.3	22.5	16	15	24.5	31.6	61.8	65.1	0.00	0.22	0.00	0.15
El Salvador	635.5	647.8	675.9	789.7	19.1	19.1	21	21	19.5	14.0	89.6	91.2	0.00	0.05	0.00	0.04
Eritrea	10.1	10.9	0.4	0.4	12.0	9.0	5	6	14.7	14.7	34.6	34.6	0.00	0.00	0.00	0.00
Estonia	1,757.0	2,099.7	9,180.4	12,056.6	29.3	27.6	15	17	42.2	48.7	87.6	84.9	0.00	0.03	0.00	0.12
Ethiopia	9.6	13.3	2.1	6.8	7.7	9.4	5	5	23.3	13.3	10.7	15.7	0.00	0.01	0.00	0.01
Fiji	425.1	445.3	545.3	457.6	6.8	6.8	12	12	6.1	5.5	74.7	76.2	0.00	0.00	0.00	0.00
Finland	8,572.3	6,168.4	17,058.4	12,407.3	50.7	37.5	21	16	57.1	44.5	93.6	90.4	0.01	0.37	0.01	0.53
France	3,834.8	3,568.3	8,371.4	7,724.4	47.2	47.1	11	10	64.3	65.3	88.1	87.6	0.03	2.54	0.05	3.92
Gabon	255.9	274.5	788.8	642.8	5.4	5.4	4	4	6.6	10.1	12.2	18.2	0.00	0.01	0.00	0.01
Gambia	26.1	22.5	1.9	0.6	4.7	4.7	6	5	10.7	5.6	28.8	10.3	0.00	0.00	0.00	0.00
Georgia	238.6	342.8	287.4	317.9	15.4	16.2	13	15	48.9	43.1	84.7	74.4	0.00	0.02	0.00	0.01
Germany	7,342.2	7,655.8	15,427.8	15,504.2	60.3	59.9	20	21	71.1	72.7	87.7	87.9	0.07	7.02	0.11	10.11
Ghana	42.9	53.0	32.3	79.8	0.8	0.8	8	7	18.1	33.3	19.6	16.4	0.00	0.02	0.00	0.02
Greece	1,593.1	1,395.6	2,205.6	2,623.3	24.1	20.4	7	8	31.5	21.7	78.5	80.5	0.00	0.17	0.00	0.23
Guatemala	406.7	404.6	391.3	448.0	16.3	16.3	18	17	21.6	20.9	69.0	68.8	0.00	0.07	0.00	0.06
Haiti	44.6	50.6	6.2	6.2	5.3	5.3	10	11	3.8	3.8	83.0	83.0	0.00	0.01	0.00	0.00
Honduras	286.2	270.1	185.4	301.3	7.2	7.2	18	17	31.1	36.8	55.7	50.9	0.00		0.00	0.02
Hungary	2,375.1	2,365.7	9,227.5	9,634.3	52.0	56.2	21	21	77.4	73.5	85.7	89.0	0.00	0.26	0.01	0.76
Iceland	5,227.7	5,393.9	5,495.0	3,934.2	9.5	9.5	9	10	52.7	38.6	31.7	26.0	0.00	0.02	0.00	0.01
India	131.9	161.7	132.9	223.3	38.5	40.8	15	14	27.3	28.7	85.8	83.1	0.02	2.25	0.01	2.20
Indonesia	379.5	451.3	353.9	438.8	39.5	37.7	26	25	29.5	30.3	60.5	60.1	0.01	1.25	0.01	0.86
Iran, Islamic Rep. of	338.0	324.9	131.7	340.7	41.2	43.9	11	10	25.2	31.7	14.8	19.7	0.00	0.28	0.00	0.21
Iraq	27.4	36.9	1.7	14.9	7.7	7.3	2	2	5.2	5.2	0.1	0.6	0.00	0.01	0.00	0.00
Ireland	7,930.5	6,736.3	26,978.4	23,133.4	58.3	60.5	16	15	55.2	51.2	92.4	92.8	0.00	0.35	0.01	0.84
Israel	3,162.4	3,232.5	8,194.6	8,265.7	59.5	37.6	14	13	49.9	54.8	94.9	95.7	0.00	0.28	0.01	0.50
Italy	4,953.1	4,151.6	8,318.7	7,540.9	40.1	42.2	16	15	54.1	53.7	91.9	91.5	0.04	2.81	0.04	3.63
Jamaica	310.2	274.2	840.9	487.0	18.8	18.8	7	7	7.5	7.5	96.2	91.1	0.00	0.01	0.00	0.01
Japan	7,951.4	7,820.7	5,675.6	5,163.5	55.4	54.9	22	21	79.6	78.1	92.5	91.8	0.12	11.02	0.06	5.18
Jordan	434.5	398.7	846.7	730.9	26.6	26.3	16	16	47.8	43.4	81.0	78.5	0.00	0.03	0.00	0.04
Kazakhstan	497.6	605.9	1,150.5	1,042.8	12.6	16.2	11	11	35.4	33.2	25.2	20.8	0.00	0.11	0.00	0.14
Kenya	59.5	61.1	59.8	58.1	10.0	10.4	11	10	17.7	21.6	49.6	48.7	0.00	0.03	0.00	0.02

Country	MVA per capita (2005 $)		Manufactured exports per capita (current $)		Medium- and high-tech MVA share in total manufacturing (percent)		Share of MVA in GDP (percent)		Medium- and high-tech manufactured exports share in total manufactured exports (percent)		Manufactured exports share in total exports (percent)		Impact of a country on world MVA (percent)		Impact of a country on world manufactures trade (percent)	
	2008	2013	2008	2013	2008	2013	2008	2013	2008	2013	2008	2013	2008	2013	2008	2013
Korea, Rep. of	5,801.4	7,180.7	8,552.6	11,043.4	82.8	63.1	27	29	72.7	72.4	97.0	97.2	0.03	3.92	0.04	4.29
Kuwait	2,397.3	1,785.7	10,332.0	10,400.6	27.4	29.6	7	6	8.2	13.4	32.6	30.6	0.00	0.07	0.00	0.28
Kyrgyzstan	67.3	54.0	86.1	94.9	6.1	3.5	12	8	23.1	18.1	27.7	34.1	0.00	0.00	0.00	0.00
Latvia	838.4	1,057.1	3,619.2	5,028.2	20.7	22.1	9	12	34.2	36.0	83.4	77.4	0.00	0.02	0.00	0.08
Lebanon	474.4	480.7	675.1	620.7	19.9	20.0	8	7	38.3	32.9	81.2	76.0	0.00	0.03	0.00	0.02
Lithuania	1,812.4	2,102.8	6,605.2	9,209.2	31.3	21.6	18	20	40.0	36.8	87.3	85.2	0.00	0.07	0.00	0.22
Luxembourg	5,327.5	3,719.1	31,440.0	22,502.8	8.6	11.0	6	5	35.8	39.2	88.5	87.0	0.00	0.02	0.00	0.09
Macedonia, Former Yugoslav Rep. of	474.0	415.5	1,468.7	1,828.1	13.7	15.5	14	12	29.6	48.1	91.8	90.3	0.00	0.01	0.00	0.03
Madagascar	44.6	37.2	61.8	42.5	3.6	3.6	15	14	2.1	4.4	79.6	55.6	0.00	0.01	0.00	0.01
Malawi	24.6	22.6	10.8	22.2	11.4	11.3	10	9	15.3	12.8	17.3	30.1	0.00	0.00	0.00	0.00
Malaysia	1,616.6	1,717.0	5,148.3	6,201.9	43.1	42.1	26	25	57.9	58.4	70.7	80.7	0.01	0.57	0.01	1.45
Malta	2,178.3	1,842.9	6,369.5	11,318.0	50.5	28.8	14	11	78.1	40.2	88.5	93.3	0.00	0.01	0.00	0.04
Mauritius	982.6	1,065.9	1,583.3	1,468.5	3.2	8.3	16	15	12.4	3.5	80.6	96.3	0.00	0.02	0.00	0.01
Mexico	1,300.0	1,340.9	1,991.8	2,514.4	37.6	43.7	16	16	76.9	78.3	78.6	81.0	0.02	1.82	0.02	2.43
Moldova, Rep. of	99.1	118.0	187.0	312.1	9.8	17.4	10	10	14.2	25.5	69.0	67.2	0.00	0.01	0.00	0.01
Mongolia	81.2	91.9	457.5	680.1	5.4	6.2	6	5	1.9	3.7	62.9	45.2	0.00	0.00	0.00	0.02
Morocco	309.3	323.6	491.4	534.8	29.1	27.4	14	13	31.6	44.5	74.9	80.4	0.00	0.12	0.00	0.14
Mozambique	45.2	51.0	14.2	44.6	11.0	11.0	13	11	35.9	44.8	12.2	28.7	0.00	0.02	0.00	0.01
Namibia	504.1	491.1	1,117.3	1,713.0	7.9	8.0	13	11	13.8	25.8	49.9	69.1	0.00	0.01	0.00	0.03
Nepal	24.5	26.3	23.9	24.0	1.9	8.6	7	6	20.8	20.3	71.7	77.2	0.00	0.01	0.00	0.01
Netherlands	4,995.2	4,813.2	24,148.3	27,818.9	43.8	48.7	12	12	53.9	50.8	73.0	81.6	0.01	0.89	0.03	3.68
New Zealand	3,504.8	3,574.8	3,176.9	3,844.8	17.2	18.3	13	12	24.6	19.1	46.3	45.5	0.00	0.18	0.00	0.14
Niger	14.9	18.4	35.8	67.0	26.4	26.4	5	6	3.9	6.0	51.4	89.3	0.00	0.00	0.00	0.01
Nigeria	27.1	44.6	30.3	35.1	33.4	33.4	3	4	73.2	19.4	5.6	6.7	0.00	0.09	0.00	0.05
Norway	5,544.4	5,211.5	9,172.9	8,101.9	58.6	46.3	8	8	50.6	46.8	25.3	26.5	0.00	0.29	0.00	0.32
Oman	1,328.4	1,297.8	2,355.0	2,308.2	14.2	48.2	9	10	28.9	40.3	18.1	18.1	0.00	0.05	0.00	0.07
Pakistan	144.1	139.1	98.6	112.1	24.6	24.6	19	17	9.1	10.4	82.8	81.6	0.00	0.28	0.00	0.16
Panama	341.2	338.2	60.0	90.2	7.6	7.6	6	4	14.0	7.9	18.9	41.3	0.00	0.01	0.00	0.00
Papua New Guinea	53.4	71.8	343.9	324.0	12.6	12.6	6	6	9.1	8.1	43.9	51.4	0.00	0.01	0.00	0.02
Paraguay	170.3	179.0	189.8	233.9	21.3	21.8	10	9	10.5	15.1	26.5	16.9	0.00	0.01	0.00	0.01
Peru	533.4	604.7	575.0	714.6	15.6	13.5	16	15	4.8	6.1	52.6	51.8	0.00	0.20	0.00	0.17
Philippines	308.5	353.4	500.0	495.6	35.0	41.5	23	22	77.6	68.6	92.1	90.3	0.00	0.39	0.00	0.38
Poland	1,850.3	2,323.6	3,938.6	4,656.8	36.5	32.9	19	21	57.6	55.2	87.5	87.3	0.01	0.98	0.01	1.40
Portugal	2,343.9	2,280.2	4,541.1	5,489.6	26.8	26.4	12	13	43.9	37.8	85.8	92.8	0.00	0.27	0.00	0.46
Qatar	4,297.7	4,595.1	2,450.6	5,693.4	33.0	25.8	7	8	43.7	38.8	5.0	9.1	0.00	0.11	0.00	0.10
Romania	828.9	854.6	2,046.0	2,625.9	28.3	38.2	15	15	47.0	55.8	90.7	86.5	0.00	0.21	0.00	0.45
Russian Federation	931.1	968.1	1,228.1	1,532.1	24.6	27.7	14	14	28.2	22.8	37.7	41.5	0.02	1.53	0.02	1.73
Rwanda	21.3	22.0	11.2	27.0	6.7	6.7	7	6	3.6	4.4	57.2	63.4	0.00	0.00	0.00	0.00
Saint Lucia	268.1	276.3	244.2	244.3	7.8	7.8	4	5	30.0	30.0	61.6	61.6	0.00	0.00	0.00	0.00
Saudi Arabia	1,599.4	2,046.1	1,909.3	2,429.9	36.3	35.9	11	11	26.0	35.7	16.4	19.1	0.01	0.65	0.00	0.55
Senegal	93.4	98.9	144.5	117.2	22.5	21.7	12	12	17.3	15.2	81.5	62.1	0.00	0.02	0.00	0.01

Country	MVA per capita (2005 $)		Manufactured exports per capita (current $)		Medium- and high-tech MVA share in total manufacturing (percent)		Share of MVA in GDP (percent)		Medium- and high-tech manufactured exports share in total manufactured exports (percent)		Manufactured exports share in total exports (percent)		Impact of a country on world MVA (percent)		Impact of a country on world manufactures trade (percent)	
	2008	2013	2008	2013	2008	2013	2008	2013	2008	2013	2008	2013	2008	2013	2008	2013
Serbia	405.8	361.3	950.6	1,289.2	21.8	20.1	14	12	32.2	48.6	84.6	83.9	0.00	0.04	0.00	0.10
Singapore	7,921.9	9,700.0	32,535.6	32,285.9	88.0	81.2	25	26	66.7	69.0	89.8	89.8	0.00	0.58	0.01	1.52
Slovakia	3,580.3	3,125.9	12,111.4	14,745.8	41.5	58.0	25	21	65.4	67.7	93.4	94.3	0.00	0.19	0.01	0.63
Slovenia	4,264.6	3,659.3	13,182.0	12,485.4	45.8	49.2	21	20	62.0	61.8	91.6	90.4	0.00	0.08	0.00	0.20
South Africa	931.9	894.0	1,016.9	1,208.9	23.6	24.4	16	15	51.1	43.7	69.1	67.0	0.01	0.52	0.00	0.50
Spain	3,370.6	2,960.7	5,269.5	5,425.2	34.3	34.3	12	12	56.9	55.9	85.4	81.9	0.02	1.54	0.02	2.01
Sri Lanka	274.5	357.2	282.1	345.5	13.9	6.8	19	19	9.3	8.2	70.6	73.5	0.00	0.08	0.00	0.06
State of Palestine	143.5	148.9	92.7	114.1	5.6	5.5	11	9	9.5	7.7	90.8	77.6	0.00	0.01	0.00	0.00
Suriname	631.5	599.2	440.5	625.0	11.6	11.6	16	13	10.3	19.4	13.4	13.8	0.00	0.00	0.00	0.00
Swaziland	726.9	641.5	888.8	888.8	0.9	0.9	30	26	29.0	29.0	92.9	92.9	0.00	0.01	0.00	0.01
Sweden	7,133.2	6,896.7	17,803.4	15,530.8	53.4	53.7	17	16	57.9	57.5	89.4	88.7	0.01	0.73	0.01	1.17
Switzerland	10,323.9	10,147.3	23,620.4	25,700.8	64.0	64.5	19	18	69.0	71.3	90.1	90.6	0.01	0.91	0.02	1.64
Syrian Arab Rep.	60.0	65.2	365.7	231.3	21.5	21.5	4	4	25.1	22.7	51.7	43.9	0.00	0.02	0.00	0.05
Tajikistan	57.9	59.1	15.5	15.5	2.6	3.0	15	12	66.3	66.3	13.8	13.8	0.00	0.01	0.00	0.00
Tanzania, United Rep. of	34.4	43.0	31.3	33.0	11.2	8.6	8	9	22.9	22.4	42.5	36.8	0.00	0.02	0.00	0.01
Thailand	1,079.7	1,168.4	2,253.1	2,998.6	46.3	40.7	36	34	60.4	59.8	84.8	88.0	0.01	0.87	0.01	1.58
Tonga	170.0	168.7	23.1	14.9	17.3	17.3	7	6	17.7	49.5	26.9	11.3	0.00	0.00	0.00	0.00
Trinidad and Tobago	918.3	941.4	8,820.1	5,564.4	39.6	39.6	6	7	20.5	17.7	62.3	74.0	0.00	0.01	0.00	0.07
Tunisia	582.1	652.8	1,523.1	1,317.7	27.3	28.8	16	17	39.2	46.8	81.9	84.9	0.00	0.08	0.00	0.11
Turkey	1,344.8	1,548.3	1,679.9	1,778.4	32.6	32.7	17	18	42.2	40.6	89.5	87.8	0.01	1.29	0.01	1.05
Uganda	24.2	27.3	14.6	17.3	15.3	15.3	7	7	14.8	13.8	34.4	38.2	0.00	0.01	0.00	0.01
Ukraine	375.0	358.7	1,233.9	1,088.6	33.9	33.9	17	17	45.0	42.3	85.6	77.8	0.00	0.18	0.01	0.39
United Arab Emirates	3,395.4	2,612.9	4,525.1	2,314.1	12.6	12.6	11	10	7.5	21.2	18.6	10.6	0.00	0.27	0.00	0.17
United Kingdom	4,043.2	3,671.4	6,051.8	2,844.5	44.9	45.9	10	10	62.7	57.1	77.0	76.3	0.03	2.57	0.03	1.42
United States	5,459.2	5,464.5	3,178.2	3,229.0	51.3	50.6	12	12	69.0	61.7	83.3	75.3	0.20	19.39	0.09	8.15
Uruguay	945.1	979.7	754.6	994.7	14.5	13.8	15	13	22.4	26.5	42.5	37.4	0.00	0.04	0.00	0.03
Venezuela, Bolivarian Rep. of	876.0	806.9	747.3	425.4	34.3	34.3	13	13	10.3	9.7	25.2	14.7	0.00	0.27	0.00	0.10
Viet Nam	173.6	235.6	423.1	1,128.9	25.7	29.8	22	23	25.8	47.4	59.0	78.4	0.00	0.24	0.00	0.82
Yemen	57.8	59.4	46.4	36.0	2.1	2.3	6	7	2.8	5.6	14.0	12.5	0.00	0.02	0.00	0.01
Zambia	68.4	76.9	126.5	182.3	21.1	21.1	10	9	12.6	23.2	30.9	25.0	0.00	0.01	0.00	0.02

Note: MVA is manufacturing value added; GDP is gross domestic product. For medium and high tech classification please see Table 7.4.
Source: UNIDO (2015a) and UNIDO (2015b).

Annex B3

Indicators of manufacturing value added and exports by industrialization level, region and income group

Table B3.1

Medium- and high-tech manufactured exports share in total manufactured exports, 2009–2013 (percent)

Group	2009	2010	2011	2012	2013
World	60	60	58	58	58
Industrialized countries	64	64	62	62	61
Industrializing countries	51	51	50	51	52
By industrialization level					
Emerging industrial countries	53	54	53	53	53
Other developing countries	31	26	28	32	36
Least developed countries	9	6	8	16	15
By region					
Africa	31	30	30	31	34
Asia and the Pacific	61	61	58	59	58
Europe	61	60	59	59	58
Latin America and the Caribbean	51	51	52	52	55
By income (industrializing countries)					
High income	49	48	48	46	47
Upper middle income	55	56	55	55	56
Lower middle income	33	31	30	34	35
Low income	8	7	7	15	15

Note: Tech classification based on Annex B5, Table B5.1.

Table B3.2

Manufacturing exports per capita, 2009–2013 (current $)

Group	2009	2010	2011	2012	2013
World	1,381	1,653	1,922	1,891	1,940
Industrialized countries	5,337	6,233	7,230	7,023	7,233
Industrializing countries	529	674	795	811	835
By industrialization level					
Emerging industrial countries	743	942	1,119	1,159	1,202
Others developing countries	47	61	76	46	45
Least developed countries	200	272	309	293	290
By region					
Africa	146	203	236	218	193
Asia and Pacific	502	654	779	805	851
Europe	1,702	2,003	2,476	2,486	2,559
Latin America	874	1,070	1,196	1,228	1,200
By income					
High income	5,993	7,011	8,109	7,838	8,066
Upper middle income	892	1,147	1,359	1,425	1,471
Lower middle income	186	235	290	296	308
Low income	44	56	65	32	38

Table B3.3
Impact of a country on world manufactures trade, 2009-2013 (percent)

Group	2009	2010	2011	2012	2013
World	100	100	100	100	100
Industrialized countries	68	66	66	65	64
Industrializing countries	32	34	34	35	36
By industrialization level					
Emerging industrial countries	29	30	31	32	33
Others developing countries	0	0	0	0	0
Least developed countries	3	3	3	3	3
By region					
Africa	2	2	2	2	2
Asia and Pacific	20	22	23	24	24
Europe	4	4	4	5	4
Latin America	5	6	5	6	5
By income					
High income	71	69	69	67	67
Upper middle income	24	25	26	27	27
Lower middle income	5	5	5	6	6
Low income	0	0	0	0	0

Table B3.4
Impact of a country on world manufacturing value added, 2009-2013 (percent)

Group	2009	2010	2011	2012	2013
World	100	100	100	100	100
Industrialized countries	67	67	68	65	65
Industrializing countries	33	33	32	35	35
By industrialization level					
Emerging industrial countries	29	29	31	31	32
Others developing countries	3	3	3	3	3
Least developed countries	0	0	0	0	0
By region					
Africa	1	1	1	1	1
Asia and the Pacific	22	22	23	23	24
Europe	3	3	3	3	3
Latin America and the Caribbean	6	6	6	6	6
By income (industrializing countries)					
High income	2	2	2	2	2
Upper middle income	25	24	25	26	27
Lower middle income	5	5	5	5	5
Low income	0	0	0	0	0

Table B3.5

Medium- and high-tech manufacturing value added share in total manufacturing, 2009–2013 (percent)

Group	2009	2010	2011	2012	2013
World	47	46	46	47	47
Industrialized countries	52	51	51	51	50
Industrializing countries	38	39	39	34	34
By industrialization level					
Emerging industrial countries	39	39	40	36	36
Others developing countries	8	8	9	9	9
Least developed countries	30	29	28	29	30
By region					
Africa	22	24	23	23	22
Asia and the Pacific	49	47	45	50	49
Europe	46	47	48	47	47
Latin America and the Caribbean	34	35	34	33	33
By income (industrializing countries)					
High income	36	36	37	38	38
Upper middle income	39	39	40	34	34
Lower middle income	35	35	35	35	35
Low income	11	11	11	11	9

Note: For medium and high tech classification please see Table 7.4.

Table B3.6

Share of manufacturing value added in GDP, 2009–2013 (percent)

Group	2009	2010	2011	2012	2013
World	15	16	16	16	16
Industrialized countries	14	15	15	15	15
Industrializing countries	20	20	20	20	20
By industrialization level					
Emerging industrial countries	22	22	22	22	22
Others developing countries	12	12	12	12	12
Least developed countries	11	11	11	12	12
By region					
Africa	10	10	10	10	10
Asia and the Pacific	25	25	25	25	25
Europe	16	16	17	17	17
Latin America and the Caribbean	14	15	15	14	14
By income					
High income	13	14	15	15	15
Upper middle income	21	22	22	22	22
Lower middle income	16	16	16	16	15
Low income	12	12	13	13	13

Note: GDP is gross domestic product.

Table B3.7
Manufactured exports share in total exports, 2009–2013 (percent)

Group	2009	2010	2011	2012	2013
World	86	84	83	83	83
Industrialized countries	90	89	88	89	87
Industrializing countries	78	76	75	75	76
By industrialization level					
Emerging industrial countries	82	82	79	80	80
Others developing countries	69	66	74	64	64
Least developed countries	52	45	46	45	48
By region					
Africa	43	46	46	45	47
Asia and Pacific	84	81	79	79	81
Europe	91	91	90	90	90
Latin America	69	67	63	64	64
By income					
High income	89	88	87	87	86
Upper middle income	81	79	78	80	81
Lower middle income	72	67	69	63	68
Low income	66	64	68	57	56

Table B3.8
Manufacturing value added per capita, 2009–2013 (constant 2005 $)

Group	2009	2010	2011	2012	2013
World	1,104	1,202	1,235	1,250	1,262
Industrialized countries	4,180	4,610	4,702	4,728	4,751
Industrializing countries	442	473	500	517	534
By industrialization level					
Emerging industrial countries	1,104	1,202	1,235	1,250	1,262
Others developing countries	189	197	200	204	199
Least developed countries	48	51	53	55	58
By region					
Africa	122	125	125	127	128
Asia and the Pacific	438	472	506	533	559
Europe	1,021	1,105	1,175	1,184	1,206
Latin America and the Caribbean	768	818	842	840	836
By income					
High income	4,498	4,952	5,048	5,067	5,087
Upper middle income	818	885	941	984	1,026
Lower middle income	172	181	189	192	194
Low income	48	50	53	55	54

Source: UNIDO elaboration based on UNIDO (2015b).

Annex B4

Summary of world trade, by industrialization level, region and income group

Table B4.1
Total exports, all commodities, 2009–2013 (current $, billions)

Group	2009	2010	2011	2012	2013
World	10,984	13,533	16,157	16,018	16,682
Industrialized countries	7,192	8,494	10,020	9,704	10,219
Industrializing countries	3,791	5,039	6,137	6,314	6,463
By industrialization level					
Emerging industrial countries	3,273	4,229	5,229	5,427	5,623
Others developing countries	54	75	84	61	61
Least developed countries	464	735	824	826	779
By region					
Asia and Pacific	2,267	3,116	3,812	3,972	4,170
Europe	451	531	661	666	693
Latin America	737	943	1,132	1,157	1,147
Africa	336	448	533	519	453
By income					
High income	7,571	8,978	10,621	10,267	10,795
Upper middle income	2,735	3,619	4,407	4,512	4,654
Lower middle income	625	866	1,051	1,192	1,175
Low income	53	70	78	47	58

Table B4.2
Primary exports, 2009–2013 (current $, billions)

Group	2009	2010	2011	2012	2013
World	1,422	1,939	2,511	2,442	2,620
Industrialized countries	626	764	998	904	1,136
Industrializing countries	795	1,175	1,513	1,537	1,484
By industrialization level					
Emerging industrial countries	553	745	1,049	1,062	1,055
Others developing countries	16	24	20	21	21
Least developed countries	227	405	444	455	409
By region (developing and emerging industrial)					
Asia and Pacific	352	596	772	798	777
Europe	32	36	45	50	57
Latin America	222	303	410	407	409
Africa	190	240	285	283	241
By income					
High income	734	913	1,226	1,140	1,372
Upper middle income	501	723	948	860	856
Lower middle income	170	279	313	423	367
Low income	17	24	24	19	25

Table B4.3

Resource-based manufactured exports, 2009-2013 (current $, billions)

Group	2009	2010	2011	2012	2013
World	2,031	2,564	3,203	3,166	3,290
Industrialized countries	1,384	1,676	2,112	2,080	2,204
Industrializing countries	647	891	1,091	1,082	1,082
By industrialization level					
Emerging industrial countries	550	728	911	922	938
Others developing countries	8	13	13	19	19
Least developed countries	89	148	165	144	127
By region (industrializing)					
Asia and Pacific	339	455	593	578	646
Europe	89	109	143	148	152
Latin America	184	236	257	271	248
Africa	36	89	98	85	36
By income					
High income	1,466	1,781	2,236	2,204	2,328
Upper middle income	403	542	657	668	655
Lower middle income	156	229	300	282	293
Low income	7	11	10	12	14

Table B4.4

Low-tech manufactured exports, 2009-2013 (current $, billions)

Group	2009	2010	2011	2012	2013
World	1,723	2,010	2,419	2,402	2,535
Industrialized countries	947	1,063	1,257	1,196	1,245
Industrializing countries	808	984	1,197	1,239	1,309
By industrialization level					
Emerging industrial countries	708	860	1,050	1,121	1,184
Others developing countries	25	33	44	14	14
Least developed countries	76	95	110	105	112
By region (industrializing)					
Asia and Pacific	557	696	866	916	982
Europe	105	118	143	141	147
Latin America	105	117	133	132	132
Africa	41	53	55	51	48
By income					
High income	1,049	1,177	1,397	1,318	1,397
Upper middle income	519	648	797	874	921
Lower middle income	131	156	187	199	205
Low income	24	30	38	10	13

Table B4.5
Medium-tech manufactured exports, 2009–2013 (current $, billions)

Group	2009	2010	2011	2012	2013
World	3,598	4,406	5,213	5,168	5,285
Industrialized countries	2,672	3,192	3,758	3,680	3,754
Industrializing countries	898	1,178	1,421	1,456	1,514
By industrialization level					
Emerging industrial countries	856	1,120	1,350	1,387	1,444
Others developing countries	3	3	4	5	5
Least developed countries	39	54	65	62	65
By region (industrializing)					
Asia and Pacific	547	744	900	930	958
Europe	149	175	218	215	224
Latin America	168	214	250	260	279
Africa	33	45	53	50	54
By income					
High income	2,811	3,327	3,916	3,731	3,820
Upper middle income	691	948	1,137	1,252	1,276
Lower middle income	93	128	157	181	184
Low income	2	3	3	4	4

Table B4.6
High-tech manufactured exports, 2009–2013 (current $, billions)

Group	2009	2010	2011	2012	2013
World	2,069	2,430	2,587	2,628	2,756
Industrialized countries	1,446	1,647	1,714	1,674	1,726
Industrializing countries	619	778	872	956	1,032
By industrialization level					
Emerging industrial countries	583	744	827	895	960
Others developing countries	1	0	1	1	1
Least developed countries	35	33	43	57	68
By region (industrializing)					
Asia and Pacific	485	624	687	761	834
Europe	72	83	99	103	105
Latin America	59	67	80	85	85
Africa	3	4	6	7	7
By income					
High income	1,197	1,346	1,375	1,328	1,391
Upper middle income	764	967	1,074	1,143	1,192
Lower middle income	103	111	132	154	169
Low income	4	5	5	3	4

Note: Tech classification of Table B4.3-B4.6 based on Annex B5, Table B5.1.
Source: UNIDO elaboration based on UNIDO (2015a) and United Nations Comtrade database (UNSD 2015).

Annex B5

Technological classification of international trade data

Table B5.1
Technology classification of exports, Standard International Trade Classification, Rev. 3

Type of export	SITC sections
Resource-based exports	016, 017, 023, 024, 035, 037, 046, 047, 048, 056, 058, 059, 061, 062, 073, 098, 111, 112, 122, 232, 247, 248, 251, 264, 265, 281, 282, 283, 284, 285, 286, 287, 288, 289, 322, 334, 335, 342, 344, 345, 411, 421, 422, 431, 511, 514, 515, 516, 522, 523, 524, 531, 532, 551, 592, 621, 625, 629, 633, 634, 635, 641, 661, 662, 663, 664, 667,689
Low-tech exports	611, 612, 613, 642, 651, 652, 654, 655, 656, 657, 658, 659, 665, 666, 673, 674, 675, 676, 677, 679, 691, 692, 693, 694, 695, 696, 697, 699, 821, 831, 841, 842, 843, 844, 845, 846, 848, 851, 893, 894, 895, 897, 898, 899
Medium-tech exports	266, 267, 512, 513, 533, 553, 554, 562, 571, 572, 573, 574, 575, 579, 581, 582, 583, 591, 593, 597, 598, 653, 671, 672, 678, 711, 712,713 ,714, 721, 722, 723, 724, 725, 726, 727, 728, 731, 733, 735, 737, 741, 742, 743, 744, 745, 746, 747, 748, 749, 761, 762, 763, 772, 773, 775, 778, 781, 782, 783, 784, 785, 786, 791, 793, 811, 812, 813, 872, 873, 882, 884, 885
High-tech exports	525, 541, 542, 716, 718, 751, 752, 759, 764, 771, 774, 776, 792, 871, 874, 881, 891

Source: UNIDO 2011a.

Annex C1

Data appendix

Table C1.1

Background papers used for producing Industrial Development Report 2016 (*IDR 2016*) figures and tables and their datasets

Citation	Link (if already available online)	Datasets used to produce the background papers, not necessarily the figures and tables for the *IDR 2016*	Corresponding figures or tables in *IDR 2016*
Foster-McGregor, Kaulich and Stehrer 2015	www.unido.org//fileadmin/user_media/Services/PSD/WP_2015_04_v2.pdf	• UN Comtrade database • UNCTAD Eora GVC database • World Development Indicators database • UNIDO Industrial Development Statistics database (INDSTAT4)	Figures 2.21
Foster-McGregor, Kaba and Szirmai 2015		• UN National Accounts database • World Development Indicators database • Maddison Project database	Figures 3.2–3.4, 3.8, 3.9
Jacob and Sasso 2015	www.unido.org//fileadmin/user_media/Services/PSD/WP_2015_07_FDI.pdf	• Financial Times FDI Markets database • OECD ANBERN database	Tables 3.2, 3.3
Kaltenberg and Verspagen 2015		• Maddison Project database • BACI database	Figures 1.2, 1.3, 2.1–2.8 Tables 1.1, 3.1
Lavopa and Szirmai 2015	www.unido.org//fileadmin/user_media/Services/PSD/WP_10_FB.pdf	• UN's main aggregate database • World Development Indicators database • ICOP database • UNIDO Industrial Development Statistics database (INDSTAT2) • Conference Board Total Economy database • National Accounts Main Aggregates (NAMAD) Database • GGDC 10 Sector database • Asia, EU and World KLEMS database • WIOD database	Box 1.4, Figures 1.8, 1.9
Mazzanti and others 2015		• EORA database • IEA database • UNIDO Industrial Development Statistics database (INDSTAT2)	Figures 5.3, 5.7 Tables 5.2, 5.6
Tregenna 2015		• ILO database • Penn World Tables • UN database	Figure 3.5
Zhong 2015		• WIOD database	Figures 5.5, 5.6

Table C1.2
Classifications used for producing *IDR 2016* figures and tables and their datasets

Classification	Further information	Citation to data sources used in the *IDR 2016*	Type of classification used in the *IDR 2016*	Corresponding figures or tables in *IDR 2016*
Value added technological classification	For further information on technological classification used, please see Annex 6.5 in UNIDO (2010b).	UNIDO 2010b	Medium- and high-technology classification	Annexes B2.1, B3.5 Tables 7.4, 7.5
	For further information on technological classification used, please see Annex A in OECD (2005).	UNIDO 2012	Low, medium and high tech classification	Annex A2 Figures 1.8, 1.9, 2.2, 2.4
	For further information on technological classification used, please see Appendix A in Foster-McGregor, Kaulich and Stehrer (2015).	Foster-McGregor, Kaulich and Stehrer 2015	Primary, low-tech and high-tech manufacturing and services classification	Figure 2.21
Trade technological classification	For detailed structure and explanatory notes of Standard International Trade Classification Rev. 3, please see http://unstats.un.org/unsd/cr/registry/regcst.asp?Cl=14	UNIDO 2011a	Resource based, low tech, medium tech, high tech sectors sections 016–899	Annexes B3.1, B4.3–B4.6, B5 Figures 7.10, 7.11, 7.13, 8.2, 8.3 Table 7.5
Regional classification	For further information on regional classification used, please see Annex A1.2 in *IDR 2016*.	UNIDO elaboration	Industrialized countries and economies (5 regions) and industrializing countries and economies (12 regions)	Figures 1.5–1.8, 1.12–1.15 Tables 1.3, 1.4
	For further information on regional classification used, please see UNIDO (2015c).	UNIDO 2015c	Industrialized countries and economies (4 regions) and industrializing countries and economies (4 regions)	Annexes B1.1, B3, B4 Figures 7.7, 7.8 Tables 7.1, 7.5, 7.7–7.9
Development level classification	For further information on development level classification used, please see UNIDO (2015c).	UNIDO 2015c	Industrialized countries and economies and industrializing countries and economies (Emerging, Other and Least developed)	Annexes B1.2, B3, B4 Figures 7.1, 7.3–7.5, 7.9, 7.11, 7.12, 7.19 Tables 7.1–7.3, 7.5, 7.7–7.9
Income classification	For further information on *World Bank Atlas* method—World Bank gross national income per capita operational guidelines and analytical classifications, please see https://datahelpdesk.worldbank.org/knowledgebase/articles/378834-how-does-the-world-bank-classify-countries	UNIDO 2015c	Low-income, upper middle-income, lower middle-income and high-income classification	Annexes B1.3, B3, B4 Table 7.5, 7.7–7.9
		Lavopa and Szirmai 2014	Low-income, upper middle-income, lower middle-income and high-income classification (meta-analysis of World Bank classifications in different years)	Figure 3.10–3.12
		World Bank 2015b	Low-income, upper middle-income, lower middle-income and high-income classification (per year 2013)	Annex A1.1 Figure 5.1, 5.2, 5.12, 5.13 Table 5.2, 5.4, Box 5.6,
		World Bank 2015b	Low-income, upper middle-income, lower middle-income and high-income classification (per year 1990)	Annex A1.2 Figure 1.5–1.8, 1.12–1.15, 2.11–2.20, 4.2 Table 1.3, 2.2, 2.3, 4.1

C1

Table C1.3
List of datasets used for producing figures and tables of the *IDR 2016*

Dataset	Further information	Datasets key references cited in *IDR 2016*	Corresponding figures and tables in *IDR 2016*
American Community Survey	The American Community Survey is the premier source for detailed information about the American people and workforce. For more information please see www.census.gov/programs-surveys/acs/	Acemoglu and Autor 2011	Figures 4.4, 4.6
BACI International Trade Database	BACI is an international trade database covering more than 200 countries and 5,000 products, between 1994 and 2007. For more information please see Gaulier and Zignago (2010).	Gaulier and Zignago 2010	Figure 2.9
CAIT Historical Emissions Data	CAIT Historic Emissions Data allows easy access, analysis and visualization of the latest available international greenhouse gas emissions data. It includes information for 186 countries, 50 U.S. states, 6 gases, multiple economic sectors, and 160 years —carbon dioxide emissions for 1850–2012 and multi-sector greenhouse gas emission for 1990–2012. For more information please see www.wri.org/resources/data-sets/cait-historical-emissions-data-countries-us-states-unfccc	WRI 2015	Figures 1.16, 1.21, 1.23
CHAT Data	The CHAT data set covers the diffusion of 104 technologies for 161 countries over the last 200 years. For more information please see Comin and Mestieri (2013).	Comin and Mestieri 2013	Figure 4.8
Competitiveness Industrial Performance Index (CIP)	The CIP index benchmarks national industrial performance of 118 countries using indicators of an economy's ability to produce and export manufactured goods competitively. For more information please see www.unido.org/data1/Statistics/Research/cip.html	UNIDO 2013c, UNIDO 2015a	Annexes B3, B2 Figures 8.1–8.4 Tables 8.1–8.5
Current Population Survey (CPS)	The CPS is used to collect data for a variety of studies on the entire U.S. population and specific population subsets from a probability-selected sample of about 60,000 occupied households. The fieldwork is conducted during the calendar week that includes the 19th of the month. For more information please see www.census.gov/cps/about/supplemental.html	Acemoglu and Autor 2011	Figures 4.4, 4.6
Institutional Characteristics of Trade Unions, Wage Setting, State Intervention and Social Pacts (ICTWSS) Database	The ICTWSS database covers institutional characteristics of trade unions, wage setting, state intervention and social pacts in 34 countries over a time period of 1960–2012. For more information please see www.uva-aias.net/208	Visser 2013	Figure 4.10
Emission Database for Global Atmospheric Research (EDGAR)	EDGAR provides global past and present-day anthropogenic emissions of greenhouse gases and air pollutants by country and on a spatial grid. The EDGARv4.2 inventory covers the time period from 1970 to 2008, with data presented for all countries, emissions provided per main source category, and spatially allocated on a 0.1° × 0.1° grid over the globe. For more information please see http://edgar.jrc.ec.europa.eu	European Commission, Joint Research Centre and Netherlands Environmental Assessment Agency 2014	Table 5.3
Eora Multiregion Input-Output (MRIO) Database	The Eora MRIO database provides a time series of high resolution input-output tables with matching environmental and social satellite accounts for 187 countries for the period 1990–2011. For more information please see http://worldmrio.com	Lenzen and others 2012; Lenzen and others 2013	Figures 2.10–2.20 Tables 2.1- 2.3

Dataset	Further information	Datasets key references cited in *IDR 2016*	Corresponding figures and tables in *IDR 2016*
EU KLEMS project	The EU KLEMS project aimed at creating a database on measures of economic growth, productivity, employment creation, capital formation and technological change at the industry level for all European Union member states from 1970 onwards. In the March 2007 release, a period is covered from 1970 to 2004 for 25 countries and a limited set of variables for 72 industries. For more information please see http://euklems.net	O'Mahony and Timmer 2009	Figures 1.16–1.18, 1.23
Fuel Combustion Statistics Database	This dataset contains a detailed set of statistics on carbon dioxide (CO_2) emissions estimates from fossil fuel combustion with years covered from 1960–2012. For more information please see http://dx.doi.org/10.1787/co2-data-en	IEA 2015b	Figures 1.22, 5.1, 5.2, 5.9
Groningen Growth and Development Centre (GGDC) 10-Sector Database	The GGDC 10-Sector database provides a long-run internationally comparable dataset on sectoral productivity performance in Africa, Asia, and Latin America. Variables covered in the data set are annual series of value added, output deflators, and persons employed for 10 broad sectors. It gives sector detail from 1950 onwards, consists of series for 11 countries in Africa, 11 countries in Asia, 2 countries in the Middle East and North Africa, and 9 in Latin-America. For comparison, also data for the United States and several European countries were added. For more information please see www.rug.nl/research/ggdc/data/10-sector-database	Timmer, de Vries and de Vries 2014	Figures 1.4, 1.7, 1.16, 1.18, 1.23 Table 1.4
ICE Futures Europe Data	ICE Futures Europe provide market data and insight into Intercontinental Exchange's global financial and commodity markets and clearing houses. For more information please see www.theice.com/market-data	Knopf and others 2014	Figure 5.15
ILOSTAT Database	The ILOSTAT database is the primary source for cross-country statistics on the labour market. The database contains over 100 indicators covering more than 230 countries and economies. For more information please see www.ilo.org/ilostat/faces/oracle/webcenter/portalapp/pagehierarchy/Page137.jspx?_afrLoop=4652569856048325&clean=true#%40%3F_afrLoop%3D4652569856048325%26clean%3Dtrue%26_adf.ctrl-state%3Dtqhjvfh3b_9	ILO 2015a	Figures 1.7, 1.16, 1.18, 1.23
International Energy Agency (IEA) Energy Flow Charts	The IEA's animated Sankey flow charts visualize energy balances on a global and country specific basis in the period 1973–2013. For more information please see www.iea.org/Sankey/index.html	IEA 2013	Figures 7.23–7.27
International Futures model	The International Futures model is a comprehensive forecasting modeling system available to the public. It uses Pardee's best understanding of global systems to produce forecasts for 186 countries to the year 2100. For more information please see http://pardee.du.edu	Frederick S. Pardee Center for International Futures 2015	Table 5.1
International Income Distribution (I2D2) Database	The I2D2 database is a global harmonized household survey database with 1,018 economies-years that represent 160 economies. For more information please see Montenegro and Patrinos (2014).	Montenegro and Patrinos 2014	Figure 4.7
Key Indicators of the Labour Market (KILM) Datbase	The KILM database is a comprehensive database of country-level data on 18 key indicators of the labour market from 1980 to the latest available year. For more information please see www.ilo.org/empelm/what/WCMS_114240/lang--en/index.htm	ILO 2015b	Figures 1.7, 1.16, 1.18, 1.23

Dataset	Further information	Datasets key references cited in *IDR 2016*	Corresponding figures and tables in *IDR 2016*
Maddison Project Database	The Maddison Project was initiated to support an effective way of cooperation between scholars to continue Maddison's work on measuring economic performance for different regions, time periods and subtopics. For more information please see: www.ggdc.net/maddison/maddison-project/home.htm	The Maddison Project 2013, Comin and Mestieri 2013	Figures 1.4, 3.2, 3.3, 3.4, 3.8 Table 1.2
Global Material Flows Database	The Global Material Flows database comprises data for more than 200 countries, the time period of 1980 to 2011, and more than 300 different materials aggregated into 12 categories of material flows. For more information please see www.materialflows.net/home	SERI and WU Vienna 2015	Figure 5.2
Combined datasets (based on multiple datasets)	Newly constructed data set with international comparable data on employment and value added by sector and unemployment. The dataset contains information for 100 countries over the period 1950–2009. Datasets used to construct a large dataset: WIOD database; Asia, EU and World KLEMS databases; OECD Structural Analysis Database; GGDC10 database, Asian Productivity Organization Database; Asian Development Bank Key Indicators for Asia and the Pacific Database; EUROSTAT Database; United Nations Economic Commission for Latin America and the Caribbean Statistical Database; United Nations Economic Commission for Africa: UNECA; United Nations Statistical Division, National Accounts Database and International Labour Organization, KILM Database For more information please see: Lavopa and Szirmai (2014).	Lavopa and Szirmai 2014	Annex A3 Figures 3.10–3.12
	Newly constructed data set with information for almost 100 countries over the period 1950–2010. Datasets used to construct a large dataset: WDI Database, Human Development Index (UNDP), Gini index (SWIID 5.0), Inclusive industrialization development index (UNSD, EUKLEMS, WIOD, INDSTAT), CLIMATE (Gallup and others 1999), Natural Resources (World Bank 2010), PWT 8.1, Education (Barro and Lee 2013), GGDC10, ILO For more information please see Lavopa (2015b).	Lavopa 2015b	Figures 4.2, 4.3 Table 4.1
	A regression framework was elaborated, using a dataset of 88 countries, including 21 advanced economies and 67 developing countries, covering the period 1950–2005. Datasets used to construct a large dataset: GGDC 10 sector database, World Development Indicators database, UN national accounts statistics, EU KLEMS database, UNIDO industrial statistics database, Maddison project dataset, Penn World Tables (Vs. 6.3) dataset. For more information on data please see Szirmai and Verspagen (2015).	Szirmai and Verspagen 2015	Figures 3.6, 3.7
OECD Social and Welfare Statistics Database	OECD Social and Welfare Statistics database provides comparable data on the distribution of household income for OECD's 34 member countries and the Russian Federation in the period 1974–2011. For more information please see: www.oecd-ilibrary.org/social-issues-migration-health/data/oecd-social-and-welfare-statistics/income-distribution_data-00654-en	OECD 2014	Figure 4.9
Penn World Tables (PWT) Database	PWT version 8.0 is a database with information on relative levels of income, output, inputs and productivity, covering 167 countries between 1950 and 2011. For more information please see www.rug.nl/research/ggdc/data/pwt	Feenstra and others 2015	Figures 1.16–1.21, 1.23
PovcalNet database	PovcalNet is the compilation of a large number of household surveys stored by the World Bank research department. For more information please see http://iresearch.worldbank.org/PovcalNet	Lakner and Milanovic 2013	Box 1.1

Dataset	Further information	Datasets key references cited in *IDR 2016*	Corresponding figures and tables in *IDR 2016*
Standardized World Income Inequality (SWIID) Database	The SWIID database provides comparable Gini indices of gross and net income inequality for 174 countries for as many years as possible from 1960 to the present along with estimates of uncertainty in these statistics. For more information please see https://dataverse.harvard.edu/dataset.xhtml?persistentId=hdl:1902.1/11992	Solt 2009, 2014	Figures 4.7, 4.11
Trade Union Membership Statistics Database	The data presented in the Trade Union Membership database are official national statistics drawn mainly from national publications with a coverage of 63 countries and a time period of 1980–2013. For more information please see Visser, Hayter and Gammarano (2015).	Visser, Hayter and Gammarano 2015	Figure 4.11
Trends Econometric Models Database	Trends Econometric Models database provides estimates of labour market indicators in the countries and years for which country-reported data are unavailable. The regional estimates and projections cover 107 countries for 1995–2013. For more information please see ILO (2014).	ILO 2014	Figures 7.14–7.20 Tables 7.10, 7.11
UN National Accounts Statistics Database	The National Accounts database presents a series of analytical national accounts tables from 1970 onwards for more than 200 countries and areas of the world. For more information please see http://unstats.un.org/unsd/snaama/Introduction.asp	UN 2014b	Figures 1.4–1.6, 1.16–1.18, 1.21, 1.23, 3.2–3.4, 3.8 Table 1.3
United Nations Commodity Trade Statistics (UN Comtrade) Database	UN Comtrade Database is a repository of official trade statistics and relevant analytical tables. It contains annual trade statistics starting from 1962 and monthly trade statistics since 2010. For more information please see http://comtrade.un.org	UNSD 2015a	Annex B4 Figures 5.14, 7.9–7.13 Tables 5.4, 5.6, 7.6–7.9
UNIDO Industrial Statistics Database at the 2-digit level of ISIC code (Rev. 3) (NDSTAT2)	The INDSTAT2 database contains time series data on the manufacturing sector for the period 1963 onwards for more than 160 countries. The database contains eight principle indicators of industrial statistics, including the index numbers of industrial production, which show the real growth of the volume of production by 2-digit of ISIC Rev. 3. For more information please see www.unido.org/en/resources/statistics/statistical-databases.html	UNIDO 2012, 2014c, 2015g	Figures 1.10, 1.11, 1.16, 1.17, 1.19, 1.20, 1.22, 1.23 Tables 7.2, 7.3, 7.5
UNIDO Manufacturing Value Added (MVA) Database	The MVA database contains country data for GDP, MVA and population for the period starting with 1990 to the latest year available. The database is updated annually. For more information please see www.unido.org/en/resources/statistics/statistical-databases.html	UNIDO 2014d, UNIDO 2015e	Figures 5.1, 7.1–7.8, 7.25–7.27 Table 7.1
UTIP-UNIDO Industrial Pay Inequality Database	The UTIP-UNIDO Industrial Pay Inequality Database focuses on measuring and explaining movements of inequality in wages and earnings around the world. The dataset provides a wage inequality Theil measure for 167 countries worldwide for the period 1963–2008. For more information please see www.edac.eu/indicators_desc.cfm?v_id=209	University of Texas and UNIDO 2015	Figures 1.16–1.18, 1.23
World Income Distribution (WYD) Database	The WYD database is on global income inequality and world income distribution, with world household survey data from 1988–2005. For more information please see http://econ.worldbank.org/WBSITE/EXTERNAL/EXTDEC/EXTRESEARCH/0,,contentMDK:22261771~pagePK:64214825~piPK:64214943~theSitePK:469382,00.html	Lakner and Milanovic 2013	Box 1.1

Dataset	Further information	Datasets key references cited in *IDR 2016*	Corresponding figures and tables in *IDR 2016*
Industrial Pollution Projection System (IPPS) Datasets	The Economics of Industrial Pollution Control Research Project at the World Bank has developed a database and approach to estimating the amount of industrial pollution and cost of abatement based on easily obtainable indicators of industry scale. For more information please see http://econ.worldbank.org/WBSITE/EXTERNAL/EXTDEC/EXTRESEARCH/0,,contentMDK:20347046~pagePK:64214825~piPK:64214943~theSitePK:469382,00.html	Hettige and others 1995	Figure 5.8
World Development Indicators (WDI) Database	The WDI database presents the most current and accurate global development data available, and includes national, regional and global estimates. The database is updated quarterly and covers 1960–2015. For more information please see: http://data.worldbank.org/data-catalog/world-development-indicators	World Bank 2015a	Figures 5.1–5.3, 5.9–5.12
World Input-Output (WIOD) Database	The WIOD database provides time-series of world input-output tables for forty countries worldwide and a model for the rest-of-the-world, covering the period from 1995 to 2011. For more information please see: Timmer and others (2015) and www.wiod.org/new_site/home.htm	Timmer and others 2015	Figures 1.1, 1.4, 1.12–1.17, 1.23, 5.9, 7.21, 7.22, 7.23 Table 3.1
World Integrated Trade System (WITS) Database	The WITS database allows users to access and retrieve information on trade and tariffs from international organizations: UN Comtrade, UNCTAD Trade Analysis Information System (TRAINS), WTO's Integrated Data Base and the World Bank and the Center for International Business, Tuck School of Business at Dartmouth College Global Preferential Trade Agreements Database. For more information please see http://wits.worldbank.org/about_wits.html	World Bank 2015c	Figures 5.12, 5.13 Table 5.4
World Top Income Database	The World Top Incomes database aims to provide online access to world top income information and information on the distribution of earnings and the distribution of wealth. Around forty-five further countries are under study and will be incorporated. For more information please see Alvaredo and others (2014) and http://topincomes.parisschoolofeconomics.eu	Alvaredo and others 2014	Figure 4.10

References

A*STAR (Agency for Science, Technology and Research), 2013a. *Singapore's Biopolis: A Success Story*. Available at <www.a-star.edu.sg/Media/News/Press-Releases/ID/1893/Singapores-Biopolis-A-Success-Story.aspx>. Accessed July 2015.

———, 2013b. *Information Sheet—Annex B. The BMS Initiative: 10 Years On*. Available at <www.a-star.edu.sg/Media/News/Press-Releases/ID/1893/Singapores-Biopolis-A-Success-Story.aspx>. Accessed July 2015.

Abdullah, A.Z., Salamatinia, B., Mootabadi, H., Bhatia, S., 2009. Current Status and Policies on Biodiesel Industry in Malaysia as the World's Leading Producer of Palm Oil. *Energy Policy*, 37(12), pp. 5440–5448.

Abramovitz, M., 1986. Catching Up, Forging Ahead, and Falling Behind. *Journal of Economic History*, 46(2), pp. 385–406.

Abramovitz, M. and David, P.A., 1996. Convergence and Deferred Catch-Up: Productivity Leadership and the Waning of American Exceptionalism. In: *The Mosaic of Economic Growth*, eds. Landau, R., Taylor, T. and Wright, G., Redwood City, CA: Stanford University Press.

Acemoglu, D., 2003. Technology and Inequality. *NBER Reporter*, Winter, pp. 12–19. Cambridge, MA: The National Bureau of Economic Research.

Acemoglu, D. and Autor, D.H., 2011. Skills, Tasks and Technologies: Implications for Employment Earnings. In: *The Handbook of Labor Economics, Volume 4B*, eds. Ashenfelter, O. and Card, D.E., Amsterdam: Elsevier.

Ács, Z.J. and Naudé, W.A., 2011. *Entrepreneurship, Stages of Development, and Industrialization*. UNU-WIDER Working Paper No. 2011/80. Helsinki: United Nations University—World Institute for Development Economics Research.

Adewuyi, A.O. and Oyejide, T.A., 2012. Determinants of Backward Linkages of Oil and Gas Industry in the Nigerian Economy. *Resources Policy*, 37(4), pp. 452–460.

Afrane, G., Arvidsson, R., Baumann, H., Borg, J., Keller, E., Milà í Canals, L. and Selmer, J.K., 2013. *A Product Chain Organization Study of Certified Cocoa Supply*. The 6th International Conference on Life Cycle Management, Gothenburg, 25–28 August 2013.

Ageyi-Holmes, A., 2014. *Tilling the Soil in Tanzania: What Do Emerging Economies Have to Offer?* PhD Dissertation. Milton Keynes: The Open University.

Aho, E., Cornu, J., Georghiou, L., Subira, A., 2006. *Creating an Innovative Europe*. Report of the Independent Expert Group on R&D and Innovation. Brussels: European Commission.

Alcorta, L. 1994. The Impact of New Technologies on Scale in Manufacturing Industries: Issues and Evidence. *World Development*, 22(5), pp. 755–769.

Alderson, A.S. and Nielsen, F., 2002. Globalization and the Great U-Turn: Income Inequality Trends in 16 OECD Countries. *American Journal of Sociology*, 107(5), pp. 1244–1299.

Allen, R.C., 2012. Technology and the Great Divergence: Global Economic Development since 1820. *Explorations in Economic History*, 49(1), pp. 1–16.

Allwood, J.M., Ashby, M.F., Gutowski, T.G. and Worrell, E., 2013. Material Efficiency: Providing Material Services with Less Material Production. *Philosophical Transactions of the Royal Society*, 371(1986), pp. 1–15.

Altenburg, T., 2010. Managed Latecomer Strategies vs. Political Capture: Can Developing Countries Handle Selective Business Promotion? In: *Effective State-Business Relations, Industrial Policy and Economic Growth*, ed. te Velde, D.W., London: Overseas Development Institute.

———, 2011. *Industrial Policy in Developing Countries: Overview and Lessons from Seven Country Cases*. Discussion Paper 4/2011. Bonn: German Development Institute.

Alvaredo, F., Atkinson, A.B, Piketty, T. and Saez, E., 2013. The Top 1 Percent in International and Historical Perspective. *Journal of Economic Perspectives*, 27(3), pp. 3–20.

———, 2014. *The World Top Incomes Database*. Available at <http://topincomes.parisschoolofeconomics.eu>. Accessed November 2014.

Amsden, A.H., 1989. *Asia's Next Giant: South Korea and Late Industrialization.* New York: Oxford University Press.

Arimura, T., Hibiki, A. and Johnstone, N., 2007. An Empirical Study of Environmental R&D: What Encourages Facilities to Be Environmentally Innovative? In: *Environmental Policy and Corporate Behaviour.* ed. Johnstone, N., Cheltenham and Northampton, MA: Edgar Elgar.

Arnold, W., 2003. *Singapore Goes for Biotech.* New York Times. Available at <www.nytimes.com/2003/08/26/business/singapore-goes-for-biotech.html>. Accessed July 2015.

Atkinson, A.B., 2013. *Ensuring Social Inclusion in Changing Labour and Capital Markets.* Economic Papers 481. Brussels: European Union.

Atkinson, R.D. and Ezell, S.J., 2012. *Innovation Economics: The Race For Global Advantage.* New Haven, CT, and London: Yale University Press.

Attah-Ankomah, R., 2014. *China's Presence in Developing Countries' Technology Basket: The Case of Furniture Manufacturing in Kenya.* PhD Dissertation. Milton Keynes: The Open University.

Auriol, E., Biancini, S. and Paillacar, R., 2012. *Intellectual Property Rights Protection in Developing Countries.* Paper presented at the 2012 Fourteenth Annual of the Conference European Trade Study Group (ETSG 2012), University of Leuven, Leuven, 13–15 September, 2012.

Autor, D.H. and Dorn, D., 2013. The Growth of Low-Skill Service Jobs and the Polarization of the US Labor Market. *American Economic Review*, 103(5), pp. 1553–1597.

Autor, D.H., Katz, L.F. and Kearney, M.S., 2006. The Polarization of the U.S. Labor Market. *American Economic Review*, 96(2), pp. 189–194.

———, 2008. Trends in U.S. Wage Inequality: Revising the Revisionists. *Review of Economics and Statistics*, 90(2), pp. 300–323.

Autor, D.H., Levy, F. and Murnane, R.J., 2003. The Skill Content of Recent Technological Change: An Empirical Exploration. *The Quarterly Journal of Economics*, 118(4), pp. 1279–1333.

Ayres, R.U., Turton, H. and Casten, T., 2006. Energy Efficiency, Sustainability and Economic Growth. *Energy*, 32(5), pp. 634–648.

Bair, J. and Gereffi, G., 2014. Towards Better Work in Central America: Nicaragua and the CAFTA Context. In: *Towards Better Work: Understanding Labour in Apparel Global Value Chains*, eds. Rossi, A., Luinstra, A. and Pickles, J., Basingstoke and Geneva: Palgrave Macmillan and International Labour Office.

Baker, D., 2012. *Technology Doesn't Cause Inequality—Deliberate Policy Change Does.* The Guardian. Available at <www.theguardian.com/commentisfree/cifamerica/2012/jul/16/technology-inequality-policy-change>. Accessed July 2015.

Baldwin, R.E., 2011. *Trade and Industrialization After Globalization's 2nd Unbundling: How Building and Joining a Supply Chain Are Different and Why It Matters.* Working Paper 17716. Cambridge, MA: The National Bureau of Economic Research.

Baldwin, R.E. and Venables, A.J., 2013. Spiders and Snakes: Offshoring and Agglomeration in the Global Economy. *Journal of International Economics*, 90(2), pp. 245–254.

Bamber, P., Fernandez-Stark, K., Gereffi, G. and Guinn, A., 2014. *Connecting Local Producers in Developing Countries to Regional and Global Value Chains: Update.* OECD Trade Policy Papers No. 160. Paris: OECD Publishing.

Bank of America, 2013. *No Time to Waste—Global Waste Primer.* Thematic Investing. Available at <www.long-finance.net/images/reports/pdf/baml_waste_2013.pdf>. Accessed August 2015. Barrett, S., 1994. Self-Enforcing International Environmental Agreements. *Oxford Economic Papers*, 46(0), pp. 878–894.

Barrientos, S., Gereffi, G. and Rossi, A., 2011. Economic and Social Upgrading in Global Production Networks: A New Paradigm for a Changing World. *International Labour Review*, 150(3–4), pp. 319–340.

Bénabou, R., 2005. Inequality, Technology, and the Social Contract. In: *Handbook of Economic Growth, Volume 1B*, eds. Aghion, P. and Durlauf, S.N., Amsterdam: North-Holland.

Berns M., Townend A., Khayat Z., Balagopal B., Reeves M., Hopkins M. and Kruschwitz N., 2009. Sustainability and Competitive Advantage. *MIT Sloan Management Review*, 51(1), pp. 19–26.

Bertani, R., 2015. *Geothermal Power Generation in the World 2010–2014 Update Report.* Proceedings World

Geothermal Congress 2015. Melbourne, Australia, 19–25, April 2015.

BFTA (Belgian Foreign Trade Agency), 2014. *Biomedical Science Sector: Singapore*. Brussels.

Bigum, M., Brogaard, L.K.-S. and Christensen, T.H, 2012. Metal Recovery from High-Grade WEEE: A Life Cycle Assessment. *Journal of Hazardous Materials,* 207–208, pp. 8–14.

Bleischwitz, R., 2009. Outline of a Resource Policy and Its Economic Dimension. In: *Sustainable Resource Management. Trends, Visions and Policies for Europe and the World,* ed. Bringezu, S. and Bleischwitz, R., Sheffield: Greenleaf Publisher.

———, 2012. Towards a Resource Policy—Unleashing Productivity Dynamics and Balancing International Distortions. *Mineral Economics*, 24(2), pp. 135–144.

Blinder, A.S., 2009. How Many US Jobs Might be Offshorable? *World Economics*, 10(2), pp. 41–78.

Bloch, R. and Owusu, G., 2012. Linkages in Ghana's Gold Mining Industry: Challenging the Enclave Thesis. *Resources Policy*, 37(4), pp. 434–442.

Bluhm, R. and Szirmai, A., 2012. *Institutions and Long-Run Growth Performance: An Analytic Literature Review of the Institutional Determinants of Economic Growth.* UNU-MERIT Working Paper Series No. 2012–033. Maastricht: United Nations University, Maastricht Economic and Social Research Institute on Innovation and Technology.

Bogliacino, F., Piva, M. and Vivarelli, M., 2012. R&D and Employment: An Application of the LSDVC Estimator Using European Microdata. *Economics Letters*, 116(1), pp. 56–59.

Bolt, J. and van Zanden, J.L., 2014. The Maddison Project: Collaborative Research on Historical National Accounts. *The Economic History Review*, 67(3), pp. 627–651.

Borgave, S. and Chaudhari, J.S., 2010. Indian Auto Component Industry: Challenges Ahead. *International Journal of Economics and Business Modeling*, 1(2), pp. 1–11.

Bosetti, V., Carraro, C. and Galeotti, M., 2006. The Dynamics of Carbon and Energy Intensity in a Model of Endogenous Technical Change. *The Energy Journal*, 27, Special Issue on Endogenous Technological Change and the Economics of Atmospheric Stabilisation, pp. 191–206.

Botchie, D., 2014. *Comparative Analysis of Asian Driver and Western Garment Making Machines: A Case of the Garments Sector in Uganda.* PhD Dissertation. Milton Keynes: The Open University.

Brynjolfsson, E. and McAfee, A., 2012a. Thriving in the Automated Economy. *The Futurist,* March-April, 46(2), pp. 27–31. Available at <http://www.wfs.org/content/futurist/march-april-2012-vol-46-no-2/thriving-automated-economy>. Accessed October 2015.

———, 2012b. *Race Against the Machine: How the Digital Revolution is Accelerating Innovation, Driving Productivity, and Irreversibly Transforming Employment and the Economy.* Research Brief. Cambridge, MA: MIT Center for Digital Business.

———, 2014. *The Second Machine Age: Work, Progress, and Prosperity in a Time of Brilliant Technologies*. New York: W. W. Norton & Company.

Cafaggi, F., Swensson, L.F.J., Macedo Jr., R.P., Andreotti e Silva, T., Gross, C.P., de Almeida, L.G. and Ribeiro, T.A., 2012. *Accessing the Global Value Chain in a Changing Institutional Environment: Comparing Aeronautics and Coffee.* IDB Working Paper Series No. IDB-WP-370. Washington, DC: Inter-American Development Bank

Canidio, A., 2013. *The Determinants of Long-Run Inequality*. Department of Economics Working Paper. Budapest: Central European University. Available at < http://www.personal.ceu.hu/staff/repec/pdf/2012_10.pdf >. Accessed November 2015.

Cantore N., 2011. *Energy Efficiency in Developing Countries for the Manufacturing Sector.* WP 15/2011. Vienna: United Nations Industrial Development Organization.

Cantore, N., Antimiani, A. and Anciaes, P.R., 2012. *Energy Price Shocks: Sweet and Sour Consequences for Developing Countries.* Working Paper 355. London: Overseas Development Institute.

Cantore, N., Nussbaumer, P., Wei, M. and Kammen, D., 2015. *Promoting Renewable Energy and Energy Efficiency in Africa: Employment Generation and Cost Effectiveness.* Working paper. Vienna: United Nations Industrial Development Organization. Forthcoming.

Cantore, N. and Padilla, E., 2010. *Emissions Distribution in Post–Kyoto International Negotiations: A Policy Perspective.* Barcelona: Universitat Autònoma de Barcelona.

Card, D.E. and DiNardo, J.E., 2002. Skill-Biased Technological Change and Rising Wage Inequality: Some Problems and Puzzles. *Journal of Labor Economics*, 20(4), pp. 733–783.

Card, D.E., Lemieux, T. and Riddell, W.C., 2004. Unions and Wage Inequality. *Journal of Labor Research*, 25(4), pp. 519–559.

Carpenter, J., 2011. Impact of GM Crops on Biodiversity. *GM Crops*, 2, pp. 1–17.

Carraro, C., Eyckmans, J. and Finus, M., 2006. Optimal Transfers and Participation Decisions in International Environmental Agreements. *The Review of International Organizations*, 1(4), pp. 379–396.

Casalet, M., Buenrostro, E., Stezano, F., Oliver, R. and L. Abelenda, L., 2011. *Evolución y Complejidad en el Desarrollo de Encadenamientos Productivos en México: Los Desafíos de la Construcción del Cluster Aeroespacial en Querétaro.* Santiago: United Nations Economic Commission for Latin America and the Caribbean.

CEPII (Centre d'Etudes Prospectives et d'Informations Internationales), 2010. *BACI.* Paris. Available at <www.cepii.fr/anglaisgraph/bdd/baci.htm>. Accessed August 2015.

Chassot, O., 2012. *Servicios Medioambientales.* Personal communication with K. Fernandez-Stark. November 2, 2012.

Chataway, J., Hanlin, R. and Kaplinsky, R., 2014. Inclusive Innovation: An Architecture for Policy Development. *Innovation and Development*, 1(4), pp. 33–54.

Chertow, M. and Ehrenfeld, J., 2012. Organizing Self-Organizing Systems: Toward a Theory of Industrial Symbiosis. *Journal of Industrial Ecology*, 16(1), pp. 13–27.

Chong, L.W., n.d. *Overview of Singapore's Biomedical Sciences Initiative.* Singapore: Ministry of Health.

Chung, S., 2011. Innovation, Competitiveness and Growth: Korean Experiences. In: *Annual World Bank Conference on Development Economics—Global 2010. Lessons from East Asia and the Global Financial Crisis*, eds. Lin, J.Y. and Pleskovic, B., Washington, DC: The World Bank.

Clough, J., 2015. *Made in India: Five Companies Demonstrate Global Leadership for the Climate.* Green Futures Magazine. Available at <www.forumforthefuture.org/greenfutures/articles/made-india-five-companies-demonstrate-global-leadership-climate-0>. Accessed August 2015.

Comin, D. and Mestieri, M., 2013. *If Technology Has Arrived Everywhere, Why Has Income Diverged*? Mimeo. Cambridge: Harvard University.

Conservation International, 2012. *Costa Rica.* Available at <www.conservation.org>. Accessed June 2015.

Cornwall, J., 1977. *Modern Capitalism: Its Growth and Transformation.* New York: St. Martin's Press.

Cowen, T., 2014. *How Technology Could Help Fight Income Inequality.* The New York Times. Available at <www.nytimes.com/2014/12/07/upshot/how-technology-could-help-fight-income-inequality.html>. Accessed July 2015.

Craven, C., 2011. The Honduran Palm Oil Industry: Employing Lessons from Malaysia in the Search for Economically and Environmentally Sustainable Energy Solutions. *Energy Policy*, 39(11), pp. 6943–6950.

CREST Working Group, 2007. *Internationalisation of R&D —Facing the Challenge of Globalisation: Approaches to a Proactive International Policy in S&T.* Analytical Report. Brussels: European Union Scientific and Technical Research Committee.

Curkovic, S. and Sroufe, R., 2011. Using ISO 14001 to Promote a Sustainable Supply Chain Strategy. *Business Strategy and the Environment*, 20(2), pp. 71–93.

Dahlman, C., 2010. *Innovation Strategies in Brazil, India and China: From Imitation to Deepening Technological Capability in the South.* In: *The Rise of Technological Power in the South*, eds. Xiaolan Fu; and Luc Soete, London: Palgrave Macmillan.

DARA and the Climate Vulnerable Forum, 2012. *Climate Vulnerability Monitor, 2nd Edition. A Guide to the Cold Calculus of a Hot Planet.* Madrid and Manila.

Das, B., Prakash, S., Reddy, P.S.R. and Misra, V.N., 2007. An Overview of Utilization of Slag and Sludge from Steel Industries. *Resources, Conservation and Recycling*, 50(1), pp. 40–57.

Davies, R.B. and Vadlamannati, K.C., 2013. A Race to the Bottom in Labour Standards? An Empirical Investigation. *Journal of Development Economics*, 103, pp. 1–14.

de Haan, A., 2015. *Social Inclusion and Structural Transformation: Concepts, Measurements and Trade-Offs.*

Background paper prepared for the 2015 Industrial Development Report. Vienna: United Nations Industrial Development Organization.

De Marchi, V., Di Maria, E. and Micelli, S., 2013. Environmental Strategies, Upgrading and Competitive Advantage in Global Value Chains. *Business Strategy and the Environment*, 22(1), pp. 62–72.

De Marchi, V., Di Maria, E. and Ponte, S., 2013. The Greening of Global Value Chains: Insights from the Furniture Industry. *Competition and Change*, 17(4), pp. 299–318.

De Marchi, V., Giuliani, E. and Rabellotti, R., 2015. *Local Innovation and Global Value Chains in Developing Countries.* Inclusive and Sustainable Industrial Development Working Paper Series, WP 05/2015. Vienna: United Nations Industrial Development Organization.

De Marchi, V. and Grandinetti, R., 2013. Knowledge Strategies for Environmental Innovations: The Case of Italian Manufacturing Firms. *Journal of Knowledge Management*, 17(4), pp. 569–582.

Díaz, A. and Puch, L.A., 2013. *A Theory of Vintage Capital Investment and Energy Use*. UC3M Working Papers. Madrid: Universidad Carlos III de Madrid.

Diop, D., Blanco, M., Flammini, A., Schlaifer, M., Kropiwnicka, M.A. and Markhof, M.M., 2013. *Assessing the Impact of Biofuels Production on Developing Countries from the Point of View of Policy Coherence for Development*. Final Report to the European Commission. Brussels.

Dustmann, C., Ludsteck, J. and Schönberg, U., 2009. Revisiting the German Wage Structure. *The Quarterly Journal of Economics*, 124(2), pp. 843–881.

Dutz, M.A., Kessides, I., O'Connell, S. and Willig, R.D., 2011. *Competition and Innovation-Driven Inclusive Growth*. Policy Research Working Paper 5852. Washington, DC: World Bank.

Eco-Innovation Observatory, 2013. *Europe in Transition: Paving the Way to a Green Economy Through Eco-Innovation*. Funded by the European Commission, DG Environment, Brussels.

EDB (Singapore Economic Development Board), 2012. *Biomedical Sciences. Singapore—The Biopolis of Asia*. Singapore.

———, 2014. *Pharmaceuticals and Biotechnology*. Available at <www.edb.gov.sg/content/edb/en/industries/industries/pharma-biotech.html>. Accessed July 2015.

Edler, J., 2010. *International Policy Coordination for Collaboration in S&T.* Manchester Business School Working Paper No. 590. Manchester: Manchester Business School.

Elharouni, F., Benmoussa, R. and Ouahman, A.A., 2013. Incentives for the Development of the Automotive Sector in Morocco and the Truth of Local Stakeholders. *International Journal of Science and Research*, 4(5), pp. 2157–2158.

El-Sebaii, A.A. and Shalaby, S.M., 2012. Solar Drying of Agricultural Products: A Review. *Renewable and Sustainable Energy Reviews*, 16(1), pp. 37–43.

Enevoldsen, M.K., Ryelund, A.V. and Andersen, M.S., 2007. Decoupling of Industrial Energy Consumption and CO_2 Emissions in Energy-Intensive Industries in Scandinavia. *Energy Economics*, 29(4), pp. 665–692.

Enright, M.J., Scott, E.E. and Chang, K.-M., 2005. *Regional Powerhouse: The Greater Pearl River Delta and the Rise of China*. Singapore: John Wiley & Sons.

Environmental Compliance Assistance Centre, 2015. *Success Story Tannery*. Available at <http://www.ecacwb.org/editor_upload/files/Success_story_tannery.pdf>. Accessed: July 2015.

EPU (Economic Planning Unit), 2010. *Tenth Malaysia Plan 2011–2015*. Putrajaya: Prime Minister's Department.

European Commission, 2012a. *Communication from the Commission to the European Parliament, the Council, the European Economic and Social Committee and the Committee of the Regions on 'Enhancing and Focusing EU International Cooperation in Research and Innovation: A Strategic Approach'.* Commission Staff Working Document. Brussels.

———, 2012b. *International Cooperation in Science, Technology and Innovation: Strategies for a Changing World. Report of the Expert Group Established to Support the Further Development of an EU International STI Cooperation Strategy*, eds. Schwaag Serger, S. and Remoe, S. Luxembourg.

European Commission, Joint Research Centre and Netherlands Environmental Assessment Agency (PBL), 2014.

Emission Database for Global Atmospheric Research (EDGAR), release CO_2 time series 1970–2013. Available at <http://edgar.jrc.ec.europa.eu, 2014>.

———, 2015. *Number of Organisations with Registered Environmental Management Systems According to EMAS and ISO 14001.* Available at <http://www.eea.europa.eu/data-and-maps/indicators/number-of-organisations-with-registered/assessment>. Accessed August 2015.

European Science Foundation, 2012. *Indicators of Internationalisation for Research Institutions: A New Approach.* A report by the ESF Member Organisation Forum on Evaluation: Indicators of Internationalisation. Strasbourg.

European Union, 2015. *The Principle of Subsidiarity.* EUR-Lex. Available at < http://eur-lex.europa.eu/legal-content/EN/TXT/HTML/?uri=URISERV:ai0017&from=EN>. Accessed November 2015.

Fagerberg, J., 1987. A Technology Gap Approach to Why Growth Rates Differ. *Research Policy*, 16(2–4), pp. 87–99.

———, 2006. Innovation: A Guide to the Literature. In: *The Oxford Handbook of Innovation*, eds. Fagerberg, J., Mowery, D.C. and Nelson, R.R., Oxford: Oxford University Press.

Fagerberg, J. and Srholec, M., 2008. National Innovation Systems, Capabilities and Economic Development. *Research Policy,* 37(9), pp. 1417–1435.

Fagerberg, J., Srholec, M. and Verspagen, B., 2010. Innovation and Economic Development. In: *Handbook of the Economics of Innovation,* Volume 2, eds. Hall, B. and Rosenberg, N., Oxford: North-Holland.

FAO (Food and Agriculture Organization of the United Nations), 2006a. *Dryer Construction for Solar-dried Fruit and Vegetables Production.* Available at <http://teca.fao.org/technology/dryer-construction-solar-dried-fruit-and-vegetables-production>. Accessed July 2015.

———, 2006b. *Introducing a Mechanical Press for Making Shea Butter in Northern Ghana.* Available at <http://teca.fao.org/read/4500>. Accessed September 2015.

———, 2008. *The State of Food and Agriculture 2008. Biofuels: Prospects, Risks and Opportunities.* Rome.

———, 2015. *FAOSTAT: Country Profile. Uganda.* Available at <http://faostat.fao.org/CountryProfiles/Country_Profile/Direct.aspx?lang=en&area=226>. Accessed July 2015.

FAO (Food and Agriculture Organization of the United Nations), IFAD (International Fund for Agricultural Development), IMF (International Monetary Fund), OECD (Organization for Economic Co-Operation and Development), UNCTAD (United Nations Conference on Trade and Development), WFP (World Food Programme), World Bank, WTO (World Trade Organization), IFPRI (International Food Policy Research Institute) and UN HLTF (United Nations High Level Task Force), 2011. *Price Volatility in Food and Agricultural Markets: Policy Responses.* Rome: Food and Agriculture Organization

Farole, T. and Winkler, D. eds., 2014. *Making Foreign Direct Investment Work for Sub-Saharan Africa: Local Spillovers and Competitiveness in Global Value Chains.* Directions in Development: Trade. Washington, DC: World Bank.

Feenstra, R.C., Inklaar, R. and Timmer, M.P., 2015. The Next Generation of the Penn World Table. *American Economic Review*, forthcoming. Available at <www.ggdc.net/pwt>.

Fine, D., van Wamelen, A., Lund, S., Cabral, A., Taoufiki, M., Dörr, N., Leke, A., Roxburgh, C., Schubert, J. and Cook, P., 2012. *Africa at Work: Job Creation and Inclusive Growth.* Washington, DC: McKinsey Global Institute.

Foster-McGregor, N., Kaulich, F. and Stehrer, R., 2015. *Global Value Chains in Africa.* Inclusive and Sustainable Industrial Development Working Paper Series, WP 04/2015. Vienna: United Nations Industrial Development Organization.

Foster-McGregor, N., Kaba, I. and Szirmai, A., 2015. *Structural Change and the Ability to Sustain Growth.* Background paper prepared for the 2015 Industrial Development Report. Vienna: United Nations Industrial Development Organization.

Frederick S. Pardee Center for International Futures, 2015. *International Futures Modeling System, Version 7.0.* Denver, CO: Josef Korbel School of International Studies, University of Denver. Available at <http://pardee.du.edu/access-ifs>.

Freeman, C., 1987. *Technology Policy and Economic Performance: Lessons from Japan.* London and New York: Frances Printer Publishers.

Freeman, C. and Soete, L., 1997. *The Economics of Industrial Innovation.* Cambridge, MA: MIT Press.

Frey, T., 2011. *55 Jobs of the Future.* Futurist Speaker. Available at <www.futuristspeaker.com/2011/11/55-jobs-of-the-future>. Accessed August 2015..

Frey, C.B. and Osborne, M.A., 2013. *The Future of Employment: How Susceptible Are Jobs to Computerization?* University of Oxford.

Frondel, M., Horbach, J. and Rennings, K., 2007. End-of-Pipe or Cleaner Production? An Empirical Comparison of Environmental Innovation Decisions Across OECD Countries. *Business Strategy and the Environment,* 16(8), pp. 571–584.

Fu, X., 2015. *China's Path to Innovation.* Cambridge: Cambridge University Press.

Fu, X., Pietrobelli, C. and Soete, L., 2011. The Role of Foreign Technology and Indigenous Innovation in the Emerging Economies: Technological Change and Catching-up. *World Development,* 39(7), pp. 1204–1212.

Galor, O., 2012. The Demographic Transition: Causes and Consequences. *Cliometrica,* 6(1), pp. 1–28.

Gaulier, G. and Zignago, S., 2010. *BACI: International Trade Database at the Product-Level. The 1994–2007 Version.* CEPII Working Paper No 2010–23. Paris: Centre d'Études Prospectives et d'Informations Internationales.

Georghiou, L. and Roessner, D., 2000. Evaluating Technology Programs: Tools and Methods. *Research Policy,* 29(4), pp. 657–678.

Gereffi, G., 1999. International Trade and Industrial Upgrading in the Apparel Commodity Chain. *Journal of International Economics,* 48(1), pp. 37–70.

———, 2015. *Global Value Chains and Development.* Background paper prepared for the 2015 Industrial Development Report. Vienna: United Nations Industrial Development Organization.

Gereffi, G. and Fernandez-Stark, K., 2011. *Global Value Chain Analysis: A Primer.* Durham, NC: Duke University, Center on Globalization, Governance & Competitiveness.

Gereffi, G., Humphrey, J. and Kaplinsky, R., 2001. Introduction: Globalisation, Value Chains and Development. *IDS bulletin,* 32(3), pp. 1–8.

Gereffi, G., Humphrey, J. and Sturgeon, T., 2005. The Governance of Global Value Chains. *Review of International Political Economy,* 12(1), pp. 78–104.

GGDC (Groningen Growth and Development Centre), 2015. *10-Sector Database.* Available at <www.rug.nl/research/ggdc/data/10-sector-database>. Accessed July 2015.

Ghana Ministry of Food and Agriculture, 2012. *Pilot Program Based Budget for 2013–2015.* Final draft. Rwanda.

Ghani, E. and O'Connell, S.D., 2014. *Can Service be a Growth Escalator in Low Income Countries?* Policy Research Working Paper 6971. Washington, DC: World Bank.

Ghisetti, C., Marzucchi, A. and Montresor, S., 2013. *Does External Knowledge Affect Environmental Innovations? An Empirical Investigation of Eleven European Countries.* NGENIO (CSIC-UPV) Working Paper Series No. 01. Valencia: Ciudad Politécnica de la Innovación.

Ghisetti, C. and Quatraro, F., 2013. Beyond Inducement in Climate Change: Does Environmental Performance Spur Environmental Technologies? A Regional Analysis of Cross-Sectoral Differences. *Ecological Economics,* 96, pp. 99–113.

Giuliani, E. and Macchi, C., 2014. Multinational Corporations' Economic and Human Rights Impacts on Developing Countries: A Review and Research Agenda. *Cambridge Journal of Economics,* 38(2), pp. 479–517.

Goldin, C. and Katz, L.F., 2009. *The Race between Education and Technology: The Evolution of U.S. Educational Wage Differentials, 1890 to 2005.* NBER Working Paper No. 12984. Cambridge, MA: The National Bureau of Economic Research.

Goldstein, A., 2002. EMBRAER: From National Champion to Global Player. *CEPAL Review,* 77, pp. 97–115.

Goos, M. and Manning, A., 2007. Lousy and Lovely Jobs: The Rising Polarization of Work in Britain. *The Review of Economics and Statistics,* 89(1), pp. 118–133.

Goos, M., Manning, A. and Salomons, A., 2014. Explaining Job Polarization: Routine-Biased Technological Change and Offshoring. *The American Economic Review,* 104(8), pp. 2509–2526.

Goyal, M., 2013. SMEs Employ Close to 40% of India's Workforce, but Contribute Only 17% to GDP.

The Economic Times. Available at <http://articles.economictimes.indiatimes.com/2013-06-09/news/39834857_1_smes-workforce-small-and-medium-enterprises>. Accessed June 2015.

Gries, T. and Grundmann, R., 2014. Trade and Fertility in the Developing World: The Impact of Trade and Trade Structure. *Journal of Population Economics*, 27(4), pp. 1165–1186.

Gries, T., Grundmann, R., Palnau, I. and Redlin, M., 2015. *Can Technological Change Drive Socially Inclusive Industrialization ? A Review of Major Concepts and Findings.* Background paper prepared for the 2015 Industrial Development Report. Vienna: United Nations Industrial Development Organization.

Grindrod, M., 2015. *Vibrant Growth in Singapore Pharma Industry Being Driven by Innovation.* ProClinical Life Sciences Recruitment Blog. Available at <http://blog.proclinical.jobs/vibrant-growth-in-singapore-pharma-industry-being-driven-by-innovation>. Accessed July 2015.

Guadagno, F., 2013. *Towards a Quantification of Industrial Policy in Korea, Taiwan, Argentina, Brazil, and Mexico.* Paper presented at the 25th Annual EAEPE Conference for the European Association for Evolutionary Political Economy, Université Paris Nord, Paris, 7–9 November 2013.

Guerrieri P. and Meliciani, V., 2005. Technology and International Competitiveness: The Interdependence between Manufacturing and Producer Services. *Structural Change and Economic Dynamics*, 16(4), pp. 489–502.

Gupta, N., Healy, D.W., Stein, A.M. and Shipp, S.S., 2013. *Innovation Policies of South Korea.* Alexandria, VA: Institute for Defense Analyses.

Hall, B.H. and Jaffe, A.B., 2012. *Measuring Science, Technology, and Innovation: A Review.* A report prepared for the Panel on Developing Science, Technology, and Innovation Indicators for the Future. Washington, DC: National Academy of Sciences.

Haraguchi, N. and Rezonja, G., 2011. *Emerging Patterns of Manufacturing Structural Change.* UNU-WIDER Working Paper No. 2011/43. Maastricht: United Nations University, Maastricht Economic and Social Research Institute on Innovation and Technology.

Haseltine, W.A., 2013. *Affordable Excellence: The Singapore Healthcare Story.* Brookings Institution.

Hatzichronoglou, T., 1997. *Revision of the High-Technology Sector and Product Classification.* OECD Science, Technology and Industry Working Papers, 1997/02, Paris: OECD Publishing.

Hausmann, R., Pritchett, L. and Rodrik, D., 2005. Growth Accelerations. *Journal of Economic Growth,* 10(4), pp. 303–329.

Hausmann, R. and Rodrik, D., 2003. Economic Development as Self-Discovery. *Journal of Development Economics*, 72(2), pp. 603–633.

———, 2006. *Doomed to Choose: Industrial Policy as Predicament.* Prepared for the Blue Sky seminar organized by the Center for International Development at Harvard University. Cambridge, MA: Harvard University.

Hettige, H., Martin, P., Singh, M. and Wheeler, D., 1995. *The Industrial Pollution Projection System.* World Bank Policy Research Working Paper Number 1431, March 1995. Washington, DC: World Bank.

Hirsch-Kreinsen, H., 2008. "Low-Technology": A Forgotten Sector in Innovation Policy. *Journal of Technology Management and Innovation*, 3(3), pp. 11–20.

Holbrook, J.A. and Godin, B., 2011. *A Proposal for Measuring Science, Technology and Innovation Activities.* Burnaby, BC: Simon Fraser University, Centre for Policy Research on Science and Technology.

Horbach, J., Rammer, C. and Rennings, K., 2012. Determinants of Eco-Innovations by Type of Environmental Impact—The Role of Regulatory Push/Pull, Technology Push and Market Pull. *Ecological Economics*, 78, pp. 112–122.

Humphrey, J. and Schmitz, H., 2000. *Governance and Upgrading: Linking Industrial Cluster and Global Value Chain Research.* IDS Working Paper 120. Brighton: Institute of Development Studies.

———, 2001. *Governance in Global Value Chains.* IDS Bulletin, 32(3), pp. 19–29. Brighton: University of Sussex: Institute of Development Studies.

———, 2002. How Does Insertion in Global Value Chains Affect Upgrading in Industrial Clusters? *Regional Studies, 36*(9), pp. 1017–1027.

Hussain, I.S., 2014. *Indian Case Study on ZLD—The Tirupur Textile cluster experience.* Presentation, 28 January,

2014. Available at http://gpcb.gov.in/images/pdf/zld_presentation_11.pdf>. Accessed July 2015.

IEA (International Energy Agency), 2008. *Energy Efficiency Policies and Measures.* Paris. Available at <www.iea.org/policiesandmeasures/energyefficiency/>. Accessed May 2015.

———, 2013a. *Technology Roadmap: Carbon Capture and Storage, 2013 Edition.* Paris.

———, 2013b. Energy Flow Charts. Paris. Available at http://www.iea.org/Sankey/index.html. Accessed July 2015.

———, 2013. *Energy Flow Charts.* Paris. Available at <www.iea.org/Sankey/>. Accessed July 2015.

———, 2014. *Technology Roadmap: Solar Photovoltaic Energy.* Paris.

———, 2015a. *Carbon Capture and Storage.* Available at <www.iea.org/topics/ccs/>. Accessed November 2015.

———, 2015b. Detailed CO_2 Estimates. *IEA CO_2 Emissions from Fuel Combustion Statistics* (database). Available at <http://dx.doi.org/10.1787/data-00429-en>. Accessed July/August 2015.

IEA (International Energy Agency), ICCA (International Council of Chemical Associations) and DECHEMA, 2013. *Technology Roadmap: Energy and GHG Reductions in the Chemical Industry via Catalytic Processes.* Paris: International Energy Agency.

———, 2013b. Energy Flow Charts. Paris. Available at http://www.iea.org/Sankey/index.html. Accessed July 2015.

ILO (International Labor Organization), 1991. *The Creation of Technological Capability in Developing Countries.* London: Pinter Publishers Limited.

———, 2014. *World Employment and Social Outlook. Trends 2015. Data Set Trends Econometric Models.* Available at <www.ilo.org/global/research/global-reports/weso/2015/lang--en/index.htm>. Accessed September 2015.

———, 2015a. *ILOSTAT Database.* Available at <www.ilo.org/ilostat>. Accessed July 2015. Geneva.

———, 2015b. *Key Indicators of the Labour Market (KILM). Database.* Available at <www.ilo.org/empelm/what/WCMS_114240/lang--en/index.htm>. Accessed July 2015. Geneva.

ILO (International Labor Organization) and IFC (International Finance Corporation), 2010. *Better Work.* Available at: <www.betterwork.org/EN/countries/Pages/Lesotho.aspx>. Accessed November 2010.

Imbs, J.M. and Wacziarg, R., 2003. Stages of Diversification. *American Economic Review*, 93(1), pp. 63–86.

Inklaar, R. and Timmer, M.P., 2012. The Relative Price of Services. *Review of Income and Wealth,* 60(4), pp. 1–16.

IRENA (International Renewable Energy Agency), 2014a. *Renewable Power Generation Costs in 2014.* Bonn.

———, 2014b. *Renewable Energy Prospects: China, Remap 2030 Analysis.* Abu Dhabi. Available at: <www.irena.org/remap>.

———, 2014c. *Renewable Energy and Jobs—Annual Review 2014.* Abu Dhabi.

Jacob, J. and Sasso, S., 2015. *Foreign Direct Investment and Technology Spillovers in Low and Middle-Income Countries: A Comparative Cross-Sectoral Analysis.* Inclusive and Sustainable Industrial Development Working Paper Series, WP 07/2015. Vienna: United Nations Industrial Development Organization.

Jenne, C.A. and Cattell, R.K., 1983. Structural Change and Energy Efficiency in Industry. *Energy Economics,* 5(2), pp. 114–123.

Jéquier, N., 1976. *Appropriate Technology: Problems and Promises.* Paris: Organization of Economic Co-Operation and Development.

Jiang, X. and Milberg, W., 2012. *Vertical Specialization and Industrial Upgrading: A Preliminary Note.* Working Paper 10. Manchester: Capturing the Gains.

Jibreel, M.B., Mumuni, E., Al-Hassan, S. and Baba, N.M., 2013. Shea Butter and Its Processing Impacts on the Environment in the Tamale Metropolis of Ghana. *International Journal of Development and Sustainability,* 2(3), pp. 2008–2019.

Jovanovic, B. and Rousseau, P.L., 2005. General Purpose Technologies. In: *Handbook of Economic Growth, Volume 1B,* eds. Aghion, P. and Durlauf, S., Amsterdam: Elsevier.

Kaba, I.S. and Szirmai, A., 2015. *Indicators of Structural Diversification and Characteristics of Economic Sustainability: an Empirical Investigation.* Background paper prepared for the 2015 Industrial Development Report. Vienna: United Nations Industrial Development Organization.

Kaldor, N., 1957. A Model of Economic Growth. *Economic Journal*, 69(268), pp. 591–624.

———, 1966. *Causes of the Slow Rate of Growth of the United Kingdom. An Inaugural Lecture.* Cambridge, MA: Cambridge University Press.

———, 1967. *Strategic Factors in Economic Development.* Ithaca, NY: Cornell University, New York State School of Industrial and Labor Relations.

Kaltenberg, M. and Verspagen, B., 2015. *Healthy Development: Uncovering Capabilities.* Background paper prepared for the 2015 Industrial Development Report. Vienna: United Nations Industrial Development Organization.

Kaphengst, T. and Smith, L., 2013. *The Impact of Biotechnology on Developing Countries.* Berlin: Ecological Institute.

Kaplan, D., 2012. South African Mining Equipment and Specialist Services: Technological Capacity, Export Performance and Policy. *Resource Policy*, 37(4), pp. 425–433.

Kaplinsky, R., 1990. *The Economies of Small: Appropriate Technology in a Changing World.* London: Intermediate Technology Publications.

———, 2015a. *Technological Upgrading in Global Value Chains and Clusters and Their Contribution to Sustaining Economic Growth in Low and Middle Income Economies.* Inclusive and Sustainable Industrial Development Working Paper Series, WP 03/2015. Vienna: United Nations Industrial Development Organization.

———, 2015b. Global Value Chains in Manufacturing Industry: Where They Came from, Where They Are Going and Why This Is Important. In: *Routledge Handbook of Industry and Development*, eds. Weiss, J. and Tribe, M., London and New York: Routledge

Kaplinsky, R. and Morris, M., 2014. *Developing Industrial Clusters and Supply Chains to Support Diversification and Sustainable Development of Exports in Africa.* Addis Ababa: African Export-Import Bank.

Kaplinsky, R. and Wamae, W., 2010. *The Determinants of Upgrading and Value Added in the African Clothing Sector: The Contrasting Experiences of Kenya and Madagascar.* IKD Working Paper No. 59. Milton Keynes: The Open University.

Katz, J., 1995.*Technology and Industrial Restructuring in Latin America: The New Evidence.* Paper presented in the seminar Transferencia de Tecnologia, Comercio y Desarrollo, University of Venecia.

Katz, L.F. and Autor, D.H., 1999. Changes in the Wage Structure and Earnings Inequality. In: *Handbook of Labor Economics, Volume 3A*, eds. Ashenfelter, O. and Card, D., Amsterdam: Elsevier.

Kaulich, F., 2012. *Diversification vs. Specialization as Alternative Strategies for Economic Development: Can We Settle a Debate by Looking at the Empirical Evidence?* Development Policy, Statistics and Research Branch Working Paper 3/2012. Vienna: United Nations Industrial Development Organization.

Keller, W., 2010. International Trade, Foreign Direct Investment, and Technology Spillovers. In: *Handbook of the Economics of Innovation, Volume 2*, eds. Hall, B.H. and Rosenberg, N., Amsterdam: Elsevier.

Ketels, C. and Memedovic, O., 2008. From Clusters to Cluster-Based Economic Development. *International Journal of Technological Learning, Innovation and Development*, 1(3), pp. 375–392.

Khanna, M., Deltas, G. and Harrington, D.R., 2009. Adoption of Pollution Prevention Techniques: The Role of Management Systems and Regulatory Pressures. *Environmental and Resource Economics*, 44(1), pp. 85–106.

Kim, L., 1980. Stages of Development of Industrial Technology in a Developing Country: A Model. *Research Policy*, 9(3), pp. 254–277.

KIPO (Korean Intellectual Property Office), 2015. *Annual Report 2014.* Daejeon.

Knopf, B., Koch, N., Grosjean, G., Fuss, S., Flachsland, C., Pahle, M., Jakob, M. and, Edenhofer, O., 2014. *The European Emissions Trading System (EU ETS): Ex-Post Analysis, the Market Stability Reserve and Options for a Comprehensive Reform.* Fondazione Eni Enrico Mattei Working Paper Series No. 79. Milan.

KOICA (Korean International Cooperation Agency), 2014. *Implementation Strategies to Activate Appropriate Technology in Development Cooperation Program.* Seongnam.

Koh, L.P. and Wilcove, D.S., 2008. Is Oil Palm Agriculture Really Destroying Tropical Biodiversity? *Conservation Letters*, 1, pp. 60–64.

Krishnaswamy, K.N., Mathirajan, M. and Subrahmanya, M.H.B., 2014. Technological Innovations and Its

Influence on the Growth of Auto Component SMEs of Bangalore: A Case Study Approach. *Technology in Society*, 38, pp. 18–31.

Krugman, P.R., 2008. Trade and Wages, Reconsidered. *Brookings Papers on Economic Activity*, 39(1), pp. 103–154.

Kuznets, S., 1955. Economic Growth and Income Inequality. *The American Economic Review*, 45(1), pp. 1–28.

Lakner, C. and Milanovic, B., 2013. *Global Income Distribution: From the Fall of the Berlin Wall to the Great Recession*. World Bank Policy Research Working Paper 6719. Washington, DC: World Bank.

Lall, S., 1992. Technological Capabilities and Industrialization. *World Development*, 20(2), pp. 165–186.

———, 2000. Technological Change and Industrialization in the Asian Newly Industrializing Economies: Achievements and Challenges. In: *Technology, Learning and Innovation: Experiences of Newly Industrializing Economies*, eds. Kim, L. and Nelson, R.R., Cambridge: Cambridge University Press.

———, 2002. Linking FDI and Technology Development for Capacity Building and Strategic Competitiveness. *Transnational Corporations*, 11(3), pp. 39–88.

Lall, S. and Narula, R., 2004. Foreign Direct Investment and its Role in Economic Development: Do We Need a New Agenda? *The European Journal of Development Research*, 16(3), pp. 447–464.

Lang-Koetz, C., Pastewski, N. and Rohn, H., 2010. Identifying New Technologies, Products and Strategies for Resource Efficiency. *Chemical Engineering and Technology*, 33(4), pp. 559–566.

Lavopa, A., 2015a. *Structural Transformation and Economic Development Can Development Traps be Avoided?* PhD thesis. Maastricht: Maastricht University.

———, 2015b. *Technology-Driven Structural Change and Inclusiveness: The Role of Manufacturing*. Inclusive and Sustainable Development Working Paper Series WP 14/2015. Vienna: United Nations Industrial Development Organization.

Lavopa, A. and Szirmai, A., 2012. *Industrialization, Employment and Poverty*. Report prepared for the International Finance Corporation, November 2012. UNU-MERIT Working Paper No. 2012–81. Maastricht: United Nations University, Maastricht Economic and Social Research Institute on Innovation and Technology.

———, 2014. *Structural Modernization and Development Traps: An Empirical Approach*. UNU-MERIT Working Paper Series No. 2014–076. Maastricht: United Nations University, Maastricht Economic and Social Research Institute on Innovation and Technology.

———, 2015. *Industrialization in Time and Space*. Inclusive and Sustainable Industrial Development Working Paper Series, WP 10/2015. Vienna: United Nations Industrial Development Organization.

Lazonick, W., Moss, P., Salzman, H. and Tulum, Ö., 2014. *Skill Development and Sustainable Prosperity: Cumulative and Collective Careers versus Skill-Biased Technical Change*. Working Paper Series No. 15. New York: Institute for New Economic Thinking.

Lee, K., 2013a. How Can Korea be a Role Model for Catch-up Development? A "Capability-based View." In: *Achieving Development Success: Strategies and Lessons from the Developing World*, ed. Fosu, A., Oxford: Oxford University Press.

———, 2013b. Knowledge Regimes and Technological Catch Up. In: *Learning, Capability Building and Innovation for Development*, eds. Dutrénit, G., Lee, K., Nelson, R.R., Vera-Cruz, A.O. and Soete, L., London: Palgrave Macmillan.

———, 2013c. *Schumpeterian Analysis of Economic Catch-up: Knowledge, Path-creation, and the Middle-Income Trap*. Cambridge: Cambridge University Press.

Lema, R., Berger, A. and Schmitz, H., 2013. China's Impact on the Global Wind Power Industry. *Journal of Current Chinese Affairs*, 42(1), pp. 37–69.

Lennon, C., Cantore, N. and Clara, M., 2015. *Fast and Furious: Kaldorian Analysis of Dynamic Sectors*. Background paper prepared for the 2015 Industrial Development Report. Vienna: United Nations Industrial Development Organization.

Lenzen, M., Kanemoto, K., Moran, D. and Geschke, A., 2012. Mapping the Structure of the World Economy. *Environmental Science and Technology*, 46(15), pp. 8374–8381.

Lenzen, M., Moran, D., Kanemoto, K. and Geschke, A., 2013. Building Eora: A Global Multi-Regional Input-Output

Database at High Country and Sector Resolution. *Economic Systems Research*, 25(1), pp. 20–49.

Liew, F.C., 2008. *Publication on Pyrometallurgy versus Hydrometallurgy.* Singapore: Engineering Department, TES-AMM.

Lin, J.Y. and Zhang, P., 2009. *Industrial Structure, Appropriate Technology and Economic Growth in Less Developed Countries.* Policy Research Working Paper 4905. Washington, DC: World Bank.

Llorens, M., Puelles, M. and Manzano, R., 2011. Consumer Behavior and Brand Preferences in Organic Grocery Products. Store Brands vs. Manufacturing Brands. *Innovation Marketing*, 7(3), pp. 109–115.

Los, B., Timmer, M.P. and de Vries, G.J., 2014. *The Demand for Skills 1995–2008: A Global Supply Chain Perspective.* OECD Economics Department Working Papers No. 1141. Paris: OECD Publishing.

Luken, R., Van Rompaey, F. and Zigová, K., 2008. The Determinants of EST Adoption by Manufacturing Plants in Developing Countries. *Ecological Economics*, 66, pp. 141–152.

Lundvall, B.-Å. and Johnson, B., 1994. The Learning Economy. *Journal of Industry Studies,* 1(2), pp. 23–42.

The Maddison Project. 2013. *Maddison Project Database, 2013 version.* Available at <www.ggdc.net/maddison/maddison-project/data.htm>.

Makki, S.S. and Somwaru, A., 2004. Impact of Foreign Direct Investment and Trade on Economic Growth: Evidence from Developing Countries. *American Journal of Agricultural Economics,* 86(3), pp. 795–801.

Malaysia Ministry of Human Resources, 2008. *Occupational Structure: Oil Palm-Based Industry.* Kuala Lumpur.

Mansour, S. and Castel, V., 2015. *Country Note: Morocco.* African Economic Outlook. Available at <www.africaneconomicoutlook.org/en/country-notes/north-africa/morocco>. Accessed August 2015.

Manyika, J., Chui, M., Bughin, J., Dobbs, R., Bisson, P. and Marrs, A., 2013. *Disruptive Technologies: Advances That Will Transform Life, Business, and the Global Economy.* Washington, DC: McKinsey Global Institute.

Marsh, P., 2012. *The New Industrial Revolution: Consumers, Globalization and the End of Mass Production.* New Haven, CT: Yale University Press.

Martínez, J.R., 2011. Centripetal Forces in Aerospace Clusters in Mexico. *Innovation and Development,* 1(2), pp. 303–318.

Massa, I., 2015. *Technological Change in Developing Countries: Trade-Offs Between Economic, Social and Environmental Sustainability.* Background paper prepared for the 2015 Industrial Development Report. Vienna: United Nations Industrial Development Organization.

Matus, K., Timmer, V. and Appleby, D., 2013. *Manufacturing Sector Discussion Paper: Background Paper for the Project on Innovation and Access to Technologies for Sustainable Development.* Sustainability Science Program Working Paper No. 2013–04. Cambridge, MA Harvard University, John F. Kennedy School of Government.

Mazzanti, M., Marin, G., Nicolli, F. and Gilli, M., 2015. *Sustainable Development and Industrial Development: Manufacturing Environmental Performance, Technology and Consumption/Production Perspectives.* Background paper prepared for the 2015 Industrial Development Report. Vienna: United Nations Industrial Development Organization.

Mazzoleni, R. and Nelson, R.R., 1998. The Benefits and Costs of Strong Patent Protection: A Contribution to the Current Debate. *Research Policy*, 27(3), pp. 273–284.

Mazzanti M. and Zoboli R., 2005. The Drivers of Environmental Innovation. Empirical Evidence for a Manufacturing System. *Economia Politica,* 3/2005, pp. 399–438.

Mbayi, L., 2011. *Linkages in Botswana's Diamond Cutting and Polishing Industry.* Making the Most of Commodities Programme Discussion Paper No. 6. Cape Town and Milton Keynes: University of Cape Town and Open University.

McKinsey & Company, 2014. *India: Towards Energy Independence 2030.* New Delhi.

McMillan, M.S., Rodrik, D. and Verduzco-Gallo, Í., 2014. Globalization, Structural Change, and Productivity Growth, with an Update on Africa. *World Development*, 63, pp. 11–32.

Michaels, G., Natraj, A. and van Reenen, J., 2014. Has ICT Polarized Skill Demand? Evidence from Eleven Countries over Twenty-Five Years. *The Review of Economics and Statistics*, 96(1), pp. 60–77.

Milanovic, B., 2012. *Global Income Inequality by the Numbers: In History and Now—An Overview*. Policy Research Working Paper Series 6259. Washington, DC: World Bank.

Milberg, W., Jiang, X. and Gereffi, G., 2014. Industrial Policy in the Era of Vertically Specialized Industrialization. In: *Transforming Economies: Making Industrial Policy Work for Growth, Jobs and Development*, eds. Salazar-Xirinachs, J.M., Nübler, I. and Kozul-Wright, R., Geneva: International Labour Organization.

Mishel, L., Shierholz, H. and Schmitt, J., 2013. *Don't Blame the Robots: Assessing the Job Polarization Explanation of Growing Wage Inequality*. EPI-CEPR Working Paper, November. Washington, DC: Economic Policy Institute.

Mishkin, S. 2013. Chinese Companies Move into Supply Chain for Apple Components. *Financial Times*. Available at <www.ft.com/cms/s/0/d70fca52-2691-11e3-9dc0-00144feab7de.html#axzz3qynmG2eW>. Accessed November 2015.

MOFCOM (Ministry of Commerce People's Republic of China), 2012. National Programs for Science and Technology. Available at <http://english.mofcom.gov.cn/aarticle/zm/201205/20120508139141.html>. Accessed November 2015.

Mohammed, F., Boateng, S. and Al-Hassan, S., 2013. Effects of Adoption of Improved Sheabutter Processing Technology on Women's Livelihoods and Their Microenterprise Growth. *American Journal of Humanities and Social Science*, 1(4), pp. 244–250.

Montenegro, C.E. and Patrinos, H.A., 2014. *Comparable Estimates of Returns to Schooling Around the World*. Policy Research Working Paper 7020. Washington, DC: World Bank.

Morris, M., Kaplinsky, R. and Kaplan, D., 2012. *One Thing Leads to Another: Promoting Industrialisation by Making the Most of the Commodity Boom in Sub-Saharan Africa*. Cape Town: University of Cape Town—Policy Research on International Services and Manufacturing.

Morrison, A., Pietrobelli, C. and Rabellotti, R., 2008. Global Value Chains and Technological Capabilities: A Framework to Study Learning and Innovation in Developing Countries. *Oxford Development Studies*, 36(1), pp. 39–58.

Mulgan, G., Tucker, S., Ali, R. and Sanders, B., 2007. *Social Innovation: What It Is, Why It Matters and How It Can be Accelerated*. London: The Young Foundation.

Narula, R., 2004. *Understanding Absorptive Capacities in an "Innovation Systems" Context: Consequences for Economic and Employment Growth*. MERIT-Infonomics Research Memorandum Series, 003.

Naudé, W.A, 2010. *Promoting Entrepreneurship in Developing Countries: Policy Challenges*. UNU-WIDER Policy Brief No. 4. Helsinki: United Nations University—World Institute for Development Economic Research.

Naudé, W.A., Santos-Paulino, A. and McGillivray, M., 2009. *Vulnerability in Developing Countries*. Tokyo: United Nations University Press.

Naudé, W.A, Szirmai, A. and Haraguchi, N. eds., 2015. *Structural Change and Industrial Development in the BRICS*. Oxford: Oxford University Press.

Nemet, G.F., 2009. Demand-Pull, Technology-Push, and Government-Led Incentives for Non-Incremental Technical Change. *Research Policy*, 38(5), pp. 700–709.

Nixson, F., 2014. The Dynamics of Global Value Chain Development: A BRICS Perspective. In: *Structural Change and Industrial Development in the BRICS*, eds. Naudé, W.A., Szirmai, A. and Haraguchi, N. Oxford and New York: Oxford University Press.

Njoroge, N., 2014. Kenya Power Rates to Drop. *Busiweek*. Available at <www.busiweek.com/index.php/index1.php?Ctp=2&pI=2097&pLv=3&srI=53&spI=20&cI=11>. Accessed August 2015.

North, D.C., Wallis, J.J. and Weingast, B.R., 2009. *Violence and Social Orders: A Conceptual Framework for Interpreting Recorded Human History*. New York: Cambridge University Press.

Ocampo, J.A, 2005. The Quest for Dynamic Efficiency: Structural Dynamics and Economic Growth in Developing Countries. In: *Beyond Reforms: Structural Dynamics and Macroeconomic Vulnerability*, ed. Ocampo, J.A., Santiago: Economic Commission for Latin America and the Caribbean and Stanford University Press.

Ocampo, J.A. and Vos, R., 2008. Structural Change and Economic Growth. In: *Uneven Economic Development*, eds. Ocampo, J.A. and Vos, R., London: Zed Books.

OECD (Organization for Economic Co-Operation and Development), 2001. *Policies to Enhance Sustainable Development. Meeting of the OECD Council at Ministerial Level, 2001.* Paris.

———, 2002. *Frascati Manual 2002: Proposed Standard Practice for Surveys on Research and Experimental Development. The Measurement of Scientific and Technological Activities.* Paris: OECD Publishing.

———, 2005. *Science, Technology and Industry Scoreboard.* Paris: OECD Publishing.

———, 2008. *Biofuels Support Policies: An Economic Assessment.* Paris: OECD Publishing.

———, 2009. *OECD Reviews of Innovation Policy: Korea 2009.* Paris: OECD Publishing.

———, 2010. *Measuring Innovation: A New Perspective—online version.* Available at <http://www.oecd.org/site/innovationstrategy/measuringinnovationanewperspective-onlineversion.htm#agenda>. Accessed April 2015.

———, 2011. *OECD Science, Technology and Industry Scoreboard 2011.* Paris: OECD Publishing.

———, 2012a. *OECD Science, Technology and Industry Outlook 2012.* Paris: OECD Publishing.

———, 2012b. *Meeting Global Challenges through Better Governance: International Co-Operation in Science, Technology and Innovation.* Paris: OECD Publishing.

———, 2012c. *Green Growth and Developing Countries: A Summary for Policy Makers.* Paris: OECD Publishing.

———, 2013. *OECD Science, Technology and Industry Scoreboard 2013: Innovation for Growth.* Paris: OECD Publishing.

———, 2014. *Income Distribution. OECD Social and Welfare Statistics (database).* Available at <http://dx.doi.org/10.1787/data-00654-en>. Accessed November 2014.

———, 2015. *OECD Statistics: Main Science and Technology Indicators.* Available at <http://stats.oecd.org/index.aspx?DatasetCode=MSTI_PUB>. Accessed July 2015.

OECD (Organization for Economic Co-Operation and Development) and Eurostat, 2005. *The Measurement of Scientific and Technological Activities: Proposed Guidelines for Collecting and Interpreting Innovation Data. Oslo Manual, 3rd edition.* Paris: OECD Publishing.

OECD (Organization for Economic Co-Operation and Development) and World Bank, n.d. *Radical and Incremental Innovation.* The Innovation Policy Platform. Available at <https://innovationpolicyplatform.org/content/radical-and-incremental-innovation>. Accessed September 2015.

OECD (Organization for Economic Co-Operation and Development) and WTO (World Trade Organization), 2013. *Aid for Trade at a Glance 2013: Connecting to Value Chains.* Paris and Geneva.

OECD (Organization for Economic Co-Operation and Development) and WTO (World Trade Organization), and World Bank, 2014. *Global Value Chains: Challenges, Opportunities, and Implications for Policy.* Report prepared for submission to the G20 Trade Ministers Meeting Sydney, Australia, 19 July 2014. Paris, Geneva and Washington, DC.

O'Mahony, M. and Timmer, M.P., 2009. Output, Input and Productivity Measures at the Industry Level: the EU KLEMS Database. *Economic Journal*, 119(538), pp. F374–F403.

Osakwe, P.N., 2007. *Foreign Aid, Resources and Export Diversification in Africa: A New Test of Existing Theories.* MPRA Paper No. 2228. Munich: Munich Personal RePEc Archive.

Panwar, N.L., Kaushik, S.C. and Kothari, S., 2011. Role of Renewable Energy Sources in Environmental Protection: A Review. *Renewable and Sustainable Energy Reviews*, 15, pp. 1513–1524.

Park, S.-H., 1989. Linkages between Industry and Services and their Implications for Urban Employment Generation in Developing Countries. *Journal of Development Economics*, 30(2), pp. 359–379.

Park, S.-H. and Chan, K.S., 1989. A Cross-Country Input-Output Analysis of Intersectoral Relationships between Manufacturing and Services and their Employment Implications. *World Development*, 17(2), pp. 199–212.

Pedersen, E., 2011. Health Aspects Associated With Wind Turbine Noise. Results from Three Field Studies. *Noise Control Engineering Journal*, 59(1), pp. 38–46.

Pereira, A.A, 2006. Biotechnology Foreign Direct Investment in Singapore. *Transactional Corporations*, 15(2), pp. 99–124.

Pfaltzgraff, L.A., De Bruyn, M., Cooper, E.C., Budarin, V. and Clark, J.H., 2013. Food Waste Biomass: A Resource for High-Value Chemicals. *Green Chemistry*, 15(2), pp. 307–314.

Pietrobelli, C. and Rabellotti, R. eds., 2007. *Upgrading to Compete: SMEs, Clusters and Value Chains in Latin America*. Cambridge, MA: Harvard University Press.

———, 2011. Global Value Chains Meet Innovation Systems: Are There Learning Opportunities for Developing Countries? *World Development*, 39(7), pp. 1261–1269.

Piketty, T., 2014. *Capital in the Twenty-First Century*. Cambridge, MA: Harvard University Press.

Piketty, T., Saez, E. and Stantcheva, S., 2011. *Taxing the 1%: Why the Top Tax Rate Could Be Over 80%*. VOX CEPR's Policy Portal. Available at <www.voxeu.org/article/taxing-1-why-top-tax-rate-could-be-over-80>. Accessed August 2015.

Popp, D., 2006. International Innovation and Diffusion of Air Pollution Control Technologies: The Effects of NOX and SO2 Regulation in the US, Japan, and Germany. *Journal of Environmental Economics and Management*, 51(1), pp. 46–71.

Porter, M.E. and van der Linde C., 1995. Toward a New Conception of the Environment-Competitiveness Relationship. *Journal of Economic Perspectives*, 9(4), pp. 97–118.

Portocarrero Lacayo, A.V., 2010. *El Sector Textil y Confección y el Desarrollo Sostenible en Nicaragua*. Geneva: International Centre for Trade and Sustainable Development. January.

Posen, A., 2014. The Economic Inequality Debate Avoids Asking Who is Harmed. *Financial Times*. Available at <www.ft.com/cms/s/0/b46a586e-1be6–11e4–9db1–00144feabdc0.html>. Accessed July 2015.

Pritchard, D.J., 2012. *Brazil, Russia, India and China Governments' Aerospace Strategies and National Policies: Implications to Canada's Aerospace Industry*. Williamsville, NY: Pravco Aviation Review L.L.C.

Pritchett, L., 1998. *Patterns of Economic Growth: Hills, Plateaus, Mountains, and Plains*. Policy Research Working Paper 1947. Washington, DC: World Bank.

Puga, D. and Trefler, D., 2010. Wake Up and Smell the Ginseng: International Trade and the Rise of Incremental Innovation in Low-Wage Countries. *Journal of Development Economics,* 91(1), pp. 64–76.

Punch, J., James, R. and Pabst, D., 2010. Wind Turbine Noise: What Audiologists Should Know. *Audiology Today*, 8, pp. 20–31.

Ramani, S.V. and Szirmai, A., 2014. Innovation in India: The Challenge of Combining Economic Growth with Inclusive Development. In: *Innovation in India: Combining Economic Growth with Inclusive Development*, ed. Ramani, S.V., New Delhi: Cambridge University Press.

Rankin, W.J., 2011. *Minerals, Metals and Sustainability: Meeting Future Material Needs*. Collingwood, Victoria: CSIRO Publishing.

Rasiah, R. and Azmi, S., 2006. *Development of Palm Oil and Related Products in Malaysia and Indonesia*. Kuala Lumpur: University of Malaya.

Raunio, M., Kautonen, M. and Saarinen, J.P., 2013. *Models for International Innovation Policy: Transnational Channels and Regional Platforms: Fostering Globalizing Innovation Communities in Finland and Abroad.* Working Paper No 9, Tampere: University of Tampere, Research Centre for Knowledge, Science, Technology and Innovation Studies.

Reck, B.K. and Graedel, T.E., 2012. Challenges in Metal Recycling. *Science*, 337(6095), pp. 690–695.

Rehfeld, K.-M., Rennings, K. and Ziegler, A., 2007. Integrated Product Policy and Environmental Product Innovations: An Empirical Analysis. *Ecological Economics*, 61(1), pp. 91–100.

Ren, T., Patel, M.K. and Blok, K., 2006. Olefins from Conventional and Heavy Feedstocks: Energy Use in Steam Cracking and Alternative Processes. *Energy*, 31(4), pp. 425–451.

Reuters, 2015. *EU Politicians Set to Back Carbon Market 2019 Reform Start*. Available at <http://uk.reuters.com/article/2015/07/07/eu-carbon-vote-idUKL-8N0ZN19G20150707>. Accessed August 2015.

Ribbink, G. Nyabuntu P. and Kumar S., 2005. *Successful Supply Chains in Uganda. A Study of Three Successful Chains in the Coffee, Dried Fruit and Fresh Vegetables Sector.* Kampala: The UK Department for International Development.

Richards, T.J., Acharya, R.N. and Molina, I., 2010. Retail and Wholesale Market Power in Organic Apples. *Agribusiness*, 27(1), pp. 62–81.

Richardson, J., 2015. Electricity Cost Decreases by 30% in Kenya Due to Geothermal. *Clean Technica*. Available at <http://cleantechnica.com/2015/03/02/electricity-cost-decreases-30-kenya-due-geothermal>. Accessed August 2015.

Robinson, J.A., 2009. *Industrial Policy and Development: A Political Economy Perspective.* Paper prepared for the 2009 World Bank ABCDE Conference, Seoul, 22–24 June 2009.

Rodriguez, C.M., 2012. *Servicios Medioambientales.* Personal communication with K. Fernandez-Stark. August 22.

Rodrik, D., 2004. *Industrial Policy for the Twenty-First Century.* KSG Working Paper RWP04-047. Cambridge, MA: Harvard University, John F. Kennedy School of Government.

———, 2014. *An African Growth Miracle?* NBER Working Paper No. 20188. Cambridge, MA: National Bureau of Economic Research.

———, 2015. *Premature Deindustrialization.* Economics Working Papers No. 107. Princeton, NJ: IAS School of Social Sciences.

Rood, S., 2013. *Monitoring and Evaluation for Innovation Policy.* World Bank Policy Brief. Washington, DC.

Rosenberg, N., 2006. Innovation and Economic Growth. In: *Innovation and Growth in Tourism*, ed. Organization for Economic Co-Operation and Development. Paris: OECD Publishing.

Santiago, F., 2015. *Innovation Policy and Industrial Policy at the Cross-Roads: A Review of Recent Experiences in Advanced Developing Countries.* Inclusive and Sustainable Industrial Development Working Paper Series, WP 09/2015. Vienna: United Nations Development Organization.

———, 1995b. Small Shoemakers and Fordist Giants: Tale of a Supercluster. *World Development*, 23(1), pp. 9–28.

Schnitzer, M., Monreal, C.M. and Powell, E.E., 2014. Wheat Straw Biomass: A Resource for High-Value Chemicals. *Journal of Environmental Science and Health, Part. B: Pesticides, Food Contaminants, and Agricultural Wastes*, 49(1), pp. 51–67.

Schumacher, E.F., 1973. *Small Is Beautiful: Economics As If People Mattered.* New York: Harper & Row.

SERI (Sustainable Europe Research Institute) and WU Vienna (Vienna University of Economics and Business), 2015. *Global Material Flow Database.* Available at <http://www.materialflows.net/home>.

Seuring S., 2013. A Review of Modeling Approaches for Sustainable Supply Chain Management. *Decision Support Systems*, 54(4), pp. 1513–1520.

Sherburne, A., 2009. Sustainability through the Supply Chain. In: *Sustainable Textiles: Life Cycle and Environmental Impact*, ed. Blackburn, R.S., Cambridge: Woodhead Publishing.

Siddiqui, K., 2010. The Political Economy of Development in Singapore. *Research in Applied Economics*, 2(2), pp. 1–31.

Sinclair, P.J.N., 1981. When Will Technical Progress Destroy Jobs? *Oxford Economic Papers*, 31, pp. 1–18.

SMBA (Small and Medium Business Administration), 2012. *Statistics: Venture Businesses and Inno-Biz.* Available at <www.smba.go.kr/eng/smes/statistics_01.do?mc=usr0001148>. Accessed July 2015.

Solt, F., 2009. *The Standardized World Income Inequality Database, Harvard Dataverse, V05.* Available at <http://hdl.handle.net/1902.1/11992>.

———, 2014. *The Standardized World Income Inequality Database.* Working paper. SWIID Version 5.0, October 2014. Iowa City: University of Iowa.

Spitz-Oener, A., 2006. Technical Change, Job Tasks, and Rising Educational Demands: Looking Outside the Wage Structure. *Journal of Labor Economics*, 24(2), pp. 235–270.

Staritz, C., 2012. *Value Chains for Development? Potentials and Limitations of Global Value Chain Approaches in Donor Interventions.* Working Paper No. 31. Vienna: Austrian Foundation for Development Research.

Steenblik, R., 2005. *Environmental Goods: A Comparison of the APEC and OECD Lists.* OECD Trade and Environment Working Paper No. 2005–04. Paris: OECD Publications.

Steinmueller, W.E., 2010. Economics of Technology Policy. In: *Handbook of The Economics of Innovation,* eds. Hall, B.H. and Rosenberg, N., Amsterdam: Elsevier.

Stern, N., 2006. *The Economics of Climate Change: The Stern Review.* Cambridge: Cambridge University Press.

Stewart, F. and Ranis, G., 1990. Macro-Policies for Appropriate Technology: A Synthesis of Findings. In: *The Other Policy:*

The Influence of Policies on Technology Choice and Small Enterprise Development, eds. Stewart, F., Thomas, H. and de Wilde, T., London: Intermediate Technology Publications.

Sturgeon, T., Van Biesebroeck, J. and Gereffi, G., 2008. Value Chains, Networks and Clusters: Reframing the Global Automotive Industry. *Journal of Economic Geography*, 8(3), pp. 297–321.

Swetman, T. and Hammond, L., 1997. *Rural Women Earn Cash for "Butter" from Nuts in Ghana.* Chatham: Natural Resources Institute.

Szirmai, A., 2012a. Industrialisation as an Engine of Growth in Developing Countries, 1950–2005. *Structural Change and Economic Dynamics*, 23(4), pp. 406–420.

———, 2012b. *Proximate, Intermediate and Ultimate Causality: Theories and Experiences of Growth and Development.* UNU-MERIT Working Paper Series, No 2012–32. Maastricht: United Nations University, Maastricht Economic and Social Research Institute on Innovation and Technology.

———, 2015. *Socio-Economic Development, Second Edition.* Cambridge, MA: Cambridge University Press.

Szirmai, A., Naudé, W.A. and Alcorta, L., 2013. Introduction and Overview: The Past, Present and Future of Industrialization. In: *Pathways to Industrialization in the 21st Century. New Challenges and Emerging Paradigms*, eds. Szirmai, A., Naudé, W.A. and Alcorta, D., Oxford: Oxford University Press.

Szirmai, A., Naudé, W.A. and Goedhuys, M. eds., 2011. *Entrepreneurship, Innovation and Development*. Oxford: Oxford University Press.

———, 2015. Manufacturing and Economic Growth in Developing Countries, 1950–2005. *Structural Change and Economic Dynamics,* 34, pp. 46–59.

Tacchella, A., Cristelli, M., Caldarelli, G., Gabrielli, A. and Pietronero, L., 2012. A New Metrics for Countries' Fitness and Products' Complexity. *Scientific Reports,* 2, 723.

Tanzi, V., 2014. The Challenges of Taxing the Big. *Revista de Economía Mundial*, 37, pp. 23–40.

Tacchella A, Cristelli M, Caldarelli G, Gabrielli A, Pietronero L., 2012. A new metrics for countries' fitness and products' complexity. *Scientific Reports*, 2:723.

te Velde, D.W. and Leftwich, A., 2010. *State-Business Relations and Economic Growth in Sub-Saharan Africa: A Review of Case Studies in Ghana, Mauritius, South Africa and Zambia.* Manchester, UK: University of Manchester, Research Programme Consortium for Improving Institutions for Pro-Poor Growth.

Timilsina, G.R. and Shrestha, A., 2010. *Biofuels: Markets, Targets and Impacts.* World Bank Policy Research Working Paper 5364. Washington, DC: World Bank.

Timmer M.P. and de Vries, G.J., 2009. Structural Change and Growth Accelerations in Asia and Latin America: A New Sectoral Dataset. *Cliometrica*, 3(2), pp. 165–190.

Timmer, M.P., de Vries, G. and de Vries, K., 2014. *Patterns of Structural Change in Developing Countries.* GGDC Research Memorandum 149. Groningen: University of Groningen and Groningen Growth and Development Centre.

Timmer, M.P., Dietzenbacher, E., Los, B., Stehrer, R. and de Vries, G.J., 2015. An Illustrated User Guide to the World Input-Output Database: The Case of Global Automotive Production. *Review of International Economics*, 23(3), pp. 575–605.

Tregenna, F., 2015. *Deindustrialisation, Structural Change and Sustainable Economic Growth.* Inclusive and Sustainable Industrial Development Working Paper Series, WP 02/2015. Vienna: United Nations Industrial Development Organization.

UN (United Nations), 2014a. *The Road to Dignity by 2030; Ending Poverty, Transforming All Lives and Protecting the Planet.* Synthesis Report of the Secretary-General on the Post-2015 Sustainable Development Agenda, A/69/700.

———, 2014b. *National Accounts Statistics: Analysis of Main Aggregates, 2013.* New York, NY. Available at <http://unstats.un.org/unsd/snaama/Introduction.asp>. Accessed July 2015.

———, 2015a. *Transforming Our World: The 2030 Agenda for Sustainable Development.* Finalized Text for Adoption. Available at <https://sustainabledevelopment.un.org/content/documents/7891TRANSFORMING%20OUR%20WORLD.pdf>. Accessed August 2015.

———, 2015b. *Water and Sanitation.* Available at <https://sustainabledevelopment.un.org/topics/waterandsanitation>. Accessed September 2015.

———, 2015c. *Trade Capacity Building. Resource Guide. Volume 1.* Vienna.

UNDESA (United Nations Department of Economic and Social Affairs), 2006. *World Economic and Social Survey 2006: Diverging Growth and Development.* New York: United Nations.

———, 2008. *Report of the Expert Group Meeting "Promoting Social Integration."* Helsinki, Finland, 8–10 July 2008.

———, 2009. *Creating an Inclusive Society: Practical Strategies to Promote Social Integration.* New York, NY.

UNDP (United Nations Development Programme), 2015. *Empowering Rural Women and Alleviating Poverty by Strengthening the Local Shea Butter Industry.* Available at <www.gh.undp.org/content/ghana/en/home/operations/projects/poverty_reduction/all-projects2.html>. Accessed September 2015.

UNECA (United Nations Economic Commission for Africa), 2013. *Economic Transformation in Africa: Drivers, Challenges and Options.* Issues Paper. Prepared for the Third Meeting of the High Level Panel of Eminent Persons, 30 January to 1 February, 2013. Addis Ababa.

———, 2014. *Sustainability of Supply Chains and Sustainable Public Procurement: A Pre-Study.* Nairobi.

———, n.d. *Resource Efficiency.* Paris.

UNEP (United Nations Environment Programme) and GRID-Arendal, 2005. *Vital Climate Change Graphics for Latin America and the Caribbean.* Nairobi.

UNESCO (United Nations Educational, Scientific and Cultural Organization), 2014. *Information and Communications Technologies for Inclusive Social and Economic Development.* Commission on Science and Technology for Development, 12–16 May 2014. Geneva.

UNIDO (United Nations Development Organization), 2007. *Africa Foreign Investor Survey 2005. Understanding the Contributions of Different Investor Categories to Development. Implications for Targeting Strategies.* Vienna.

———, 2010a. *Making Private Standards Work for You: A Guide to Private Standards in the Garments, Footwear and Furniture Sectors.* Vienna.

———, 2010b. *Industrial Statistics: Guidelines and Methodology.* Vienna.

———, 2011a. *Industrial Development Report 2011. Industrial Energy Efficiency for Sustainable Wealth Creation—Capturing Environmental, Economic and Social Dividends.* Vienna.

———, 2011b. *UNIDO Green Industry. Policies for Supporting Green Industry. Vienna.*

———, 2012. *Industrial Statistics Database 2-Digit Level, ISIC Revision 3 (INDSTAT2), 2012.* Vienna.

———, 2013a. *Industrial Development Report 2013. Sustaining Employment Growth—The Role of Manufacturing and Structural Change.* Vienna.

———, 2013b. *Sun and Solar Drying, Techniques and Equipment.* Vienna. Available at <www.unido.org/fileadmin/import/32146_33SUNANDSOLARDRYING.16.pdf>. Accessed July 2015.

———, 2013c. *The Industrial Competitiveness of Nations. Looking Back, Forging Ahead. Competitiveness Industrial Performance Report 2012/2013.* Vienna.

———, 2014a. *Emerging Green Technologies for the Manufacturing Sector.* Vienna.

———, 2014b. *Introduction to UNIDO: Inclusive and Sustainable Industrial Development.* Vienna.

———, 2014c. *Industrial Statistics Database 2-Digit Level, ISIC Revision 3 (INDSTAT2), 2014.* Vienna.

———, 2014d. *Manufacturing Value Added 2014.* Vienna.

———, 2015a. *Competitive Industrial Performance Index 2015.* Vienna. Forthcoming.

———, 2015b. *Industrial Prosumers of Renewable Energy. Contribution to Inclusive and Sustainable Industrial Development.* Vienna.

———, 2015c. *International Yearbook of Industrial Statistics 2015.* Edward Elgar Publishing Limited. Cheltenham.

———, 2015d. *ISID Operation Platform.* Available at <https://isid.unido.org>. Accessed August 2015.

———, 2015e. *Manufacturing Value Added 2015.* Vienna. Forthcoming.

———, 2015f. *Meeting Standards, Winning Markets.* Available at <http://www.unido.org/tradestandardscompliance.html>. Accessed November 2015.

———, 2015g. Industrial Statistics Database 2-Digit Level, ISIC Revision 3 (INDSTAT2), 2015. Vienna.

UNIDO (United Nations Industrial Development Organization) and ICSHP (International Center on Small Hydro Power), 2013. *World Small Hydropower*

Development Report 2013. Available at <www.smallhydroworld.org>. Accessed August 2015.

UNIDO (United Nations Industrial Development Organization) and UNU-MERIT (United Nations University, Maastricht Economic and Social Research Institute on Innovation and Technology) 2014. *Diffusion Strategy of Green Technology and Green Industry in Africa: A Study of Renewable Energy Technology Market and Energy Efficiency Adoption in Maize and Cassava Processing Industries in Kenya and Nigeria.* Vienna and Maastricht.

University of Texas and UNIDO (United Nations Industrial Development Organization), 2015. *UTIP-UNIDO Industrial Pay Inequality Dataset.* Austin, TX. Available at <http://utip.gov.utexas.edu/data.html>.

UNOSSC (United Nations Office for South-South Cooperation), 2015. *South-South Gateway to Sustainable Prosperity. Annual Reports 2009–2014.* South-South Global Assets and Technology Exchange. New York.

UNSD (United Nations Statistics Division), 2015a. *United Nations Commodity Trade Statistics (Comtrade) Database.* New York, NY. Available at <http://comtrade.un.org>. Accessed September 2015.

———, 2015b. *Correspondence between ISIC Rev. 2 and ISIC Rev. 3.1.* Available at <http://unstats.un.org/unsd/cr/registry/regso.asp?Ci=25&Lg=1&Co=&T=0&p=8>. Accessed November 2015.

USAID (United States Agency for International Development), 2009. *Biofuels in Asia: An Analysis of Sustainability Options.* Washington, DC.

van Epps, H.L., 2006. Singapore's Multibillion Dollar Gamble. *Journal of Experimental Medicine,* 203(5), pp. 1139–1142.

van Reenen, J., 2011. Wage Inequality, Technology and Trade: 21st Century Evidence. *Labour Economics,* 18(6), pp. 730–741.

van Zanden, J.L., Baten, J., d'Ercole, M.M., Rijpma, A., Smith, C. and Timmer, M. eds., 2014. *How Was Life? Global Well-Being Since 1820.* Paris: OECD Publishing.

van Zanden, J. L., Baten, J., Foldvari, P. and van Leeuwen, B., 2014. The Changing Shape of Global Inequality 1820–2000; Exploring a New Dataset. *Review of Income and Wealth,* 60(2), pp. 279–297.

Verdolini, E. and Galeotti, M., 2011. At Home and Abroad: An Empirical Analysis of Innovation and Diffusion in Energy Technologies. *Journal of Environmental Economics and Management,* 61(2), pp. 119–134.

Verspagen, B., 1991. A New Empirical Approach to Catching Up or Falling Behind. *Structural Change and Economic Dynamics,* 2(2), pp. 359–380.

———, 2004. Innovation and Economic Growth. In: *The Oxford Handbook of Innovation,* eds. Fagerberg, J., Mowery, D.C. and Nelson, R.R., Oxford: Oxford University Press.

Vértesy, D., 2011. *Interrupted Innovation: Emerging Economies in the Structure of the Global Aerospace Industry.* Doctoral Thesis. Maastricht: Maastricht University.

Visser, J., 2013. *Data Base on Institutional Characteristics of Trade Unions, Wage Setting, State Intervention and Social Pacts, 1960–2011 (ICTWSS), Version 4.0, April 2013.* Amsterdam: Institute for Advanced Labour Studies AIAS University of Amsterdam.

Visser, J. Hayter, S. and Gammarano, R. 2015. *Trends in Collective Bargaining Coverage: Stability, Erosion or Decline?* Labour Relations and Collective Bargaining Policy Brief No. 1. ILO, Geneva.

Vivarelli, M., 1995. *The Economics of Technology and Employment: Theory and Empirical Evidence.* Aldershot: Edward Elgar Publishing.

———, 2013. *The Employment and Skill Impacts of Multifaceted Innovation.* Background paper prepared for the Industrial Development Report 2013. Vienna: United Nations Industrial Development Organization.

———, 2015. Innovation and Employment. *IZA World of Labor,* 154(May), pp. 1–10. Bonn: Institute for the Study of Labor.

von Weizsäcker, E.U., 1989. *Erdpolitik: Ökologische Realpolitik an der Schwelle zum Jahrhundert der Umwelt.* Darmstadt: Wissenschaftliche Buchgesellschaft.

Wagner, C.S., 2006. International Collaboration in Science and Technology: Promises and Pitfalls. In: *Science and Technology Policy for Development, Dialogues at the Interface,* eds. Box, L. and Engelhard, R., London: Anthem Press.

Wang, Z.-J., Lin, H., Huang, J., Hu, R., Rozelle, S. and Pray, C., 2009. *Bt* Cotton in China: Are Secondary Insect

Infestations Offsetting the Benefits in Farmer Fields? *Agricultural Sciences in China*, 8(1), pp. 83–90.

Warwick, K., 2013. *Beyond Industrial Policy, Emerging Issues and New Trends.* OECD Science Technology and Industry Policy Papers No. 2. Paris: OECD Publishing.

Watch, D.D., 2010. *Research for the Global Good: Supporting a Better World for All.* Melbourne: Images Publishing.

Weiss, J., 2015. *Taxonomy of Industrial Policy.* Inclusive and Sustainable Industrial Development Working Paper Series, WP 08/2015. Vienna: United Nations Development Organization.

Westphal, L.E., 2002. Technology Strategies for Economic Development in a Fast Changing Global Economy. *Economics of Innovation and New Technology*, 11(4/5), pp. 275–320.

Wild, S., 2014. SA joins 3D printing revolution. *Mail & Guardian Online.* Available at <http://mg.co.za/article/2014-09-26-sa-joins-3d-printing-revolution>. Accessed August 2015.

WIOD (World Input-Output Database), 2015. *World Input-Output Database.* Available at <www.wiod.org/database/index.htm>. Accessed September 2015.

WIPO (World Intellectual Property Organization), 2004. *WIPO Intellectual Property Handbook, Second Edition.* Geneva.

Witt, H., Patel, R. and Schnurr, M., 2006. Can the Poor Help GM Crops? Technology, Representation & Cotton in the Makhathini Flats, South Africa. *Review of African Political Economy*, 109, pp. 497–513.

Wood, A., 1998. Globalisation and the Rise in Labour Market Inequalities. *The Economic Journal*, 108(450), pp. 1463–1482.

World Bank, 2010. *Innovation Policy: A Guide for Developing Countries.* Washington, DC.

———, 2015a. *Data—World Development Indicators.* Washington, DC. Available at <http://data.worldbank.org/data-catalog/world-development-indicators>. Accessed June 2015.

———, 2015b. *Historical Classifications by Income.* Washington, DC. Available at <https://datahelpdesk.worldbank.org/knowledgebase/articles/378834-how-does-the-world-bank-classify-countries>. Accessed February 2015.

———, 2015c. *World Integrated Trade Solution (WITS). International Merchandise Trade, Tariff and Non-tariff Measures.* Washington, DC. Available at: <http://wits.worldbank.org/Default.aspx>.

WRI (World Resources Institute), 2015. *CAIT Climate Data Explorer.* Washington, DC. Available at <http://cait.wri.org>.

WTO (World Trade Organization), 2014. *World Trade Report 2014. Trade and Development: Recent Trends and the Role of the WTO.* Geneva.

Wright, G. and Czelusta, J., 2004. Why Economies Slow: The Myth of the Resource Curse. *Challenge*, 47 (2), pp. 6–38.

Yüksek, I., 2012. *Technical, Economical and Environmental Benefits of Blastfurnace Cement.* Available at <www.worldcement.com/europe-cis/09072012/Benefits_of_blastfurnace_cement_1159/>. Accessed July 2015.

Zeng, D.Z., 2014. *China's Special Economic Zones and Industrial Clusters: Success and Challenges.* Available at <http://blogs.worldbank.org/developmenttalk/china-s-special-economic-zones-and-industrial-clusters-success-and-challenges>. Accessed February 2014.

Zhong, S., 2015. *Structural Decomposition of Energy Consumption, Energy Intensity, Emissions, and Emissions Intensity.* Background paper prepared for the 2015 Industrial Development Report. Vienna: United Nations Industrial Development Organization.

www.ingramcontent.com/pod-product-compliance
Lightning Source LLC
Chambersburg PA
CBHW082347230926
51846CB00004B/939
* 9 7 8 9 2 1 1 0 6 4 5 4 4 *